AF521978

# Macromolecular Crystallography

# Macromolecular Crystallography conventional and high-throughput methods

EDITED BY

**Mark Sanderson**
**and**
**Jane Skelly**

OXFORD
UNIVERSITY PRESS

OXFORD
UNIVERSITY PRESS

Great Clarendon Street, Oxford OX2 6DP

Oxford University Press is a department of the University of Oxford.
It furthers the University's objective of excellence in research, scholarship, and education by publishing worldwide in

Oxford New York

Auckland Cape Town Dar es Salaam Hong Kong Karachi
Kuala Lumpur Madrid Melbourne Mexico City Nairobi
New Delhi Shanghai Taipei Toronto

With offices in

Argentina Austria Brazil Chile Czech Republic France Greece
Guatemala Hungary Italy Japan Poland Portugal Singapore
South Korea Switzerland Thailand Turkey Ukraine Vietnam

Oxford is a registered trade mark of Oxford University Press
in the UK and in certain other countries

Published in the United States
by Oxford University Press Inc., New York



First published 2007



British Library Cataloguing in Publication Data
Data available

Library of Congress Cataloging in Publication Data
Data available

Typeset by Newgen Imaging Systems (P) Ltd., Chennai, India
Printed in Great Britain
on acid-free paper by
Antony Rowe, Chippenham, Wiltshire

ISBN 978–0–19–852097–9

10 9 8 7 6 5 4 3 2 1

# Preface

The nature of macromolecular crystallography has changed greatly over the past 10 years. Increasingly, the field is developing into two groupings. One grouping are those who continue to work along traditional lines and solve structures of single macromolecules and their complexes within a laboratory setting, where usually there is also extensive accompanying biochemical, biophysical, and genetic studies being undertaken, either in the same laboratory or by collaboration. The other grouping consists of 'high-throughput' research whose aim is take an organism and solve the structure of all proteins which it encodes. This is achieved by trying to express in large amounts all the constituent proteins, crystallizing them, and solving their structures. This volume covers aspects of the X-ray crystallography of both of these groupings.

Clearly, macromolecular crystallographers wonder what will be the role in the future of the single research group in the context of the increasing numbers of 'high-throughput' crystallography consortia. Certainly there will be a need for both enterprises as macromolecular crystallography is not always a straightforward process and an interesting structural problem can be snared by many pitfalls along the way, be they problems of protein expression, folding (Chapters 1 and 2), crystallization, diffractibility of crystals, crystal pathologies (such as twinning), and difficulties in structure solution (Chapters 3 and 4). The success of a project requires being able to intervene and solve problems *en route* in order to take it to its successful conclusion. As the 'high-throughput' crystallographic consortia solve more single proteins, the traditional crystallographic groups are moving away from similar studies towards studying protein–protein, protein–DNA, and protein–RNA complexes (Chapters 14 and 15), viruses, and membrane proteins (Chapter 16). Our ability to crystallize these larger assemblies and membrane proteins is increasingly challenging and in turn helped by robotic crystallization whose development was greatly spurred by the needs of 'high-throughput' crystallography.

In this volume has been included a wide range of topics pertinent to the conventional and high-throughput crystallography of proteins, RNA, protein–DNA complexes, protein expression and purification, crystallization, data collection, and techniques of structure solution and refinement. Other select topics that have been covered are protein–DNA complexes, RNA crystallization, and virus crystallography. In this book we have not covered the basic aspects of X-ray diffraction as these are well covered in a range of texts. One which we very strongly recommend is that written by Professor David Blow, Outline of Crystallography for Biologists, Oxford University Press, 2002.

**Safety: it must be stressed that X-ray equipment should under no circumstances be used by an untrained operator. Training in its use must be received from an experienced worker.**

It remains for us as editors to thank all the contributors for all their hard work in preparing the material for this volume. We should like to thank the commissioning team at OUP, Ian Sherman, Christine Rode, Abbie Headon, Helen Eaton (for cover design preparation), Elizabeth Paul and Melissa Dixon for all their hard work and advice in bringing this edited volume to completion.

M. R. Sanderson and J. V. Skelly

# Contents

# Contributors

**S. S. Abdel-Meguid**, ProXyChem, 6 Canal Park, # 210 Cambridge, MA 02141, USA. sherin.s.abdel-meguid@proxychem.com

**J. P. Abrahams**, Biophysical Structural Chemistry, Leiden Institute of Chemistry, Einsteinweg 55, 2333 CC Leiden, The Netherlands. Abrahams@chem.leidenuniv.nl

**N. G. A. Abrescia**, Division of Structural Biology, Henry Wellcome Building for Genome Medicine, University of Oxford, UK. nicola@strubi.ox.ac.uk

**A. Radu Aricescu**, Division of Structural Biology, Henry Wellcome Building of Genome Medicine, University of Oxford, UK. radu@strubi.ox.ac.uk

**T. Batuwangala**, Domantis Ltd, 315 Cambridge Science Park, Cambridge, CB4 OWG, UK. Thil.Batuwangala@Domantis.com

**N. S. Berrow**, The Protein Production Facility, Henry Wellcome Building of Genome Medicine, University of Oxford, UK. nick@strubi.ox.ac.uk

**S. K. Burley**, SGX Pharmaceuticals, Inc., 10505 Roselle Ave., San Diego, CA 92121 and 9700 S. Cass Ave., Building 438, Argonne, IL 60439, USA. sburley@sgxpharma.com

**W. M. Carson**, Center for Biophysical Sciences and Engineering, University of Alabama at Birmingham, 251 CBSE, 1025 18th Street South, Birmingham, AL 35294–4400, USA. carson@uab.edu

**N. E. Chayen**, Department of BioMolecular Medicine, Division of Surgery, Oncology, Reproductive Biology and Anaesthetics, Faculty of Medicine, Imperial College, London SW7 2AZ, UK. n.chayen@imperial.ac.uk

**K. L. D'Amico**, SGX Pharmaceuticals, Inc., 10505 Roselle Ave., San Diego, CA 92121 and 9700 S. Cass Ave., Building 438, Argonne, IL 60439, USA. Kevin_damico@sgxpharma.com

**M. Delarue**, Unite de Biochimie Structurale, Institut Pasteur, URA 2185 du CNRS, 25 rue du Dr. Roux, 75015 Paris, France. delarue@pasteur.fr

**B. Francois**, IBMC-CNRS-ULP, UPR9002, 15 rue René Descartes, 67084 Strasbourg, France.

**E. E. Fry**, Division of Structural Biology, Henry Wellcome Building of Genome Medicine, University of Oxford, UK. liz@strubi.ox.ac.uk

**W. Furey**, Biocrystallography Laboratory, VA Medical Center, University Drive C, Pittsburgh, PA 15240, USA and Department of Pharmacology, University of Pittsburgh, Pittsburgh, PA 15261, USA. fureyw@pitt.edu

**D. Jeruzalmi**, Department of Molecular and Cellular Biology, Harvard University, 7 Divinity Avenue, Cambridge, MA 02138, USA. dj@mcb.harvard.edu

**J. W. Koss**, SGX Pharmaceuticals, Inc., 10505 Roselle Ave., San Diego, CA 92121 and 9700 S. Cass Ave., Building 438, Argonne, IL 60439, USA. John_koss@sgxpharma.com

**V. S. Lamzin**, European Molecular Biology Laboratory (EMBL), c/o DESY, Notkestrasse 85, 22603 Hamburg, Germany. victor@embl-hamburg.de

**I. Laponogov**, Randall Division of Cell and Molecular Biophysics, Kings College London, UK. ivan.laponogov@kcl.ac.uk

**B. Masquida**, IBMC-CNRS-ULP, UPR9002, 15 rue René Descartes, 67084 Strasbourg, France. B.Masquida@ibmc.u-strasbg.fr

**R. J. Morris**, John Innes Centre, Norwich Research Park, Colney, Norwich NR4 7UH, UK.
Richard.Morris@bbsrc.ac.uk

**H. M. K. Murthy**, Center for Biophysical Sciences and Engineering University of Alabama at Birmingham, CBSE 100, 1530, 3rd Ave. South, Birmingham, AL 35294–4400, USA.
murthy@cbse.uab.edu

**S. Ness**, Biophysical Structural Chemistry, Leiden Institute of Chemistry, Einsteinweg 55, 2333 CC Leiden, The Netherlands.
sness@sness.net

**J. E. Nettleship**, The Protein Production Facility, Henry Wellcome Building of Genome Medicine, University of Oxford, UK.
joanne@strubi.ox.ac.uk

**R. J. Owens**, The Protein Production Facility, Henry Wellcome Building of Genome Medicine, University of Oxford, UK.
ray@strubi.ox.ac.uk

**N. S. Pannu**, Biophysical Structural Chemistry, Leiden Institute of Chemistry, Einsteinweg 55, 2333 CC Leiden, The Netherlands.
raj@chem.leidenuniv.nl

**A. Perrakis**, Netherlands Cancer Institute, Department of Molecular Carcinogenesis, Plesmanlaan 21, 1066 CX, Amsterdam, Netherlands.
a.perrakis@nki.nl

**J. S. Plaisier**, Biophysical Structural Chemistry, Leiden Institute of Chemistry, Einsteinweg 55, 2333 CC Leiden, The Netherlands.
plaisier@chem.leidenuniv.nl

**Sarah Sainsbury**, The Oxford Protein Facility, Henry Wellcome Building for Genomic Medicine, University of Oxford, UK OX3 7BN
sarah@strubi.ox.ac.uk

**M. R. Sanderson**, Randall Division of Cell and Molecular Biophysics, Kings College London, UK.
mark.sanderson@kcl.ac.uk

**J. V. Skelly**, School of Chemical and Life Sciences, University of Greenwich, Central Avenue, Chatham Maritime, Kent, ME4 4TB UK.
jvskelly@yahoo.com

**D. W. Smith**, SGX Pharmaceuticals, Inc., 10505 Roselle Ave., San Diego, CA 92121 and 9700 S. Cass Ave., Building 438, Argonne, IL 60439,USA.
David_smith@sgxpharma.com

**M. K. Sohi**, Randall Division of Cell and Molecular Biophysics, New Hunt's House, Guy's Campus, Kings College London, SE1 1UL, UK.
maninder.sohi@kcl.ac.uk

**D. K. Stammers**, Division of Structural Biology, Henry Wellcome Building of Genome Medicine, University of Oxford, UK.
daves@strubi.ox.ac.uk

**D. I. Stuart**, Division of Structural Biology, Henry Wellcome Building of Genome Medicine, University of Oxford, UK.
dave@strubi.ox.ac.uk

**S. R. Wasserman**, SGX Pharmaceuticals, Inc., 10505 Roselle Ave., San Diego, CA 92121 and 9700 S. Cass Ave., Building 438, Argonne, IL 60439, USA.
Stephen_wasserman@sgxpharma.com

**C. M. Weeks**, Hauptman-Woodward Medical Research Institute, Inc., 700 Ellicott Street, Buffalo, NY 14203–1102, USA.
weeks@hwi.buffalo.edu

**A. Werner**, IBMC-CNRS-ULP, UPR9002, 15 rue René Descartes, 67084 Strasbourg, France.

**E. Westhof**, IBMC-CNRS-ULP, UPR9002, 15 rue René Descartes, 67084 Strasbourg, France.
e.westhof@ibmc.u-strasbg.fr

CHAPTER 1

# Classical cloning, expression, and purification

Jane Skelly, Maninder K. Sohi, and Thil Batuwangala

## 1.1 Introduction

The ideal protein-expression strategy for X-ray structural analysis should provide correctly folded, soluble, and active protein in sufficient quantities for successful crystallization. Subsequent isolation and purification must be designed to achieve a polished product as rapidly as possible, involving a minimum number of steps. The simplest and least expensive methods employ bacterial hosts such as *Escherichia coli*, *Bacillus*, and *Staphylococcus* but if the target protein is from an eukaryotic source requiring post-translational processing for full functionality, an eukaryotic vector–host system would be appropriate – although it should be noted that in many instances the lack of processing can prove an advantage in crystallization (Table 1.1). Microbial eukaryotes, such as yeast and filamentous fungi, process their gene products in a way that more closely resembles higher organisms. Yeast is non-pathogenic and its fermentation characteristics are well known. Both *Saccharomyces cerevisiae* and *Pichia pastoris* strains are used extensively for large-scale expression of heterologous proteins. Whereas yeast, unless supplied with an appropriate leader sequence, export protein to the cell vacuole the filamentous fungi, *Aspergillus nidulans* and *Aspergillus niger*, secrete their gene products directly into the growth medium. Secretion is often preferred because it facilitates recovery of the product. DNA can also be inserted into the fungal genome at a high copy number, although the genetics are less well characterized. High-level expression of cytoplasmic, secretory, or cell surface proteins can be achieved in cultured insect cells using recombinant baculovirus vectors. Furthermore, in insects the post-translational modifications are similar to those of eukaryotes. For some very large molecules, the only feasible way of obtaining correctly-folded, active protein is by expression in mammalian cells. Mammalian expression vectors are usually hybrids, containing elements derived from prokaryotic plasmids and controlling sequences from eukaryotes such as promoters and transcription enhancers required for the expression of foreign DNA. Alternatively, *in vitro* protein expression in cell-free systems is being developed specifically for structural proteomics, where only the protein of interest is expressed, improving the yield of stably-active eukaryotic proteins as well as simplifying their purification. Product size, stability, the presence of disulphide bonds, and whether the product is likely to be toxic to the host are all important considerations when choosing a suitable expression system. Levels of expressed gene product are measured as a percentage of the total soluble cell protein which can vary from $<1\%$ to $>50\%$ depending on several factors:

1. the vector-host system;
2. gene copy number;
3. transcription and translation efficiency;
4. mRNA stability;
5. stability and solubility of gene product;
6. the conditions of fermentation and induction, as detailed for each vector.

**Table 1.1** Selection of host cells for protein amplification

| Host | Advantages | Disadvantages |
|---|---|---|
| Prokaryotic expression: *E. coli* | Rapid growth with high yields (up to 50% total cell protein)<br>Extensive range of vectors<br>Simple, well-defined growth media | Lack of post-translational processing<br>Product may prove toxic to host<br>Product incorrectly folded and inactive<br>High endotoxin content |
| Eukaryotic expression: yeast *Pichia pastoris*; *Saccharomyces cerevisiae* | Rapid growth with ease of scale-up and processing<br>High yields (g/l in *Pichia pastoris*)<br>Inexpensive<br>Formation of disulphides<br>Glycosylation | Glycosylated product differs from mammalian systems |
| Insect cell expression: *Baculovirus* | High-level expression<br>Formation of disulphides<br>Glycosylation | Glycosylated product may differ from mammalian systems<br>Not necessarily fully functional |
| Mammalian expression | Fully-functional product | Expensive media<br>Slow growth rates |
| Filamentous fungi: *Aspergillus nidulans*; *Aspergillus niger* | Secretion of large quantities into growth media | Genetics not well characterized |
| Cell-free systems | Only protein of interest expressed<br>Simple purification | Expensive reagents |

## 1.2 Cloning and expression

PCR cloning is usually the preferred route to constructing an expression vector containing the gene of interest. The choice of vector will inevitably depend on the source and characteristics of the gene product, the quantity of product required, and its purification strategy. Detection and purification can be simplified by using a fusion partner, such as glutathione-*S*-transferase (GST), a histidine tag, or a recognition motif sequence such as the c-*myc* epitope (see Section 1.2.5).

### 1.2.1 Construction of a recombinant *E. coli* expression vector by PCR

Once a suitable vector has been selected the coding sequence of the target protein to be cloned must first be amplified from either genomic or a cDNA template by PCR, for which suitable forward and reverse oligonucleotide primers are needed. A range of web-based software is available for designing primers (www.clcbio.com). Important considerations in primer design include: the primer length, which for most applications is between 18 and 30 bases; the chosen 5′ and 3′-end primer sequences; their melting temperatures (Tm) which should not be lower than 60°C; and the GC content, which should range between 40% and 60%. The 5′-end primer which overlaps with the 5′ end of the coding sequence is designed to contain: a suitable restriction endonuclease recognition site for cloning into the expression vector; a 5′ extension to the restriction site; a start codon; and an overlapping sequence. The 3′-end primer overlaps the complementary DNA strand and should supply: a second restriction site; a 5′ extension; a stop codon; and an overlapping sequence. If tags or fusion partners are appended, additional bases may be required in the antisense primer to ensure sequences are in frame. The primers are normally synthesized using a commercial synthesizer. It is not usually necessary for oligonucleotides to be purified for routine PCR. A typical amplification protocol consists of 25 cycles each of a denaturing step at 95°C for 1 min, an annealing step to be calculated from the melting temperatures of the primers used, and an extension at 72°C for 1 min followed by a final 10 min extension step at 72°C. The reaction conditions can be optimized, with the number of cycles being increased for mammalian genomic DNA. Purification of the fragment is not

## Protocol 1.1 Construction of recombinant vector by PCR

**1.** Digest the vector with specific restriction enzymes to generate ends compatible for ligation with the coding sequence to be cloned.
**2.** Purify using preparative agarose gel electrophoresis.
**3.** Extract from agarose using a commercial gel extraction kit.
**4.** Amplify the insert sequence using suitable oligonucleotide primers.
To a PCR tube on ice, add:

5 μl 10× PCR reaction buffer
5 μl dNTP mix (2 mM each dATP, dCTP, dGTP, dTTP)
5 μl of each forward and reverse primers (10 pmol/μl)
0.5 μl DNA template (100–250 ng for mammalian genomic DNA,
20 ng for linearized plasmid DNA)
1.0 μl *Taq* DNA polymerase
2–8 μl 25 mM $MgCl_2$ solution
28.5 μl sterile water

Set up negative control reactions omitting primers or DNA substrate.

Amplify DNA for 25 cycles with the appropriate sequence of melting (95°C for 1 min), annealing temperature (to be calculated from the melting temperatures of the primers used), and replication (72°C for 1 min), followed by a final 10 min extension step at 72°C.

Purify amplified DNA using a commercial kit (Qiagen QiaQuick).

Determine the concentration of the insert.

**5.** Digest the insert with specific restriction enzymes.
To a microcentrifuge tube add:

5 μl of appropriate 10× restriction enzyme buffer
0.5 μl 100× BSA
0.2 μg DNA
2.5 μl of restriction enzyme(s)
Sterile water to a volume of 50 μl

Incubate the reaction mixture for 2–4 h at the temperature appropriate for the restriction enzyme used.

Purify using either agarose or a commercial kit.

(If the two enzymes do not have a compatible buffer, perform the digestion in two steps, purifying the insert after each step.)

**6.** Ligate vector and gene product
To a microcentrifuge tube add:

100 ng of digested vector DNA
Insert fragment (1:1 to 3:1 molar ratio of the insert to the vector)
4 μl 5× ligation buffer
1 μl T4 DNA ligase

Incubate at 16°C for 4–16 h or at 4°C overnight.
**7.** Transform competent *E. coli* host with recombinant vector and select for recombinants by antibiotic resistance appropriate for the plasmid.
**8.** Identify colonies by PCR or plasmid mini-preps.
**9.** DNA sequence the construct.

usually necessary unless PCR introduces contaminating sequences.

The PCR product is then digested with appropriate restriction enzymes, unless it is to be first ligated into a TA cloning vector. (TA cloning involves two stages – firstly cloning into a TA vector followed by subcloning into an expression vector.) It is important to ensure that the insert is properly digested, and when carrying out a simultaneous double digestion the enzymes must be compatible with the buffer supplied. Before the ligation step the insert should be purified, either by electrophoresis on agarose or with a commercial kit. It is always good practice to carry out a controlled ligation with the vector alone. The recombinant ligated vector is next introduced into the selected host strain (Section 1.2.2.1). The cells are first made competent for transformation by treatment with calcium chloride using a standard procedure (Appelbaum and Shatzman, 1999). Recombinants containing the inserted gene can be conveniently screened by PCR, using vector-specific and gene-specific primers.

Directional cloning (by ligation into two different restriction sites) is usually the preferred option, having the advantage of not requiring dephosphorylation of the vector and also avoiding possibility of the product ending up in the wrong orientation. Finally, it is important to sequence the construct in order to identify any mutations that may have been generated during the PCR reaction. Protocol 1.1 outlines the sequence of steps involved in the construction of a recombinant vector. Specific details of materials, including preparations of buffer reagents, may be found in standard laboratory manuals and on manufacturers' web sites.

For high-throughput (HTP) the gene of interest can be cloned in parallel into a variety of expression vectors containing different tags and/or fusion partners, and into vectors for a variety of expression systems. Gateway™ (www.invitrogen.com) cloning technology is discussed in Chapter 2.

### 1.2.2 Prokaryotic expression systems

The most effective way to maximize transcription is to clone the gene of interest downstream from a strong, regulatable promoter. In *E. coli* the promoter providing the transcription signal consists of two consensus sequences situated –1 and –35 bases upstream from the initiation codon. High-level expression vectors contain promoter regions situated before unique restriction sites where the desired gene is to be inserted, placing the gene under the direct control of the promoter. Differences between consensus promoter sequences influence transcription levels, which depend on the frequency with which RNA polymerase initiates transcription. In addition to these regulatory elements, expression vectors possess a selectable marker – invariably an antibiotic resistance gene.

When choosing an *E. coli* expression system for production of eukaryotic cDNA, the differences between prokaryotic and eukaryotic gene control mechanisms must be addressed. In *E. coli*, the ribosome binding site (RBS) consists (in most cases) of the initiation codon AUG and the purine-rich Shine–Dalgano sequences located several bases upstream. Vectors have been constructed which provide all the necessary signals for gene expression including ribosome binding sites, strong regulatable promoters and termination sequences, derived from *E. coli* genes with the reading frame removed. Multiple cloning sites (MCS) are provided in these vectors to facilitate insertion of the target gene.

Eukaryotic DNA contains sequences recognized as termination signals in *E. coli*, resulting in premature termination of transcription and a truncated protein. Also, there are differences in codon preference affecting translation, which may ultimately result in low levels of expression or even premature termination. Not all of the 61 mRNA codons are used equally (Kane, 1995). Rare codons tend to occur in genes expressed at low level and their usage depends on the organism. (The codon usage per organism can be found in the Codon Usage Database (www.kazusa.or.jp/codon). To overcome this, site-directed mutagenesis may be carried out to replace the rare codons by more commonly-occurring ones, or alternatively by coexpression of the genes encoding rare tRNAs. *E. coli* strains that encode for a number of rare codon genes are now commercially available (see Section 1.2.2.1). There is also a possibility that expression of high levels of foreign protein may prove toxic to the *E. coli* host inducing cell fragility, therefore placing the recombinant cell at a disadvantage. Specific post-translational modifications, such as *N*- and *O*-glycosylation, phosphorylation and specific cleavage (e.g. removal of the N-terminal methionine residue) required for full functionality of the recombinant protein, will not be carried out in bacteria.

Probably the best example of regulatory gene expression in bacteria is the *lac* operon, which is extensively used in the construction of expression vectors (Jacob and Monod, 1961). The *lac* promoter contains the sequence controlling transcription of the *lacZ* gene coding for β-galactosidase, one of the enzymes that converts lactose to glucose and galactose. It also controls transcription of lacZ′, which encodes a peptide fragment of β-galactosidase. Strains of *E. coli* lacking this fragment are only able to synthesize the complete and functional enzyme when harbouring vectors carrying the lacZ′ sequence, for example pUC and M13. This is useful for screening recombinants. The *lac* promoter is induced by allolactose, an isomeric form of lactose, or more commonly, isopropyl β-D-thiogalactoside (IPTG), a non-degradable substrate, at a concentration of up to 1 mM in the growth medium. Basal expression (expression in the absence of inducer) may be reduced by addition of glucose to the media. The *lac*UV5, *tac*, and *trc* promoters are all repressed by the *lac* repressor.

The *trp* promoter is located upstream of a group of genes responsible for the biosynthesis of tryptophan. It is repressed in the presence of tryptophan but induced by either 3-indolyacetic acid or the absence of tryptophan in the growth medium (i.e. a defined minimal medium such as M9CA). The *tac* promoter is a synthetic hybrid containing the –35 sequence derived from the trp promoter and –10 from *lac*. It is several times stronger than either *lac* or *trp* (Amann *et al.*, 1983).

pBAD expression (www.invitrogen.com) utilizes the regulatory elements of the *E. coli* arabinose operon (*araBAD*), which controls the arabinose metabolic pathway. It is both positively and negatively regulated by the product of the *araC* gene, a transcriptional regulator which forms a complex with L-arabinose (Ogden *et al.*, 1980). The tight regulation provides a simple but very effective method for optimizing yields of soluble recombinant protein at levels just below the threshold at which they become insoluble. Induction is by the addition of arabinose. Again, basal expression may be repressed by the addition of 2% glucose to the growth medium, an important consideration if the protein of interest is known to be toxic to the host. Currently there are nine pBAD expression vectors with a variety of vector-specific features.

Bacteriophage lambda $P_L$ is an extremely powerful promoter responsible for the transcription of bacteriophage lambda DNA. $P_L$ expression systems offer tight control as well as high-level expression of the gene of interest. The $P_L$ promoter is under the control of the lambda cI repressor protein, which represses the lambda promoter on an adjacent operator site. Selected *E . coli* host strains synthesize a temperature-sensitive defective form of the cI repressor protein, which is inactive at temperatures greater than 32°C. Expression is induced by a rapid temperature shift. The host cells are usually grown at 28°C to 32°C to midlog phase when the temperature is rapidly adjusted to 40°C as described in Protocol 1.2. Alternatively, the cI repressor may be placed under the control of the tightly-regulated *trp* promoter, and expression is then induced by the addition of tryptophan. With no tryptophan present, the cI repressor binds the operator of $P_L$, preventing expression. However, in the presence of tryptophan the tryptophan–*trp* repressor complex forms and prevents transcription of the cI repressor gene, allowing transcription of the cloned gene. Induction can be achieved at lower temperatures although basal expression can be a problem.

The T7 RNA polymerase recognizes the bacteriophage T7 gene 10 promoter, which is carried on the vector upstream of the gene of interest. Being more efficient than the *E. coli* RNA polymerase, very high levels of expression are possible. Up to 50% of the total cell protein can be attained in a few hours after induction. First, the target gene is cloned using an *E. coli* host which does not contain the T7 polymerase gene. Once established, the plasmids are then transferred into the expression host harbouring the T7 polymerase under the control of an inducible promoter, usually *lacUV5*. Induction is by addition of IPTG. Besides high-level expression, the system offers the advantage of very tight control. Since the host-cell RNA polymerase does not recognize the T7 promoter, it prevents basal expression which might prove harmful to the host. Control can be tightened even further by coexpressing T7 lysozyme from an additional plasmid (pLysS/pLysE) in the expression strain which inactivates any spurious T7 polymerase produced under non-inducing conditions. An extensive series of derivatives of the original pET vectors constructed by Studier and Moffatt (1986) are commercially available (www.novagen.com).

### 1.2.2.1 *Bacterial hosts*

Most *E. coli* host strains used for high-level expression are descended from K12. *E. coli* strains should ideally be protease deficient, otherwise some degree of proteolysis is more or less inevitable as evident in multiple banding on SDS gels. For this reason, *E. coli* B strains deficient in the ATP-dependent *lon* (cytoplasmic) and *ompT* (periplasmic) proteases are normally used. As in the case of T7 polymerase, some vectors require host strains carrying additional regulatory elements for which a variety of derivatives of BL21 strains are commercially available. However, BL21 does not transform well so an alternative strain for cloning and maintenance of vector should be used, for example JM105. *E. coli* strains that encode for a number of rare codon genes include: BL21 (DE3) CodonPlus-RIL AGG/AGA (arginine), AUA (isoleucine), and CUA (leucine) (www.stratagene.com); and Rosetta or Rosetta (DE3) AGG/AGA (arginine), AUA (isoleucine), and CCC (proline) (www.novagen.com). For membrane-bound proteins, expression in mutant strains C41 (DE3) and C43 (DE3) could improve expression levels (Miroux and Walker, 1996). For proteins with disulphide bonds, host strains have been produced which have a more oxidizing cytoplasmic environment. For example AD494 (Novagen) has a mutation in thioredoxin (trxB) and Origami (Novagen) carries a double mutation (*trxB, gor*) in the thioredoxin and glutathione reductase genes.

*1.2.2.2 Expression method and plasmid stability*

A single colony from a freshly-streaked plate of the transformed host is used to inoculate a 30–50 ml starter culture of Luria broth (LB) medium containing the appropriate antibiotic. It is important not to allow starter cultures to grow above $OD_{600nm} > 1$. Cells should then be centrifuged and resuspended in fresh medium for inoculation of the main culture. Growth and induction conditions vary with the vector–host expression system. Usually, cells are grown to midlog phase before induction, either by a rapid shift in temperature or addition of an inducer to the medium (Protocol 1.2). It is important to maintain good aeration in the fermentation vessel.

Plasmid instability can arise when the foreign protein is toxic to the host cell. During rapid growth plasmids may be lost or the copy number reduced, allowing the non-recombinant cells to take over. As a precaution it is essential to maintain antibiotic resistance. As ampicillin is inactivated by β-lactamases secreted by *E. coli* into the medium one may spin overnight cultures and resuspend the pellet in fresh media.

*1.2.2.3 Engineering proteins for purification*

Fusion proteins containing a tag of known size and function may be engineered specifically for overexpression and detection, as well as for facilitating purification by rapid two-step affinity procedures directly from crude cell lysates (Table 1.2). Customized fusions may be constructed tailoring to the specific needs of the protein. For example an N-terminal signal sequence can be used to direct the recombinant product into the periplasm. Using an appropriate leader sequence, antibody fragments can be secreted into the periplasm and through the outer *E. coli* membrane into the culture medium where they can be effectively reconstituted. The oxidizing environment of the periplasm allows disulphide-bond formation and minimizes degradation. Expression vectors incorporating the *ompT* and *pelB* leader sequences upstream of the 5′ cloning sites are commercially available (Stader and Silhavy, 1990; Nilsson *et al.*, 1985). Proteins expressed as fusions with *Staphylococcal* protein A can be purified to near-homogeneity

**Protocol 1.2 Growth and induction of expression of a heterologous sequence from vector $P_L$ promoter**

This is an analytical scale induction to check for expression levels.

*E. coli* host strain AR58 containing defective phage lambda lysogen transformed with a recombinant vector which carries antibiotic resistance for kanamycin (Kan R).

**1.** Grow recombinant and control *E. coli* strains overnight at 32°C in LB containing kanamycin antibiotic.
**2.** Dilute the overnight culture 1 in 60–100 into fresh LB containing kanamycin.
**3.** Grow cultures at 32°C in a shaker until the $OD_{650nm}$ reaches 0.6–0.8.
**4.** Remove a 1-ml sample for analysis. Pellet samples for 30 sec at 16,000 g in a microcentrifuge, decant the medium, and place tubes on dry ice.
**5.** Move cultures to a water bath set at 40°C and continue growing the cultures at this temperature for 2 h.
**6.** Remove 1 ml aliquot from each for analysis.
**7.** Record the $OD_{650nm}$; typically it will be 1.3 or higher if the gene product is not toxic to the cell.
**8.** Harvest remaining cells by centrifugation and freeze at −70°C.
**9.** For large-scale cultures, use 2-litre flasks (with baffles). Induce by adding 1/3 volume prewarmed LB at 65°C to the culture.

This protocol is adapted from Appelbaum, E. and Shatzman, A. R. (1999). Prokaryotes in vivo expression systems. In: *Protein Expression Practical Approach*, Higgins, S. J. and Hames, B. D., eds. Oxford University Press.

by immobilization on IgG (Nilsson *et al.*, 1985). However, the drawback of using immunoaffinity procedures is that immunological detection can be made complicated. Consequently these strategies have been largely superseded by fusions based on non-immunoaffinity methods. Among the vectors that have proved popular are pTrcHis, with a tag consisting of a sequence of polyhistidines (usually 6 × His), which can be immobilized by metal chelation (Protocol 1.3), and pGEX based on *Schistosoma japonicum* glutathione S-transferase as the fusion tag (Smith and Johnson, 1988), which uses immobilized glutathione for isolation (Protocol 1.4). Both are commercially available in kit form (www.invitrogen.com; www.gehealthcare.com). pGEX vectors feature a *tac* promoter for inducible (IPTG), high-level expression and an inducible *lac* gene for use in any *E. coli* host. Thirteen pGEX vectors are available, nine with expanded MCSs. The pGEX-6P series provides all three translational reading frames linked between the GST coding region and MCS. The plasmid

**Table 1.2** *E. coli* expression systems

| Vector family | Fusion Tag | Promoter/induction | Purification |
|---|---|---|---|
| pGEX | Glutathione S-transferase | P*tac*<br>IPTG | Glutathione Sepharose Fast Flow™ |
| pET | $(His)_6$ | T7/IPTG | Chelating Sepharose Fast Flow™ |
| pBAD | $(His)_6$ | $P_{BAD}$<br>0.2% L-arabinose | Chelating Sepharose Fast Flow™ |
| pTRX<br>ThioFusion™ | Thioredoxin | $P_L$<br>Temperature shift 37°C to 42°C | Nickel-chelating resins |
| pTrcHis | $(His)_6$ | *trc* | Nickel-chelating resins |
| pEZZ18 | IgG binding domain of protein | *lacUV5* protein A | IgG Sepharose 6 Fast Flow |
| pRSET | $(His)_6$ | T7 | Chelating Sepharose Fast Flow™ |

## Protocol 1.3 Purification of soluble $His_6$-tagged protein on Ni-NTA agarose

### Materials

Sonication buffer: 50 mM sodium phosphate, 300 mM NaCl, pH 7.0–8.0

Ni-NTA agarose (Qiagen™)

Chromatography column: 20 ml bed volume

Wash buffer: 50 mM sodium phosphate, 300 mM NaCl, 30 mM imidazole, pH 7.0–8.0

Elution buffer: 50 mM sodium phosphate, 300 mM NaCl, 250–500 mM imidazole, pH 7.0–8.0

SDS-polyacrylamide gel electrophores (SDS-PAGE) system

### Method

1. Resuspend cells harvested from 1-litre culture in 10 ml sonication buffer.
2. Add lysozyme to 0.2 mg/ml and incubate at 4°C for 30 min.
3. Sonicate the cells at 4°C.
4. Draw lysate through 20-gauge syringe needle to shear the DNA and reduce viscosity if necessary.
5. Centrifuge the lysate at 40,000 g for 2–3 h and collect the supernatant.
6. Add 8 ml of 50% (v/v) slurry of Ni-NTA agarose equilibrated in the sonication buffer to the supernatant. Stir for 1 h.
7. Load the agarose into the column.
8. Wash with 20 ml of the wash buffer and collect 5 ml fractions checking $A_{280nm}$ until it is <0.01.
9. Elute the protein from the agarose with 20 ml of the elution buffer. Collect 2 ml fractions.
10. Analyse 5 μl aliquots of the fractions by SDS-PAGE after incubating the protein sample with an equal volume of the sample buffer for SDS-PAGE at 37°C instead of boiling to avoid cleavage of the protein.

Adapted from protocol supplied by QIAexpress™

provides *lac* Iq repressor and confers resistance to ampicillin. Novagen's PET system offers a wide variety of fusion tags including both N and C-terminal polyhistidines (www.novagen.com). Other widely-used tags include a calmodulin-binding peptide (www.stratagene.com), the maltose binding protein (www.westburg.nl), polyarginine, and cellulose-binding tag. These are all helpfully reviewed by Terpe (2003). Another tag utilizes the stability characteristics of *E. coli* thioredoxin, which when used as a fusion confers its heat tolerance and solubility properties upon the recombinant protein (Yasukawa *et al.*, 1995). Providing the target sequence with a C-terminal tag will ensure that only full-length protein is purified. All these fusion tags are available commercially in a variety of vectors with MCSs ensuring easy transfer of inserts.

Invariably, for crystallization it is desirable to remove the tag thus avoiding any possible interference with folding and tertiary structure. This may prove problematic, particularly if the proteolytic cleavage site introduced into the vectors for this purpose, is not unique. Shorter fragments may result, leading to microheterogeneity. Histidine-tagged protein appears to be less of a problem in this respect since successful crystallization and high-resolution structure solution has been achieved with the protein-polyhistidine sequence (His6) remaining intact. Cleavage may be carried out either on the immobilized media or after elution of the product.

**Protocol 1.4 Purification of soluble GST-tagged recombinant protein and cleavage of the GST tag using thrombin and factor Xa**

**Materials**

Binding buffer: 1 × PBS (140 mM NaCl, 2.7 mM KCl, 10 mM $Na_2HPO_4$, 1.8 mM $KH_2PO_4$, pH 7.0–8.0)
Elution buffer: 50 mM Tris-HCl, 10 mM reduced GSH, pH 7.0–8.0
Prepacked MicroSpin™GST or GSTrap FF columns (GE Healthcare)
Hi-trap Benzamidine FF column
SDS-polyacrylamide gel electrophoresis system
Thrombin: 500 units in 0.5 ml PBS (stored at –80°C)
Factor Xa: 400 units in water to give a final solution of 1 Unit/μl stored at –80°C

**Method**

Cleavage of the fusion protein off the column:

1. Add the cell lysate to a prepacked MicroSpin™GST or GSTrap FF column equilibrated with the binding buffer.
2. Wash the column with the binding buffer.
3. Elute the fusion protein with the elution buffer.
4. Cleave the eluted fusion protein with site-specific protease thrombin or Factor Xa.
5. Desalt the sample using a Hi-trap desalting column.
6. Add the sample to a MicroSpin™GST or GSTrap FF column equilibrated with the binding buffer.
7. Collect the eluate and analyse it by SDS-PAGE or by mass spectroscopy.
8. Remove the protease using a Hi-trap benzamidine column.

Cleavage of the fusion protein on the column:

1. Add the cell lysate to a prepacked MicroSpin™GST column or GSTrap FF column equilibrated with the binding buffer.
2. Wash with the binding buffer.
3. If using a GSTrap FF, connect the column directly to a Hi-trap benzamidine FF column.
4. Cleave the fusion protein with a site-specific protease (thrombin, factor Xa or any other protease).
5. Collect the flow through sample and analyse on a SDS-PAGE or by mass spectroscopy.

For scale-up:
GSTrap FF (1 ml) column binds 10–12 mg fusion protein
GSTrap FF (5 ml) column binds 50–60 mg fusion protein

1. Equilibrate the column with 5 column volumes of the binding buffer.
2. Maintain loading flow rate 0.2–1 ml /min for 1 ml column and 1–5 ml/min for 5 ml column.
3. Wash with 5–10 column volumes of binding buffer.
4. Elute with 5–10 volumes of elution buffer.

Adapted from protocol supplied by GE Healthcare.

Polyhistidine tags of other lengths (e.g. His4 or His10) may provide useful alternatives. The amount of enzyme, temperature, and length of incubation required for complete digestion varies according to the specific fusion protein. Thrombin, factor Xa, and enterokinase are the most commonly used proteases. Thrombin in particular tends to cleave promiscuously. Another disadvantage of fusions is the alteration in the sequence of the tagged protein that may be necessary in order to supply the cleavage site. For GST fusions it is advisable to use the PreScission Protease cleavage site (www.gehealthcare.com). The GST tag then can be removed and the protein purified in a single step on the column (Protocol 1.4). The PreScission Protease also has the useful property of being maximally active at +4°C thus allowing cleavage to be performed at low temperatures and so improving the stability of the target protein. The protease can be removed after cleavage using a HiTrap Benzamidine column. The GST 96 well Detection Module provides a convenient ELISA assay for testing lysates. Cloning procedures are specific for each vector and manufacturers' instructions should be closely followed.

Fusions may also be designed against which antibodies may be raised that can be used for detection. An example is the tripeptide Glu-Glu-Phe motif for the immunoaffinity of HIV enzymes which is recognized by the YL1/2 monoclonal antibody to α-tubulin (Stammers *et al.*, 1991).

To determine the optimum conditions, protein amplification should be monitored at various stages during pilot experiments before scaling-up. It should be emphasized that not all proteins are amenable to amplification in *E. coli*. Considerable time, effort, and hours of frustration can be spent in constructing a suitable expression system and optimizing yields. In particular, growth media, antibiotics, and chemical inducers can be prohibitively expensive. This is a major consideration when scaling up as large-scale fermentation involving high cell densities may simply result in the loss of vector through selection or, as mentioned above, the product may prove toxic to the host. Should the above-mentioned expression strategies fail to provide adequate levels of product in *E. coli* it is advisable to switch to yeast or insect cells.

### 1.2.3 Yeast expression systems

Being eukaryotes, yeast cells carry out some of the post-translational processes found in mammalian cell lines and rarely give rise to inclusion bodies. *Pichia pastoris* and *Saccharomyces cerevisiae* are both well characterized, easy to handle, and grow relatively quickly to high densities in defined medium. *Pichia* is particularly suited for large-scale production, being capable of yielding tens of milligrams per litre of fully functional recombinant material without loss of yield. *P. pastoris* is widely chosen for analytical and structural studies. *Pichia* expression vectors are commercially available for inducible and constitutive expression, as well as the production of secreted proteins coupled with a fusion partner for rapid purification and immunological detection. *Pichia*, being a methyltrophic yeast, can utilize methanol as its sole carbon source in the absence of glucose. Expression is regulated by the AOX1 promoter, which controls the expression of alcohol oxidase, the enzyme involved in the first stage of methanol production (Cregg *et al.*, 1993). Another yeast species that has been successfully utilized for high-level expression of heterologous proteins is *Candida utilis* (Kondo *et al.*, 1997).

One potential problem is that yeast cells can acidify the culture medium and may also contain compounds that affect binding of His-tags to the resin. Detailed protocols for culturing and handling yeast cells are available from Clontech laboratories (www.clontech.com).

### 1.2.4 Baculovirus expression system

Expression in insect cells is a common method of production of recombinant proteins for structural studies. The advantages of using insect cells include relatively high expression levels compared to other eukaryotic expression systems, expression of multiple genes, capacity for expressing unspliced genes, ease of scale up, simplified cell growth, and the possibility of protein production in high density suspension cultures. In addition, post-translational processing modifications to eukaryotic proteins expressed in insect cells are similar to those of mammalian cells and this facilitates the production of biologically active eukaryotic proteins.

Baculovirus expression is the most frequently used method for expression in insect cells and employs *Autographa californica* nuclear polyhedrosis virus (AcNPV), a double stranded (ds) DNA virus that infects arthropods. The baculovirus expression system utilizes features of the viral life cycle to introduce recombinant DNA coding the gene of interest into insect cells (Miller, 1988; O'Reilly *et al.*, 1992).

The protein polyhedrin, which is produced in large amounts during the very late phase of viral life cycle, acts to occlude virus particles and protects them from proteolysis during host cell lysis. However, polyhedrin is non-essential in the viral life cycle (Smith *et al.*, 1983). Another non-essential protein expressed at high levels in the very late phase of the baculovirus life cycle is p10 which is involved in polyhedra formation (Williams, 1989; Vlak *et al.*, 1988). Baculovirus expression systems take advantage of this since the protein of interest can be produced in large amounts by generating recombinant baculovirus with the gene of interest replacing the polyhedrin or p10 genes with expression being driven by the polyhedrin or p10 promoter (Smith *et al.*, 1983).

As the Baculovirus genome is too large (134 kb) to be used as a cloning vehicle into which foreign genes can be inserted directly using standard molecular biology techniques, a transfer vector is used to insert the gene of interest into the Baculovirus genome (Ayres *et al.*, 1994). In brief, the gene of interest is cloned into a (bacterially propagated) transfer vector flanked by viral-specific sequences. The transfer vector containing the gene of interest is then mixed with wild type Baculovirus DNA and cotransfected into insect cells. The gene is introduced into the Baculovirus genome, by homologous recombination, mediated by the viral specific flanking sequences. Recombinant virus expressing the gene of interest can be produced in this manner and the absence of the polyhedrin gene allows identification of recombinant viral plaques since the viruses containing polyhedrin have a different morphology. However, the frequency of recombination during the production of recombinant virus is low (Kitts and Possee, 1993) and identification of recombinant plaques is difficult and time-consuming. Due to the obvious advantages of being able to produce large quantities of correctly folded, biologically active protein, a number of improvements have been made in order to make the production and propagation of recombinant virus more convenient and rapid. The Baculogold™(www.bdbiosciences.com) and the BacPak (www.clontech.com) systems utilize the method of linearization of Baculovirus DNA to increase the frequency of recombination. The basis of this method is the introduction of rare restriction sites for the enzyme *Bsu*391 within the polyhedrin gene locus and the ORF1629 gene that is essential for viral replication (Possee and Howard, 1987; Kitts and Possee, 1993). Thus, linearization of Baculovirus DNA with *Bsu*391 results in the excision of the essential ORF1629 gene and renders this DNA non-infective. Cotransfection of the linearized Baculovirus DNA with a transfer vector, containing the missing sequence, restores infectivity. Moreover, since the gene of interest is within the ORF1629 locus on the transfer vector, almost 100% recombinant frequency can be achieved (Kitts and Possee, 1993). It is important to remember that only the vectors that contain the entire deleted region of the polyhedrin gene can rescue the deletion by homologous recombination.

The most commonly used insect cell lines *Spodoptera frugiperda* 9, 21 (*Sf*9, *Sf*21 respectively) and HighFive (derived from *Trichoplusia ni*) (Geisse *et al.*, 1996) are commercially available from Invitrogen and BD Biosciences. Healthy insect cells adhere to the surface of the plate forming a monolayer and double every 18–24 h. Infected cells stop dividing, become enlarged and uniformly round, have enlarged nuclei, and do not attach to the surface of the plate. The possibility of cell growth in serum free media with these cell lines also has the advantage of ease of purification of secreted proteins. Insect cells can be grown both as monolayers and in suspension. A culture is usually started as a monolayer (Protocol 1.5) and then transferred to suspension into a spinner flask for large-scale protein production. Healthy cells from a log-phase

## Protocol 1.5 Starting an insect cell culture

### Materials

*Sf*9 insect cells (BD Biosciences or Invitrogen)
EX-CEL 405™serum-free medium for insect cells (RJH Biosciences) or Sf 900 II SFM (Invitrogen-GIBCO)
Fetal calf serum (LabClinics SA, Barcelona)
Gentamycin sulphate, 10 mg/ml stock (BD Biosciences)
Amphotericin B, 250 μg/ml stock (BD Biosciences)
Haemocytometer
27°C incubator
Tissue culture flasks
Sterile 10 ml pipettes
Sterile Spinner flasks (200 ml and 1 litre), spinner apparatus
Sterile 25 ml and 50 ml plastic tubes
Sterile plastic Pasteur pipettes
Trypan Blue Stain (0.2%) solution in PBS

### Method

1. Equilibrate the medium at room temperature.
2. Add gentamycin sulphate (50 μg/ml), amphotericin B (2.5 μg/ml), and fetal calf serum (10%).
3. Remove a batch of frozen cells from liquid nitrogen storage dewar.
4. Thaw the cells quickly by dipping the vial in a water bath at 37°C for about 30 sec.
5. Spray the outside of the vial with 70% ethanol and place it in a sterile hood.
6. Transfer the cells to a 25 ml sterile universal vial using a sterile plastic Pasteur pipette.
7. Add 20 ml medium drop wise.
8. Centrifuge at 600 g, at room temperature, for 2–5 min.
9. Aspirate the supernatant using a 25 ml sterile pipette taking care not to disturb the cell pellet.
10. Resuspend the cells in 10 ml fresh medium.
11. Mix 10 μl cell suspension with 10 μl of Trypan Blue solution and estimate the viable cell density using a haemocytometer. Non-viable cells turn blue.
12. Adjust cell density to 250,000 viable cells/ml medium.
13. Transfer the cell suspension to a 50 ml tissue culture flask and incubate at 27°C or room temperature for 48 h.
14. After 48 h, examine the flask using a light microscope and reincubate until the cells become confluent.
15. Dislodge the cells by tapping the flask gently on a bench and transfer the cell suspension to a 25 ml universal tube.
16. Pellet the cells by centrifugation at 1000 g for 2–5 min.
17. Transfer the supernatant to a 50 ml sterile tube and add two volumes of fresh medium.
18. Resuspend the cell pellet in 10 ml of the medium in Step 17 and determine cell density.
19. Seed the cells into a new tissue culture or spinner flask at a density of 250,000 cells/ml using the medium in Step 17 (some fresh medium may be added if required).
20. Incubate the flask at 27°C or room temperature for 48 h.
21. Split or scale up the culture when the cell density reaches $2 \times 10^6$ cells/ml.
22. Steps 19–21 are repeated until a required culture volume is obtained.

Insect cells can be adapted to a serum-free insect cell medium by slowly decreasing the concentration of fetal calf serum in the medium.

*Note*: Protocols 1.5 to 1.11 have been adapted from *Baculovirus Expression System Manual*, 6th edn, May 1999 (www. bdbiosciences.com).

culture are transferred for long-term storage in liquid nitrogen. Properly stored cells (Protocol 1.6) remain viable for several years.

The plasmid DNA used for cotransfection should be as pure as possible. Insect cells are sensitive to impurities in plasmid samples and may lyse before the recombinant virus is regenerated, resulting in very low viral titres. For good cotransfection experiments, monolayers of healthy cells with an initial confluency of 60–70% are required.

Procedures for obtaining recombinant Baculovirus using linearized BaculoGold™ DNA are simple (Protocol 1.7) and normally generate high-titre stocks. The viral titre is determined by plaque assay (Protocol 1.8) so that known amounts of the recombinant virus are used in subsequent virus amplification experiments (Protocol 1.9) to produce large viral stocks.

A small-scale titration experiment is carried out to determine the optimum amount of the recombinant viral stock required for protein production using a 6-well tissue culture plate with a monolayer of $6 \times 10^5$ cells per well. The wells are infected with 0, 10, 20, 40, 60, and 80 μl of the recombinant

## Protocol 1.6 Storing insect cells

### Materials

Insect cell culture at 1–2 × $10^6$ cells/ml

EX-CEL 405™serum-free medium for insect cells containing gentamycin sulphate (50 μg/ml), amphotericin B (2.5 μg/ml), and fetal calf serum (10%)

Sterile cryovials

Liquid nitrogen storage dewar

A small polystyrene box or a cryovial holder

Dimethyl sulphoxide

50 ml sterile centrifuge tubes

### Method

**1.** Harvest cells from a culture containing 1–2 × $10^6$ cells/ml by centrifugation at 1000 g for 10 min.

**2.** Transfer the supernatant to a sterile tube.

**3.** Keep the cell pellet on ice.

**4.** Resuspend the pellet in 10% dimethyl sulphoxide and 90% medium (2 volumes conditioned medium +1 volume fresh medium). Cell density should be about 5 × $10^6$ cells/ml.

**5.** Dispense into cryovials (1 ml/vial).

**6.** Enclose the cryovials in a vial holder or small polystyrene box.

**7.** Transfer the box to –80°C and leave overnight.

**8.** Transfer to liquid nitrogen storage dewar.

## Protocol 1.7 Cotransfection of insect cells to produce recombinant Baculovirus

### Materials

*Sf*9 cell culture at 1–2 × $10^6$cells/ml

1 μg linearized BaculoGold™ DNA (BD Biosciences)

Recombinant Baculovirus transfer vector containing the insert

60-mm tissue culture plates

EX-CEL 405™ medium for insect cells containing 10% fetal calf serum

Transfection buffer A and B set (BD Biosciences)

Sterile microcentrifuge tubes

Sterile pipettes and pipette tips

SDS-polyacrylamide gel electrophoresis system

### Method

**1.** Prepare two 60-mm tissue culture plates.

**2.** Pipette 2 × $10^6$ *Sf*9 cells onto each plate.

**3.** Place the plates on a level surface.

**4.** Allow the cells to adhere to the bottom of the plate to form a monolayer (40–45 min).

**5.** Mix 0.5 μg linearized BaculoGold™ DNA with 2–5 μg recombinant Baculovirus transfer vector containing the insert and allow the mixture to sit at room temperature for 5 min.

**6.** Aspirate the medium from the plates.

**7.** Pipette 3 ml fresh medium onto plate number 1 (control plate).

**8.** Add 1 ml of buffer B to the mixture prepared in Step 5.

**9.** Pipette 1 ml of buffer A onto plate number 2.

**10.** Add the solution from Step 8 drop wise to plate number 2 and rock the plate gently.

**11.** Incubate both plates at 27°C for 4 h.

**12.** After 4 h aspirate the medium from plate number 2, wash the cell monolayer with 3 ml of fresh medium by rocking the plate gently, and remove the medium.

**13.** Add 3 ml of fresh medium to plate number 2 and incubate the plate at 27°C for 4–5 days.

**14.** After 4 days compare the two plates for infection. Infected cells are larger than the uninfected, have enlarged nuclei, stop dividing, and become detached from the surface of the plate. The cells on plate number 1 should remain uninfected.

**15.** After 5 days transfer the supernatant to sterile centrifuge tubes. The supernatant of plate number 2 contains the recombinant virus.

**16.** Store the recombinant virus at 4°C in a dark place.

**17.** Harvest the cells from the plates.

**18.** Pellet the cells by centrifugation at 2500 g for 5 min.

**19.** Analyse the cell pellets from both plates for expression of the recombinant protein by SDS-PAGE.

## Protocol 1.8 Purification of recombinant virus and determination of viral titre by plaque assay

### Materials

*Sf*9 cell culture
100-mm and 12-well tissue culture plates
Baculovirus transfection supernatant
Agarplaque-Plus™agarose (BD Biosciences: Pharmingen)
EX-CEL 405™serum-free medium for insect cells containing 50 μg/ml gentamycin sulphate and 2.5 μg/ml amphotericin B
Sterile water containing 50 μg/ml gentamycin sulphate and 2.5 μg/ml amphotericin B
A sterile plastic box that can accommodate six 100 ml tissue culture plates
Microcentrifuge tubes
Sterile pipette tips
Microscope
SDS-polyacrylamide gel electrophoresis system

### Method

**1.** Prepare 3% agarose solution in deionized water and sterilize by autoclaving.
**2.** Label six 100-cm plates each containing a monolayer of $2.1 \times 10^7$ cells.
**3.** Allow the cells to form a monolayer by placing the plate on a level surface.
**4.** Replace the old medium with 10 ml of fresh medium in each plate.
**5.** Prepare $10^{-3}$, $10^{-4}$, $10^{-5}$, $10^{-6}$, and $10^{-7}$dilutions of the virus transfection supernatant.
**6.** Pipette 100 μl of the dilute virus transfection supernatant onto each plate except for the control.
**7.** Transfer the plates to 27°C incubator and leave for 1 h.
**8.** Cool the agarose solution to 45°C and equilibrate two volumes of the medium to room temperature.
**9.** Transfer the plates from the incubator to the hood and remove the medium.
**10.** Add the insect cell medium at room temperature to the agarose solution and mix quickly.
**11.** Add 10 ml of the agarose solution prepared in Step 10 to the side of each plate and cover the cell monolayer completely by gently tilting the plate.
**12.** Leave the plates undisturbed on a level surface until the agarose is set.
**13.** Place the plates in a sterile plastic box containing some sterile tissues sprayed with sterile water containing 50 μg/ml gentamycin sulphate and 2.5 μg/ml amphotericin B.
**14.** Incubate at 27°C for 6–10 days until visible plaques appear.
**15.** Examine the plates for plaques using a microscope and select a plate containing well separated plaques, and mark the position of each plaque with a marker pen.
**16.** Count the number of plaques and calculate the number of plaque forming units per ml virus stock.
**17.** Remove an agarose plug over the plaque using a sterile pipette tip and place in a microcentrifuge tube containing 1 ml insect medium.
**18.** Pick up 10–20 plaques as in Step 17 and place in separate tubes.
**19.** Elute the virus from the agarose plug by rotating the tubes overnight in a cold room.
**20.** Add 200 μl virus from each tube to separate wells of 12-well tissue culture plates each containing a monolayer of $2 \times 10^5$ cells per well in 1 ml insect cell medium and incubate the plates at 27°C for 3 days.
**21.** Collect the medium containing the virus from the wells, remove the cells by centrifugation at 1000 g for 5 min at 4°C, and store the virus at 4°C. Test the cells for protein expression by SDS-PAGE.

virus stock. At the end of the incubation the recombinant protein level in the wells is compared by SDS-polyacrylamide gel electrophoresis. This method is quicker and easier than the plaque assay.

For the large-scale expression of a non-secreted protein, cell monolayers in several tissue culture flasks are infected with virus and the cells containing the protein are harvested (Protocol 1.10). The more convenient way of producing large quantities of a recombinant protein is where a large volume of the cell culture is infected in a spinner flask and cells are harvested by centrifugation (Protocol 1.11). The secreted protein is usually expressed using the insect cell medium that contains either a low concentration (2%) of fetal calf serum or serum-free insect cell medium. The methods used for the purification of proteins expressed in the Baculovirus system are similar to those described in the bacterial expression system section.

### Protocol 1.9 Amplification of the recombinant Baculovirus virus stock

**Materials**

*Sf*9 cell culture
150-mm tissue culture plates
Recombinant Baculovirus low titre viral stock
Insect cell medium
Microcentrifuge tubes
Microscope

**Method**

**1.** Pipette $2.1 \times 10^7$ cells onto a plate and allow the cells to form a monolayer by placing the plate on a level surface for 40–45 min at room temperature or 27°C.
**2.** Add 100 μl of the recombinant virus stock, keeping the multiplicity of infection below one.
**3.** Incubate the plate at 27°C for 3 days.
**4.** Examine the plate for signs of infection after 2 days of incubation using a microscope.
**5.** Collect the virus supernatant and remove cell debris by centrifugation at 10,000 g for 5 min at 4°C.
**6.** Store at 4°C in a dark place or cover the tube with a piece of aluminium foil.
**7.** Determine the recombinant viral titre by plaque assay.
**8.** Amplify two or three times to obtain a high titre stock by repeating Steps 1 to 5.
**9.** Store the virus stock at 4°C in a dark place for up to 6 months or at –80°C for longer storage periods.

### Protocol 1.10 Expression of the recombinant protein in the Baculovirus system in monolayer cultures

**Materials**

*Sf*9 cell culture
150-mm tissue culture plates
High-titre recombinant viral stock
EX-CEL 405™serum-free medium containing 50 μg/ml gentamycin sulphate, 2.5 μg/ml amphotericin B, and 10% fetal calf serum
27°C incubator
Microscope
Haemocytometer
Trypan Blue Stain (0.2%) solution in PBS

**Method**

**1.** Pipette $2.0 \times 10^7$ cells onto each of the several plates and allow the cells to form a monolayer by placing the plate on a level surface.
**2.** Add fresh medium to a final volume of 30 ml in each plate without disturbing the cell monolayer.
**3.** Calculate the amount of virus stock required using the equation:

ml of virus required = multiplicity of infection (plaque forming units/cell) × number of cells/titre of virus per ml.

**4.** Add high-titre virus stock to the plates so that multiplicity of infection is between 3 and 10.
**5.** Transfer the plates to 27°C incubator and leave for 3 days.
**6.** Examine the plates for signs of infection using a light microscope.
**7.** Harvest the supernatant and cells from the plates and pellet the cells at 1000 g for 5–10 min at room temperature. Secreted proteins are found in the supernatant whereas non-secreted proteins remain in the pellet.
**8.** Store the pellets and supernatants at –80°C.
**9.** If the recombinant protein is in the cell pellet, resuspend the pellet in an appropriate lysis buffer containing inhibitors of proteases.
**10.** Purify the target protein using the same methods as described for purification of proteins expressed in the bacterial system.
**11.** If the recombinant protein is secreted it will be present in the supernatant. In that case, add an equal volume of an appropriate buffer, containing inhibitors of proteases, to the supernatant and proceed with the purification of the target protein as described for the bacterial system.

**Protocol 1.11 Expression of the recombinant protein in the Baculovirus system in suspension cultures**

**Materials**

*Sf*9 cell culture
EX-CEL 405™serum-free medium for insect cells containing gentamycin sulphate, amphotericin B, and fetal calf serum
60-mm tissue culture plates
Baculovirus high-titre virus stock
Microcentrifuge tubes
50 ml sterile centrifuge tubes
Light microscope
Spinner flask and spinner apparatus
Haemocytometer
27°C incubator
SDS-polyacrylamide gel electrophoresis system

**Method**

**1.** Pipette *Sf9* cell culture containing $2 \times 10^6$ cells/ml into a spinner flask.
**2.** Pipette 1 ml of the culture into a 60-mm tissue culture plate.
**3.** Place the plate on a level surface for 30–40 min.
**4.** Examine the plate to see if the cells are healthy.
**5.** Calculate the volume of recombinant virus required to infect at a multiplicity of infection of 3–10 by the equation:

$$\text{ml of virus required} = \text{multiplicity of infection} \times \text{number of cells/titre of recombinant virus per ml}$$

**6.** Add the required volume of the recombinant virus.
**7.** Incubate the spinner flask at 27°C spinning at 30–70 rpm for 1 h.
**8.** Equilibrate fresh medium at 27°C.
**9.** Transfer the spinner flask to the hood and adjust cell density to $1 \times 10^6$ cells/ml by adding the medium prepared in Step 8.
**10.** Incubate the spinner flask at 27°C, spinning at 30–70 rpm, for 2–4 days.
**11.** Follow the progress of the infection by removing aliquots of the culture and examining under a microscope.
**12.** Transfer $2 \times 1.5$ ml of the culture to microcentrifuge tubes and separate cells from the medium by centrifugation at 2500 g for 5 min for analysis by SDS-PAGE.
**13.** Transfer the rest of the culture to sterile centrifuge tubes and harvest the cells by centrifugation at 2500 g for 5 min at room temperature.
**14.** For non-secreted protein store the cell pellets at –80°C; for secreted protein, keep the supernatant in sterile tubes at –80°C.
**15.** Lyse the cell pellet from Step 12 in an appropriate cell lysis buffer and analyse the lysate and also the supernatant from Step 12 by SDS-PAGE for the presence of the recombinant protein.

### 1.2.5 Recombinant protein expression in mammalian systems

*1.2.5.1 Transient expression*

Expression by transient transfection methods using mammalian cell lines is a convenient and rapid method of producing recombinant proteins when *E. coli* systems fail to produce correctly folded, structurally homogeneous protein. Moreover, it is a method that is routinely used to produce proteins for crystallization.

Successful expression depends on several factors, such as efficiency of delivery of vector DNA to the host cell, transcriptional and translational control elements on the vector, mRNA stability, genetic properties of the host, chromosomal site of integration of the gene of interest, and potential toxicity of recombinant proteins to host cells. Delivery of vector DNA can be achieved by either infection of the host cell line with a virus containing the recombinant gene of interest or by direct transfection of vector DNA.

Introduction of a gene of interest into the host cell line by viral infection is a convenient method since a large number of cells can be infected simultaneously. Systems employing Semliki Forest Virus, Vaccinia Virus, and Retoviral vectors are used. However, drawbacks include the requirement for special precautions when engineering and preparing the viral stocks and limitation of expression for a short time due to cytopathic effects of viral infection.

In the case of recombinant protein production for structural biology applications and specifically for obtaining milligram quantities of protein for crystallization, direct transfection of vector DNA is a convenient option mainly due to advances in transfection methods and reagents. Vectors containing the gene of interest require the following essential features: (1) a bacterial origin of replication for vector propagation in *E. coli*; (2) a constitutive or inducible promoter; (3) mRNA cleavage and polyadenylation signal; (4) a transcription termination signal; (5) a Kozak sequence for optimal ribosome binding; (6) a translation termination signal; and (7) in the case of transient expression an SV40 (or other viral) origin of replication for maintenance of vector DNA in host cell. Additional features such as purification tags, secretion signals, fusion moieties, and protease cleavage sites can also be engineered.

Vectors containing the target gene in an expression cassette are engineered with the Kozak sequence and, if desired, an appropriate purification tag downstream of a powerful promoter. Strong viral promoters from cytomegalovirus (CMV) or SV40 are commonly used in most mammalian expression vectors. The elongation factor (EF)-1 promoter is a widely used non-viral promoter due to equivalent or even better expression levels compared to viral promoters.

Human embryonic kidney (HEK) 293, baby hamster kidney (BHK), and COS cells are commonly used in transient expression systems due to the high transfection efficiencies with these cell lines. Genetically modified HEK-293 and COS lines that express the SV40 large T antigen (HEK 293T, COS-1, -3, and -7) or the Epstein–Barr Virus nuclear antigen (HEK 293 EBNA) are particularly preferred. Plasmids carrying SV40 or EBV origins of replication are amplified and maintained by high extrachromosomal replication levels when used with these cell lines. Thus when used in combination with a powerful eukaryotic or viral promoter (e.g. SV40, CMV or EF-1) high transcription and translation levels of target genes can be achieved. Traditional chemical methods of introducing DNA into eukaryotic cell cultures are: (1) Calcium phosphate mediated transfection (Graham and van der Eb, 1973), where DNA is mixed with calcium phosphate to form a fine precipitate which is dispersed in the cultured cells; and (2) DEAE-Dextran mediated transfection (McCutchan and Pagano, 1968) where DNA is mixed with DEAE-Dextran and dispersed in the cultured cells. An alternative method is cationic lipid mediated transfection (Felgner *et al.*, 1987). Cationic headgroups of the lipid associate with the negatively charged phosphates on the DNA. The lipid-DNA complexes contact the cell membrane and fusion results in the internalization of DNA into the cell. This is a highly efficient, reproducible method of transfection with low toxicity that is ideal for large scale protein expression. There are a number of commercially available, lipid-based transfection reagents but a much cheaper alternative for large-scale transient transfections is to use polyethylenimine (PEI), an organic macromolecule which is low in toxicity and yields high transfection efficiencies.

Proteins can be expressed intracellularly or secreted into the growth medium. For example if one aims to express the extracellular domain of a cell surface receptor, the engineered gene sequence can be cloned into the expression vector with the native membrane targeting signal. Providing that the signal sequence is recognized in mammalian cells, the recombinant gene product will be secreted in into the medium. Likewise, N-terminal secretion signal sequences from other proteins can be engineered with gene or gene fragments of interest to secrete a protein or its subdomains.

After the uptake of foreign DNA, extrachromosomal replication peaks at around 48 h post transfection after which cells begin to shed the high number of plasmid copies. This is followed by cell death, probably due to the inability to endure the presence of excessive quantities of extrachromosomally replicating DNA (Gerard and Gluzman, 1985; Geisse, *et al.*, 1996). In the case of COS cells, recombinant protein expression peaks at around 72 h after transfection. However, in spite of cell deterioration and death, expression continues for a further 5–10 days (Edwards and Aruffo, 1993). Typically, with COS and HEK 293T cells protein expression can be allowed for 3–4 days post-transfection.

#### *1.2.5.2 Stable expression*

Stable lines are produced by cloning a homogeneous cell population from a heterogeneous cell pool and CHO cell lines are the most commonly used type for stable recombinant protein expression. Features of vectors used to produce stable cell lines are identical to those used in transient expression with an additional feature of having a drug resistance gene. Stable integration of foreign DNA in to the host cell genome is achieved by applying drug selection after transfection. The frequency of DNA integration is dependent on the cell line. A plasmid encoding a drug resistance marker can be cotransfected with the plasmid containing the gene of interest. Frequency of integration of foreign DNA by cotransformation depends on the cell line and efficiency of transfection. When cotransfection is inefficient it is probably best to have both the gene of interest and the drug selection marker on a single plasmid (Kaufmann, 1990).

When the gene of interest and the selection marker are on two separate plasmids they may integrate in chromosomal loci with differing transcriptional activity and, as a result, a large number of cell populations will need to be screened for expression. This can be avoided by having both the gene of interest and resistance marker on a single plasmid vector. However, there is no firm evidence that one method is better than the other.

Selection markers can be bacterial genes for which there is no mammalian equivalent (e.g. neomycin phosphotransferase, hygromycin phosphotransferase). But the most common are glutamine synthase (GS) and dihydrofolate reductase (DHFR) gene systems. GS Gene Expression Systems is licensed by Lonza (www.lonza.com). GS is an allosteric enzyme required for the production of glutamine from glutamate and this enzyme is inhibited by L-methionine sulphoximine (MSX), which is a transition state analogue of the reaction:

$$\text{Glutamate ammonia} + \text{ATP} \leftrightarrow \text{Glutamine} + \text{ADP} + \text{Pi}$$

CHO cells transfected with the GS minigene do not require glutamine in the growth media, provided that sufficient glutamate is present. However, GS is essential for cell survival when glutamate is absent from the media and full MSX inhibition in this situation is lethal. Therefore, MSX treatment encourages an increase in the copy number of the GS gene when cells are cultured in the absence of glutamine, so effectively coamplifying any associated genes.

DHFR is an enzyme in the pathway for *de novo* biosynthesis of purines and pyrimidines. In the absence of DHFR (i.e. in DHFR-CHO cells), purine and pyrimidine salvage pathways are activated. These salvage pathways can be inhibited by drugs such as methotrexate. Drug treatment encourages an increase in the copy number of the resistance gene, so effectively coamplifying any associated genes on the transfected vector. There is usually and broad variation of the level of expression of the gene of interest which is dependent on the site of integration within the host chromosome.

#### *1.2.5.3 Glycosylation*

Proteins expressed in mammalian cells undergo N-linked glycosylation which may cause problems during crystallization, particularly if there are multiple glycosylation sites. Heavy glycosylation obscures the protein surface and reduces the possibility of lattice formation mediated by the protein surface during crystallization. Microheterogeneity also prevents the formation of reproducible crystal contacts. There are different strategies to tackle the problem of glycosylation. N-linked glycosylation sites can be eliminated by site directed mutagenesis and this is a frequently used strategy. Also, glycoproteins can be treated with endoglycosidases such as endoglycosidase H (Endo H) and N-Glycosidase F (PNGase F) to cleave sugar chains or cocktails of exoglycosidases to shorten glycan chains. However, complete deglycosylation is sometimes difficult to achieve and sensitivity to deglycosylases can vary between proteins.

A number of mutant CHO cell lines have been obtained by mutagenesis and selection with lectins (Stanley, 1981). Lec3.2.8.1 (or LecR) is a mutant CHO cell line that produces truncated N-linked oligosaccharides of the endoglycosidase H (EndoH) sensitive Man5GlcNAc2 type (Stanley, 1989). Protein folding proceeds normally in the ER but *N*-glycans are not processed beyond the EndoH sensitive Man5GlcNAc2 intermediate in the Golgi (Stanley, 1981). This enables the production of correctly folded, endoglycosidase sensitive glycoproteins with defined glycosylation. The features of the CHO LecR cell line make it highly useful for preparing soluble glycoproteins that can be readily deglycosylated prior to crystallization (Davis *et al.*, 1993).

### 1.2.6 Cell-free systems

Cell-free *in vitro* expression systems are currently being developed at the Centre for Eukaryotic Structural Genomics at the University of Wisconsin. As these systems express only the protein of interest and require smaller volumes, lengthy concentration steps are avoided. The disadvantage is the expense of the reagents. Cell-free systems are available from QIAGEN (www.qiagen.com), Invtrogen (www.invitrogen.com), and CellFreeSciences.

## 1.3 Protein extraction and isolation

### 1.3.1 Cell disruption

The cells are harvested by low-speed centrifugation and resuspended in lysis buffer before disruption. Either chemical or mechanical methods may be used for disruption. The choice depends on the source of the protein (that is bacterial, yeast, insect, or mammalian, intracellular or extracellular) and the physicochemical properties of the recombinant product, as well as the scale of the extraction. For bacterial cells, enzymic digestion with hen egg white lysozyme, which specifically catalyses the hydrolysis of 1,4 glycosidic bonds in the peptideoglycan cell wall of Gram-positive bacteria, is a gentle procedure which minimizes denaturation of the product. For Gram-negative bacteria, for example *E. coli*, metal chelators such as EDTA are required to chelate cations that maintain the integrity of the outer lipopolysaccharides. Chemical disruption methods require a cocktail of anions, reducing agents, non-anionic detergents, and chaotropic agents in order to avoid irreversible denaturation of the product. Detergent-based lysis reagents are commercially available, including BugBuster™ (Merck) Fastbreak (Promega).

Mechanical disruption methods include sonication, high-speed homogenization using a French press, and bead milling, which is especially suitable for yeast cells which are difficult to break. Usually yeast cells are disrupted using a combination of physical and chemical methods. The BeadBeater™ (www.biospec.com) disrupts micro-organisms with better than 95% efficiency. Up to 80 g (wet weight) of cells can be processed in a typical 3-min run. Whereas chemical methods create contamination, the chief disadvantage of mechanical methods is the generation of heat and aerosols. For this reason all procedures should be carried out in an ice bath. Sonication should be carried out in batches of 100 ml in short bursts. Even if protease-deficient host strains such as $lon^-$ have been used it is still advisable to include protease inhibitors in the resuspension buffer. The serine protease inhibitors aprotinin and 1 mM phenylmethanesulphonyl fluoride (PMSF) are commonly used but for proteins particularly susceptible to proteolysis a cocktail of inhibitors to cope with each class of protease may be required.

### 1.3.2 The removal of cell debris and nucleic acids

Cell debris may be removed by centrifugation at 10,000 g for 30 min. The nucleic acids being the major contaminant can be removed by precipitation with a positively-charged polymer such as polyethyleneimine PEI (typically 0.5–1% of a 10% solution). Addition of magnesium to the resuspension buffer will assist in the enzyme digestion of DNA by DNAse. Some loss of protein may occur by copreciptation, which is especially the case with some DNA-binding proteins. This can usually be avoided by a 1:1 dilution of the crude extract with buffer.

### 1.3.3 Refolding strategies

High-level expression of full-length proteins in *E. coli* may result in the production of inclusion bodies. These are insoluble, inactive aggregates resulting from inappropriate folding and association via hydrophobic interactions. The proteins are functionally inactive in their aggregated state. The formation of inclusions can be advantageous for purification, provided the protein can be successfully solubilized and renatured into its active form. This involves isolation of inclusions and removal of unwanted coprecipitated proteins, followed by treatment with detergents and denaturants such as urea, and finally extensive dialysis in a suitable buffer containing refolding additives (salts, chaotropes, redox agents). For proteins containing disulphide bonds, redox systems need to be included in the renaturation buffer (Creighton, 1986). Although proved to be successful for relatively small proteins and polypeptides, it cannot be guaranteed that larger proteins will refold into their native conformations. Kits for optimizing the refolding conditions of inclusion body products are available from Takara Mirus Bio Madison Wisconsin (www.takaramirusbio.com). The iFOLD™ is a new system for determining the optimum conditions for recombinant protein refolding marketed by Novagen Merck Biosciences (www.novagen.com). The System provides a comprehensive set of conditions for target protein refolding based on the REFOLD database (http://refold.med.monash.edu.au) in a 96-well plate format amenable for high-throughput automation.

*1.3.3.1 Coproduction of the target protein with chaperones or foldases*

The formation of inclusions may be avoided in the first place by engineering the secretion of the product into the periplasm. Protein folding *in vivo* is facilitated by a group of molecular chaperones belonging to conserved families of proteins. These include the Hsp100 (ClpA, ClpB, ClpP), Hsp90 (HtpG), Hsp70 (DnaK), Hsp60 (GroEL), and α-crystalline-like small heat-shock proteins (IbpA, IbpB). Chaperones interact transiently with non-native protein substrate, GroEL and DnaK, together with their cochaperones (GroES for GroEL; DnaJ and GrpE for DnaK), maintain denatured proteins in non-aggregated states whilst assisting their refolding by a mechanism of recurrent ATPase-driven cycles of substrate binding and release.

If strong promoters such as T7 are used and the level of functional GroESL does not increase proportionally, correct folding may not occur. One effective way to increase solubility of foreign proteins in *E. coli* is by coproduction of the bacterial chaperones GroESL (Yasukawa *et al.*, 1995). Coproduction of GroESL with transcription factors and oncogene products resulted in soluble protein. A difference in the redox state between *E. coli* and eukaryotic cells may also affect protein solubility. This has been demonstrated by the fact that GST fusions produced in *E. coli* bind to the glutathione Sepharose beads with greater efficiency than similar GST fusions produced in mammalian cells. The same authors demonstrated that coproduction of bacterial thioredoxin (Trx) is more effective than the GroE system in producing soluble protein. Coexpression of one or more of the three different types of foldase (disulphide oxidoreductase (DsbA) and disulphide isomerase (DsbC); peptidyl prolyl isomerases (PPIs) and protein disulphide isomerase (PDI)) could lead to higher levels of soluble protein.

*1.3.3.2 Refolding chromatography with minichaperones*

Peptides consisting of residues from GroEL immobilized on agarose have proved effective minichaperones (Altamirano *et al.*, 1997). The procedure used both column chromatography and batch-wise methods to renature an insoluble protein from an inclusion body, refold apparently irreversibly denatured proteins, and to recondition enzymes that have lost activity on storage. Fragments were immobilized by two methods: Ni-NTA resin and CNBr-activated Sepharose 4B.

### 1.3.4 Purification

The popular use of fusions and/or generic tags obviates the need for extensive, time-consuming, multistep purification except where their use is undesirable due to the loss of structural information through tag interference, loss of solubility after cleavage, or simply prohibitive cost of proteases. In an optimally-designed purification scheme it should be possible to achieve a high-level of purity in fewer than four key stages without compromising percentage yield. To do this, the physicochemical properties of the target protein should be well defined and a rapid, reliable assay developed to monitor the progress of the purification. If the properties are unknown then a standard protocol of ion exchange (IEX), hydrophobic interaction (HIC), and gel filtration (GF) is followed. The three essential phases of any purification are: capture, followed by intermediate purification to remove the bulk of impurities (nucleic acids, proteins, and endotoxins), and a final polishing step. The programmable AKTA pilot system (GE Healthcare) is ideal for method development and small and medium scale preps. AKTA Xpress is a dedicated, high-throughput system.

#### 1.3.4.1 *Primary isolation and concentration*

The initial stage is the capture, stabilization, and subsequent enrichment of soluble product. If the pI is known, ion exchange (IEX) on a rigid matrix can be used for concentration and as a preliminary clean-up. Supernatant (see Section 1.7.2) can be applied directly to Sepharose Fast Flow™. STREAMLINE™ (www. gehealthcare.com) expanded bed adsorption is particularly suitable for secreted proteins in large volumes of crude supernatant as no preliminary clean-up is required. At this stage buffers should be relatively inexpensive and any additives required for maintaining stability should be easy to remove at a later stage if necessary. The volume of supernatant can be reduced either by ultrafiltration or selective precipitation with salts, organic solvents, or long-chain polymers, for example polyethylene glycol (PEG). Ammonium sulphate is commonly used because of its high solubility and stabilizing properties. Ethanol and acetone have proved successful in the fractionation of extracellular proteins such as plasma proteins, polypeptide hormones, and in the extraction of histones from non-recombinant sources.

#### 1.3.4.2 *Chromatography*

To avoid loss of active material the number of chromatographic steps must be minimized. This is best achieved by combining chromatographic steps in a logical sequence to maximum effect. Buffer exchange, desalting, dialysis, and ultrafiltration should be avoided where possible between chromatographic stages. Size-exclusion (GF), which desalts and dilutes the sample, should follow a concentration step such as ion-exchange (IEX), and ion-exchange would not be appropriate after ammonium sulphate fractionation because of the extensive desalting required to allow the protein to bind. Instead, hydrophobic interaction chromatography (HIC), which binds proteins preferentially at high ionic strengths, could be substituted. The choice of media is governed by scale of production and resolution. Functional properties have been combined with matrices of various strengths and porosities to optimize flow rates and selectivity. For intermediate purification Sepharose Fast Flow™ is routinely used for general and large-scale separations and Sepharose High Performance media is preferred for high-resolution applications. The small-scale High Trap columns are useful for method development. The Tricorn High Performance columns are a new generation of high-resolution of columns which come prepacked with MonoQ, MonoS, Superdex 200, and Superdex 75 from GE Healthcare. At the final polishing phase there is frequently a trade off between recovery and resolution, as peak-cutting may be necessary.

Affinity chrmomatography exploits the biological properties of the molecule by reversible absorption to an immobilized ligand coupled to an insoluble support. A wide variety of media is available for affinity applications (Table 1.3). For immunoaffinity, antibodies raised against the target protein are coupled to an activated adsorbant, for example cyanogen bromide activated Sepharose™. The high binding capacities and specificities require harsh conditions for elution, often requiring denaturing conditions.

### 1.3.5 Product analysis

For crystallization at least 95% purity is desirable. SDS PAGE stained with Coomassie™ Brilliant Blue R-250 provides a crude but reasonable primary indicator of purity and expression. Most minor contaminants (<5%) can be detected by silver staining. For enzyme isomers and proteins which differ in charge but not size, analytical isoelectric focusing and/or two-dimensional PAGE is necessary. All these electrophoretic methods can be carried out using the Phast System™ (GE Healthcare) or equivalent. Proteins which possess an optical chromophore can also be assessed using the ratio of the absorption peak in visible spectrum to the absorption at 280 nm. The Protein 200-HT2 assay of Agilent Technologies (Palo Alto California) identifies, determines size, and quantitates proteins from 14 kD to 200 kD. Microheterogeneity caused by chemical modification, partial denaturation, or incomplete

**Table 1.3** Affinity chromatography applications

| Ligands | Target molecules | Examples |
|---|---|---|
| Lectins | α-D-glucopyranosyl | Adenosine deaminase |
| Concanavalin A | α-D-mannopyranosyl | γ-Glutamyl transferase |
| Heparin | Nucleic acid binding proteins | DNA helicases<br>Restriction endonucleases<br>DNA/RNA polymerase |
| | Growth factors | Endothelial cell GF |
| Ni-NTA | $His_6$-tag | N- or C-terminus (His)6 |
| Calmodulin | Calmodulin-dependent enzymes | ATPase |
| ssDNA | DNA-dependent enzymes | DNA-polymerases alkyl transferase |
| Triazine dyes:<br>Cibacron blue | $NAD^+$, $NADP^+$-dependent enzymes | Dehydrogenases<br>Quinone reductases |
| 5′-AMP | ATP-dependent enzymes | cAMP-dependent protein kinase |
| Protein A | Immunoglobulins | Fc region of IgGs |
| MAB | Antigens | Protein A fusions |
| Glutathione | Glutathione S transferases | Glutathione S transferase fusions |
| Amylose | Maltose binding protein | MBP fusions |

Modified from Skelly and Madden (1996). In: *Crystallographic Methods and Protocols*, Jones, C., Mulloy, B. and Sanderson, M. R., eds. Humana Press.

post-translational modification can be detected by electrospray mass spectrometry (ESI-MS) which is accurate to ~1 Da (Cohen and Chait, 2001). Matrix-assisted laser desorption ionization (MALDI) is a less sensitive technique but, unlike ESI-MS, it can be used in the presence of buffers and detergents. Capillary electrophoresis (CE) is another powerful tool for the detection of chemical modifications in proteins, and is capable of separating and quantifying a single-charge difference. Dynamic light scattering (DLS) can provide evidence of aggregation and the oligomeric state of a protein, which can be a helpful indication of its potential to crystallize.

## References


Altamirano, M. M., Golbik R., Zahn, R., Buckle, A. M., and Fersht, A. (1997). Refolding chromatography with immobilised mini-chaperones *Proc. Nat. Acad. Sci. USA* **94**, 3576–3578.

Amann, E., Brosius, J., and Ptashne, M. (1983). Vectors bearing a hybrid trp-lac promoter useful for regulated expression of cloned genes in *Escherichia coli*. *Gene* **25**, 167–178.

Appelbaum, E. and Shatzman, A. R. (1999). Prokaryotes *in vivo* expression systems. In: *Protein Expression Practical Approach*, Higgins, S. J. and Hames, B. D., eds. Oxford University Press.

Ayres, M. D., Howard, S. C., Kuzio, J., Lopez-Ferber, M., and Possee, R. D. (1994). The complete DNA sequence of Autographa californica nuclear polyhedrosis virus. *Virology* **202**, 586–605.

Cohen, S. L. and Chait, B. T. (2001). Mass spectrometry as a tool for protein crystallography. *Annu. Rev. Biophys. Biomol. Struct.* **30**, 67–85.

Cregg, J. M., Vedvick, T. S., and Raschke, W. C. (1993). Recent advances in the expression of foreign genes in *Pichia pastoris*. *Bio/Technology* **11**, 905–910.

Creighton, T. E. (1986). Disulphide bonds as probes of protein folding pathways. *Method Enzymol.* **131**, 83–106.

Davis, S. J., Puklavec, M. J., Ashford, D. A., Harlos, K., Jones, E. Y., Stuart, D. I., and Williams, A. F. (1993). Expression of soluble recombinant glycoproteins with predefined glycosylation: application to the crystallization of the T-cell glycoprotein CD2. *Protein Eng.* **6**, 229–232.

Edwards, C. P. and Aruffo, A. (1993). Current applications of COS cell based transient expression systems. *Curr. Opin. Biotechnol* **4**, 558–563.

Felgner, P. L., Gadek, T. R., Holm, M., Roman, R., Chan, H. W., Wenz, M., Northrop, J. P., Ringold, G. M., and Danielsen, M. (1987). Lipofection: a highly efficient, lipid-mediated DNA-transfection procedure. *Proc. Nat. Acad. Sci. USA* **84**, 7413–7417.

Geisse, S., Gram, H., Kleuser, B., and Kocher, H. P. (1996). Eukaryotic expression systems: a comparison. *Protein Expr. Purif.* **8**, 271–282.

Gerard, R. D. and Gluzman, Y. (1985). New host cell system for regulated simian virus 40 DNA replication. *Mol. Cell Biol.* **5**, 3231–3240.

Graham, F. L. and van der Eb, A. J. (1973). A new technique for the assay of infectivity of human adenovirus 5 DNA. *Virology* **52**, 456–467.

Jacob, F. and Monod, J. (1961). Genetic regulatory mechanisms in the synthesis of proteins. *J. Mol. Biol.* **3**, 318–356.

Kane, J. F. (1995). Effects of rare codon clusters on high-level expression of heterologous proteins in *Escherichia coli*. *Curr. Opin. Biotechnol.* **6**, 494–500.

Kaufmann, R. J. (1990). Selection and coamplification of heterologous genes in mammalian cells. *Method Enzymol.* **185,** 537–566.

Kitts, P. A. and Possee, R. D. (1993). A method for producing recombinant baculovirus expression vectors at high frequency. *Biotechniques* **14,** 810–817.

Kondo, K., Miura, Y., Sone, H., Kobayashi, K., and Iijima, H (1997). High-level expression of a sweet protein, monellin, in the food yeast *Candida utilis. Nature Biotechnol.* **15,** 453.

McCutchan, J. H. and Pagano, J. S. (1968). Enchancement of the infectivity of simian virus 40 deoxyribonucleic acid with diethylaminoethyl-dextran. *J. Natl. Cancer Inst.* **41,** 351–357.

Miller, L. K. (1988). Baculoviruses as gene expression vectors. *Ann. Rev. Microbiol.* **42,** 177–199.

Miroux, B. and Walker, J. E. (1996). Over-production of proteins in *Escherichia coli*: mutant hosts that allow synthesis of some membrane proteins and globular proteins at high levels. *J. Mol. Biol.* **260,** 289–298.

Nilsson, B., Holmgren, E., Josephson, S., Gatenbeck, S., Philipson, L., and Uhlen, M. (1985). Effecienct secretion and purification of human-like growth factor I with a gene fusion vector in staphylocci. *Nucleic Acids Res.* **13,** 1151–1162.

Ogden, S., Haggety, D., Stoner, C. M., Kolodrubetz, D., and Schleif, R. (1980). The *Escherichia coli* L-arabinose operon: binding sites of the regulatory proteins and a mechanism of positive and negative regulation. *Proc. Nat. Acad. Sci. USA* **77,** 3346–3350.

O'Reilly, D. R., Miller, L. K., and Luckow, V. A. (1992). *Baculovirus Expression Vectors: A Laboratory Manual.* W. H. Freeman and Co., New York.

Possee, R. D. and Howard, S. C. (1987). Analysis of the polyhedrin gene promoter of the Autographa californica nuclear polyhedrosis virus. *Nucleic Acids Res.* **15,** 10233–10248.

Skelly, J. V. and Madden, C. B. (1996). Overexpression, isolation and crystallisation of proteins. In: *Crystallographic Methods and Protocols*, Jones, C., Mulloy, B. and Sanderson, M. R., eds. Humana Press, New Jersey, USA, p. 23.

Smith, D. B. and Johnson, K. S. (1988). Single step purification of polypeptides expressed in *Escherichia coli* as fusions with gluthatione S-transferase. *Gene* **67,** 31–40.

Smith, G. E., Summers, M. D., and Fraser, M. J. (1983). Production of human beta interferon in insect cells infected with a baculovirus expression vector. *Mol. Cell Biol.* **3,** 2156–2165.

Stadier, J. A. and Silhavy, T. J. (1990). Engineering *Escherichia coli* to secrete heterologous gene products. *Method Enzymol.* **185,** 166–187.

Stammers, D. K., Tisdale, M., Court, S., Parmar, V., Bradley, C., and Ross, C. K. (1991). Rapid purification and characterisation of HIV-1 reverse transcriptase and RNase H engineered to incorporate a C-terminal tripeptide alpha-tubilin epitope. *FEBBS Lett.* **283,** 298–302.

Stanley, P. (1981). Selection of specific wheat germ agglutinin-resistant (WgaR) phenotypes from Chinese hamster ovary cell populations containing numerous lecR genotypes. *Mol. Cell Biol.* **1,** 687–696.

Stanley, P. (1989). Chinese hamster ovary cell mutants with multiple glycosylation defects for production of glycoproteins with minimal carbohydrate heterogeneity. *Mol. Cell Biol* **9,** 377–383.

Studer, F. W. and Moffatt, B. A. (1986). Use of bacteriophage 7 RNA polymerase to direct selective high level expression of cloned genes. *J. Mol. Biol.* **189,** 113–130.

Terpe, K. (2003). An overview of tag fusions: from molecular and biochemical fundamentals to commercial systems. *App. Microbiol. Biotechnol.* **60,** 523–533

Vlak, J. M., Klinkenberg, F. A., Zaal, K. J., Usmany, M., Klinge-Roode, E. C., Geervliet, J. B., Roosien, J., and van Lent, J. W. (1988). Functional studies on the p10 gene of Autographa californica nuclear polyhedrosis virus using a recombinant expressing a p10-beta-galactosidase fusion gene. *J. Gen. Virol.* **69,** 765–776.

Williams, G. V., Rohel, D. Z., Kuzio, J., and Faulkner, P. (1989). A cytopathological investigation of Autographa californica nuclear polyhedrosis virus p10 gene function using insertion/deletion mutants. *J. Gen. Virol.* **70,** 187–202.

Yasukawa, T., Kanei-Ishii, C., Maekawa, T., Jiro Fujimoto, J., Yamamoto, T., and Ishii, S. (1995). Increase of solubility of foreign proteins in Escherichia coli by coproduction of the bacterial thioredoxin. *Biol. Chem.* **270,** 25328–25331.

### Manual

*Baculovirus Expression Vector System Manual*, 6th edition May 1999 (www. bdbiosciences.com).

CHAPTER 2

# High-throughput cloning, expression, and purification

Raymond J. Owens, Joanne E. Nettleship, Nick S. Berrow, Sarah Sainsbury, A. Radu Aricescu, David I. Stuart, and David K. Stammers

## 2.1 Introduction

High-throughput sequencing of eukaryotic, viral, and bacterial genomes is providing a huge database of proteins with potential for structure–function analysis. In response to this opportunity, structural genomics projects have been initiated world-wide with the aim of establishing high-throughput structure determination on a genome-wide scale. Crucial to this effort has been the development of protein production technologies for the high-throughput cloning, expression, and purification of proteins. Large-scale structural genomic projects were initiated in the US by the National Institute of Health (NIH) and in Japan by the Riken Laboratory from 1998 to 2000. European projects followed, including the Protein Structure Factory in Berlin (www.proteinstrukturfabrik.de), Oxford Protein Production Facility (OPPF) (www.oppf.ox.ac.uk), and the EU-sponsored Structural Proteomics In Europe (SPINE: www.spineurope.org) programme. The scale of these projects has been smaller than the US/Japan initiatives, with a focus at the outset on human and viral targets. For all projects, there has been an emphasis on parallel processing, both in terms of molecular cloning, expression, and purification, driven by the need to accommodate relatively large numbers of potential targets for structural biology at an acceptable cost. This has led to varying degrees of automation and most of the groups involved have set up semiautomated liquid handling systems to carry out some or all of their protocols. However, the protocols can equally well be carried out manually with appropriate equipment, for example multichannel pipette dispensers. The motivation to implement automation is largely to enable processes to be scaleable and sustainable as error-free operations. In this article we review the technical developments that have come from structural proteomics and provide protocols for carrying out cloning, expression, and purification procedures in a relatively high-throughput (HTP) and parallel approach.

## 2.2 Cloning

Two options are available for constructing the expression vectors required for protein production, namely ligation-dependent and ligation-independent cloning (LIC). The former makes use of standard restriction enzyme digestion in combination with DNA ligation to produce the vectors. Whereas the latter utilizes either some form of recombination event or the production and annealing of single-stranded overhangs, both of which avoid the need to restriction digest the input DNA. Typically, in both cases the starting DNA is a PCR product corresponding to the whole or part of an open reading frame (ORF) produced from either a genomic or cDNA template. The PCR primers incorporate either restriction enzyme recognition sites or the sequences required for LIC reactivity. By using rare cutting restriction enzyme sites, ligation based cloning has been used effectively for semi-automated high-throughput cloning (Lesley *et al.*, 2002). However, most projects have adopted LIC for the obvious reason that it is independent of the input sequence. The three methods that are commercially available are described below. These methods are generally carried out in 96-well format and are therefore amenable to laboratory automation using standard liquid handling systems.

### 2.2.1 Ligation-independent cloning methods

#### *2.2.1.1 LIC-PCR*

Ligation-independent cloning of PCR products (LIC-PCR) was developed over 10 years ago (Aslandis, 1990; Haun *et al.* 1992). It is based on the use of T4 DNA polymerase in the presence of a single deoxyribonucleotide to produce 12–15 bp overhangs in a PCR product that are complementary to sequences generated in the recipient vector (Fig. 2.1a). These extensions anneal sufficiently strongly to allow transformation of *E. coli* without the need to ligate the fragments, which is carried out by repair enzymes in the host. Advantages of the LIC-PCR system are that it does not require specialized vectors and the reagents are relatively inexpensive. However, the system does require the preparation of a high-quality, linearized vector, which will require batch checking to ensure high efficiency of cloning. A limitation of the LIC-PCR is that one of four bases has to be preselected as the 'lock' in the compatible overhangs and hence the base pair composition of the annealing regions is limited to using the other three bases. Consequently, the method is not entirely sequence independent and cannot be used to join any sequence to any other sequence. However, by appropriate vector design LIC-PCR has been successfully implemented in HTP mode (Stols *et al.*, 2002). A protocol for carrying out LIC-PCR using a commercially available vector system (www.novagen.com) is described in Protocol 2.1.

#### *2.2.1.2 Gateway™*

Gateway™ cloning technology is a modification of the recombination system of phage λ (Walhout *et al.*, 2000; Hartley *et al.*, 2000). The Gateway™ system utilizes a minimum set of components of the λ system for *in vitro* transfer of DNA, namely the λ integrase protein, λ excisionase, the *E. coli* protein integration host factor (IHF) and the *att* recombination sequences attached to the DNA to be cloned. Directional cloning of the DNA insert is ensured by using two nearly identical but non-compatible versions of the λ *att* recombination site (Fig. 2.1b). Expression vectors are usually constructed in two stages. In the first step an entry (a.k.a. master or capture) clone is generated using recombination between *attB* and *attP* sites in the input DNA, usually a PCR product, and the donor vector respectively (BP reaction). The inserted DNA can then be transferred to one or more destination vectors to generate expression clones (LR reaction). The ability to generate rapidly and with high efficiency multiple expression vectors with different formats (e.g. fusion tags) from the same starting vector is a unique property of the system. To select for the desired recombinants and against parental plasmids, in both BP and LR steps, the Gateway™ system uses the *E. coli* lethal gene *ccdB* in combination with differential antibiotic-resistance markers on the entry and destination plasmids. The Gateway™ method is shown schematically in Fig. 2.1b and detailed protocols are available from the manufacturer (www.invitrogen.com).

The use of this method of ligation-independent cloning has been reported by several large-scale cloning projects (Luan *et al.*, 2004; Abergel, 2003; Vincentelli *et al.*, 2003). In general, it appears that the BP reaction is largely insensitive to the concentration of input PCR product and for ORFs <2 kb, cloning efficiency yields of nearly 90% can be obtained (Marsischky and LaBaer, 2004). For larger inserts (2–3 kb) a 50% drop in yield has been reported (Marsischky and LaBaer, 2004). Ease of use comes at a price since the 28–31 bp *att* sequences add to the cost of the primers and the recombination enzymes – BP clonase (λ integrase + *E. coli* IHF) and the LR clonase (λ excisionase + λ integrase + *E. coli* IHF) – are relatively expensive compared to standard DNA-modifying enzymes. Consequently, we and others (Braun *et al.*, 2002) have modified the standard protocol by halving the recommended volume of reagents for both BP and LR steps, hence reducing the final reaction volume to 10 μl without loss in performance. In using the Gateway™ system, it is important to be aware of the effect the *att* recombination sequences may have on expression and/or solubility of the cloned DNA product if they form part of the translated sequence. This can be avoided by positioning the *att* sites outside of the ORF, but some of the flexibility of the system is lost since only a single fusion format is possible.

### 2.2.1.3 In-Fusion™

The In-Fusion™ method is both insert sequence-independent and enables cloning of PCR products directly into any cloning or expression vector (Fig. 2.1c). The mechanism of the reaction has not been fully reported but relies on the presence of homology between extensions on the PCR product and the ends of a linearized vector; the optimal length for these homologous sequences is around 15 base pairs. Once these homologous extensions have been incorporated into the PCR product no further processing of the insert is required prior to the In-Fusion™ reaction, in contrast to PCR-LIC methods (Protocol 2.2). The main advantage of the In-Fusion™ method is that the user can define the exact sequence of these primer extensions without the limitations on base and codon usage inherent in the T4 polymerase-based LIC system. With minor vector modifications (e.g. insertion in the cloning

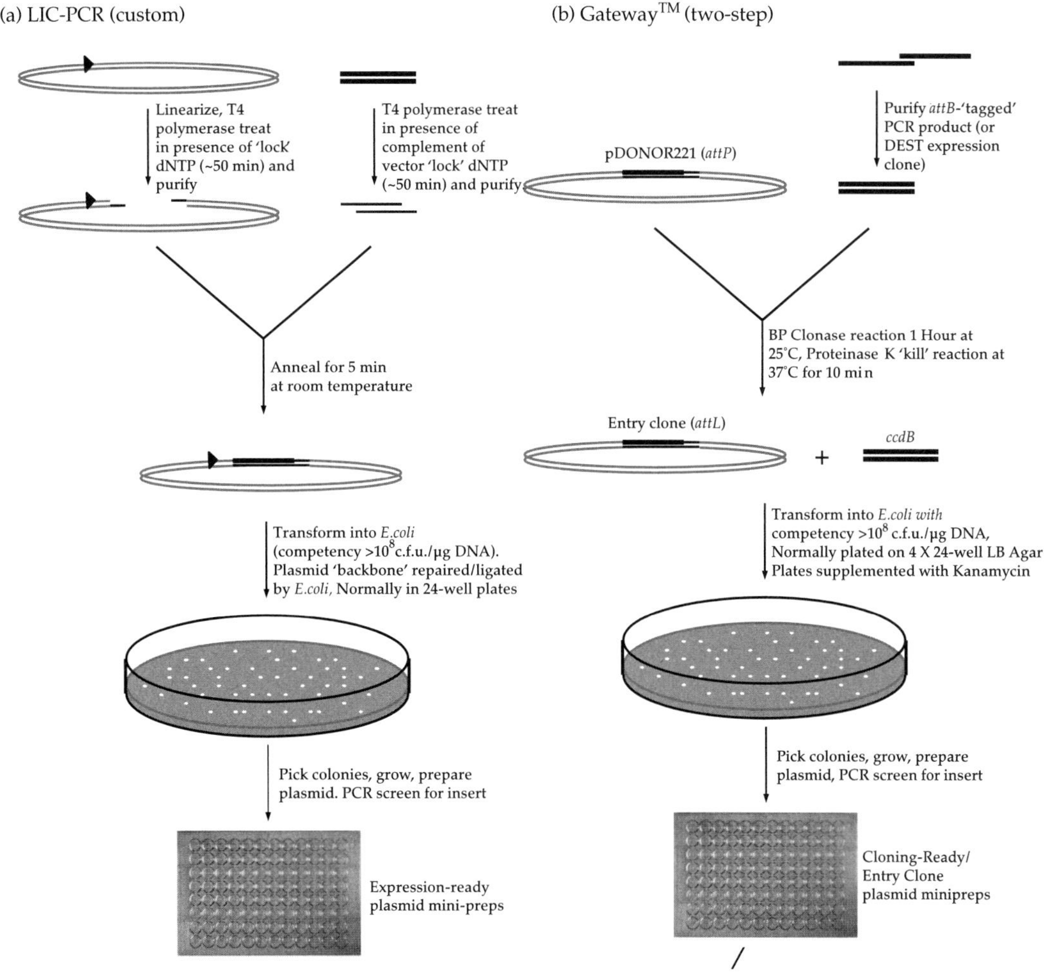


**Figure 2.1** Continued

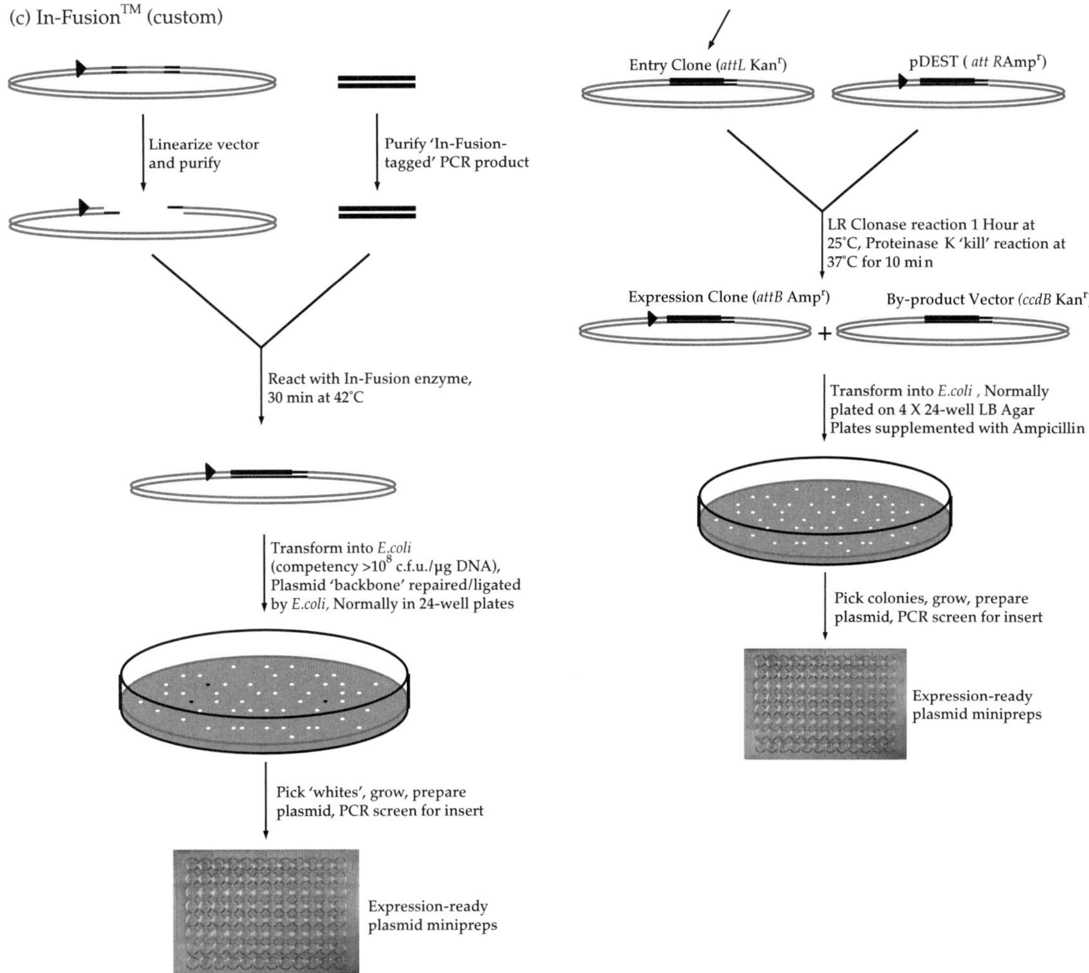


**Figure 2.1** Schematic representation of different ligation-independent cloning strategies. Grey plasmid sections represent the plasmid backbone with the origin of replication etc., hatched lines represent the gene of interest, double-hatched lines represent the lethal *ccd*B gene, black arrow-heads represent the transcription promoter and solid black lines represent the 'cloning sites' (*att* sites, In-Fusion™ sites or LIC sites for Gateway™, In-Fusion™ and LIC-PCR respectively). NB in all cases cloning is directional, i.e. 'left' and 'right' cloning sites sequences are non-identical.

site of the lacZ α-peptide for blue/white screening or a lethal gene cassette) this method becomes eminently amenable to HTP manipulations (Berrow *et al.*, 2007). In addition, In-Fusion™ enables the user to define exactly the resultant (fusion) protein sequence, using fully host-optimized codons, without incorporating undesirable vector-derived amino acids. However, the system does require the preparation of a high-quality, linearized vector which will require batch checking to ensure high efficiency of cloning.

### 2.2.2 Choice of vectors

*2.2.2.1 Promoters*

The choice of vectors obviously depends upon the expression system that will be used and there is a wide availability of optimized reagents from both commercial and academic sources. For expression in *E. coli,* the most widely used system for HTP protein production is the pET/T7 promoter vector developed by Studier which makes use of T7 RNA polymerase to direct expression of cloned

## Protocol 2.1 LIC-PCR

**1.** Amplify the desired insert sequence using appropriately designed PCR primers. The 5′-end of the primers must incorporate the following sequences:

sense primer: 5′ GAC GAC GAC AAG ATX – insert specific sequence 3′
antisense primer: 5′ GAG GAG AAG CCC GGT* – insert specific sequence 3′

**2.** Gel purify the PCR product and resuspend in TlowE buffer (10 mM Tris-HCl, 0.1 mM EDTA, pH 8.0). Treat the PCR product with T4 polymerase in the presence of dATP, the reaction may be performed in either a sterile PCR plate or microcentrifuge tubes (this generates single-stranded 5′ overhangs at both ends of the PCR product as far as the first 'T' in the primer extensions using the examples above). Incubate at 22°C for 30 min (higher temperatures are unsuitable for this reaction).
**3.** Inactivate the T4 polymerase enzyme by incubating at 75°C for 20 min and store the prepared Ek/LIC insert at −20°C if not used immediately. This prepared insert can be annealed to any of the Ek/LIC vectors.
**4.** Anneal approximately 100 ng of each insert with 50 ng of vector** at 22°C for 5 min. Add EDTA to a final concentration of approximately 6 mM, incubate at 22°C for 5 min.
**5.** Transform competent *E. coli* (with a competency greater than $10^8$ c.f.u./μg DNA) with 1 μl of the annealing reaction. Select for recombinants by plating on LB agar plates supplemented with the antibiotic appropriate for the plasmid used.
**6.** Pick colonies and prepare expression-ready plasmids.
**7.** PCR screen colonies or plasmid minipreps as usual with a vector-specific (e.g. T7) forward primer and your gene-specific reverse primer.

### Notes

The primer sequences described here are specific to the Novagen Ek/LIC system and its related suite of vectors, they cannot be used with other, unrelated vectors in the LIC protocols.
X- The first nucleotide of the insert-specific sequence must complete the codon ATX.
*If C-terminal tag sequences are desired, additional bases may be required in the antisense primer to ensure the C-terminal sequences are in frame. If a C-terminal tag is not desired a stop codon could be included in the insert-specific sequence
**Assuming that this is a commercially prelinearized and T4 polymerase-treated vector, if not, the vector must be linearized, gel purified, and T4 polymerase treated in the presence of dTTP (this generates single-stranded 5′ overhangs at both ends of the PCR product as far as the first 'A' in the ends of the linear vector that will complement the overhangs generated on the T4 polymerase-treated PCR products).

genes (Studier *et al.*, 1990). Regulated expression is achieved by using strains in which expression of a chromosomal copy of T7 polymerase is under the control of the *lacUV5* promoter and hence inducible by the addition of IPTG (DE3 strains). In addition, by incorporating the *lac* operator sequence just downstream of the start of the T7 promoter, repression of this T7/lac promoter is achieved by expression of the *lac* repressor (either in *cis* or *trans*). Further levels of control can be obtained by coexpressing T7 lysozyme (pLysS) at a low level in the expression strain. T7 lysozyme inactivates any T7 polymerase produced under non-inducing conditions, ensuring tight control of the T7 promoter vector.

For other production hosts (yeast, insect, and mammalian cells), standard promoter formats have been used in combination with HTP cloning methods to produce vectors for expression screening (see Section 2.3.2). A particularly interesting development is the use of multipromoter plasmids for expression in two or more hosts from a single vector. The construction of a dual *E.coli* (T7 promoter) and baculovirus transfer vector (*polH* promoter) for expression in insect cells has been described (Chambers *et al.*, 2004). A three-promoter vector (T7, p10, and hCMV or CAG promoter) is available from Novagen (pTriEX™) and its use reported for comparing protein expression in *E. coli* and insect cells (Xu and Jones, 2004).

**Protocol 2.2 In-Fusion™**

**1.** PCR products with approximately 15 bp extensions at each end that are homologous to the ends of the linearized vector should be gel purified for best results. To obtain the best results with lowest non-recombinant 'background' the linearized vectors should also be gel purified. Take 10–100 ng of insert and 100 ng vector in a total volume of 10 μl of either 10 mM Tris, pH 8.0 or sterile $H_2O$.
**2.** Add this to a well of the dry-down In-Fusion™ plate. Mix contents briefly by pipetting up and down.
**3.** React for 30 min at 42°C.
**4.** Dilute IMMEDIATELY with 40 μl TE Buffer (10 mM Tris, pH 8.0, 1 mM EDTA) and either transform IMMEDIATELY or freeze the reaction until you are ready to transform. *E. coli* with a transformation efficiency in excess of $1 \times 10^8$ are recommended and if the vector has been modified for blue–white screening ensure that an appropriate *E. coli* host strain is used; 5 μl of the diluted reaction should give tens to hundreds of colonies per well of a 24-well plate.
**5.** Plate on 24-well format LB agar supplemented with antibiotic and, if appropriate, X-Gal and IPTG (dilute a 20% X-Gal, in dimethyl formamide, stock 1:1000, dilute the IPTG 500 mM stock 1:500 in warm agar before pouring). Plate 10 μl of cells, shake plates well to spread the cell suspension, and allow at least 10–15 min for the plates to dry off before inverting.
**6.** Incubate overnight at 37°C.
**7.** Pick colonies for plasmid miniprepping as usual (if blue–white screening is used then blue colonies should constitute $\ll$10% if the reactions were successful. (The blue colonies are derived from inefficiently linearized parental plasmid and should not be picked as they are non-'recombinant'). Two colonies are normally sufficient to find a recombinant clone but more may be required.
**8.** PCR screen colonies or plasmid minipreps as usual with a vector-specific (e.g. T7) forward primer and your gene-specific reverse primer.

*2.2.2.2 Tags*

Several fusion protein vectors have been developed for recombinant protein expression since it is recognized that fusion vectors can enhance productivity and/or solubility of target proteins compared to non-fusion versions. Typically, the fusion partner resident in the expression vector is joined to the N-terminus of the protein of interest via a linker region containing a protease cleavage site to enable subsequent removal of the fusion protein. The most commonly used proteases are Tev (Parks *et al.*, 1994) and 3C (Cordingley *et al.*, 1989) since both have highly specific linear recognition sequences that are very rarely encountered in other sequences, thereby minimizing the risk of cleavage within the target protein (see Section 2.4 for experimental details). Some of the most commonly used tags also provide affinity purification strategies, for example glutathione S-transferase (GST) binds to immobilized glutathione and maltose binding protein binds to amylose matrices enabling selection of the fusion protein (Smith and Johnson, 1988; Alexandrov *et al.*, 2001). As an alternative to expressing proteins as relatively large hybrids, short N- or C-terminal hexahistidine tags are used to facilitate purification by metal chelate chromatography (see Section 2.4). The use of a short fusion tag, typically seven to eight amino acids, does not necessitate introduction of a cleavage site to remove the tag. However, His-tagged vectors with cleavage sites are available. Combining fusion proteins with N-His tags offers the potential benefits of improved expression levels and solubility of the fusion partner with a generic purification strategy.

There is no clear consensus as to which fusion partners give the best performance in terms of enhancing expression and solubility, though fusion proteins generally perform better than short His tags (Braun *et al.*, 2002; Dyson *et al.*, 2004; Hammarstrom *et al.*, 2002). However, cleaving off the fusion partner can lead to precipitation of the target protein. Due to these uncertainties, most structural proteomics projects have opted to work exclusively with His-tagged proteins and invest effort into selecting the targets in terms of domain definition (Folkers *et al.*, 2004). Recently, fusion vectors which combine many of the above attributes have been developed using ubiquitin-like proteins (Ubls) as the fusion partner (Baker, 1996). Ubls (e.g. SUMO) are small (approximately 100 amino acid), eukaryotic proteins that are known to exert chaperone-like effects on fused proteins in *E. coli* and yeast (Butt *et al.*, 1989). Purification of the fusion protein is enabled by adding an N-His tag to the Ubl (Catanzariti *et al.*, 2004; Malakhov *et al.*, 2004). The Ubl can be removed from the target protein *in vitro* using specific deubiquitylating enzymes now available from recombinant sources. The cleavage occurs after the final glycine residue at the carboxy-terminus of the Ubl protein irrespective of the amino acid that immediately follows, thus regenerating the native N-termini on removal of the fusion partner.

## 2.3 Expression screening

A major feature of HTP protein production pipelines is the inclusion of a screening step at a relatively small scale to identify constructs suitable in terms of soluble protein yield for scale-up and subsequent protein purification. Given the relatively high cost of the latter steps in terms of time and resources, particularly if eukaryotic expression is required, the screening stage is seen as crucial to the overall process. In this section the approaches to the evaluation of expression at small scale will be reviewed.

### 2.3.1 *E. coli*

#### *2.3.1.1 Choice of strains and media*

For the most part, B strains of *E. coli* are preferred for recombinant protein expression since they are deficient in the ATP-dependent *lon* and membrane-associated *ompT* proteases that may otherwise lead to degradation of the heterologously expressed protein. A variety of derivatives of the original BL21 strain are available, including those carrying additional plasmids, which either provide fine control of T7 vectors (pLysS see Section 2.2.2.1) or supply tRNAs for codons that are relatively rare in *E. coli* but may be commonly used in eukaryotic proteins. The presence of rare codons, particularly in the first 25 positions, is often a major cause of poor expression (Gia-Fen and Chen, 1990). Specialized BL21 strains include C41, which appears to favour production of membrane proteins, and B834, a methionine auxotroph for biosynthetic labelling with selenomethionine (Table 2.1).

The choice of culture media determines the biomass achievable in simple batch cultures and therefore the overall yield of protein. This assumes that the cell-specific productivity remains the same under different culture conditions. Most small-scale expression screens are carried out in 96- or 24-well, deep-well plates using enriched complex media, for example TB, 2YT, and GS96 (QBiogene), to ensure maximum biomass. Typically, these media support growth to optical densities (OD) of 5–10 $OD_{600}$ units compared to 2–3 $OD_{600}$ units in standard Luria broth (LB). Two options are available for inducing expression from T7 vectors either by addition of IPTG (range 0.1–1.0 mM) or by using a formulation of glucose and lactose in the media which leads to autoinduction (Studier, 2005). In the first method, the time of induction and therefore the growth state of the culture (usually midlog phase) can be predetermined, whereas in the second method induction occurs usually in late log stage and cannot be controlled. However, a major operational advantage of the autoinduction method is that once the cultures have been set up no further manipulation is required prior to harvest and expression evaluation. Further, in our experience autoinduction can lead to overall higher levels of expression. Another important parameter in the expression of recombinant proteins in *E. coli* is the culture temperature; reducing the temperature from 37°C to 20°C or even lower has been found to improve the solubility of the expressed protein. By carrying out small-scale expression tests in parallel, varying media, induction regime, and/or temperature, the optimum conditions for expression of a given target protein can be determined.

#### *2.3.1.2 Assay format*

The starting point for any expression assay is lysing the cells after harvesting by centrifugation and then separation of the soluble from insoluble fractions. Cell lysis can be carried out using standard protocols by either a freeze-thaw cycle followed by treatment with DNAse/lysozyme or by sonication with/without lysozyme. Alternatively, commercial detergent-based lysis reagents, for example BugBuster™ (Merck), FastBreak™ (Promega), CelLytic™ (Sigma), and Poppers™ (Pierce) can be used. Chemical methods lend themselves to 96-well formats, though sonicators which can accommodate a 96-well plate are available (Misonics). For HTP methods, centrifugation is generally avoided for fractionation of the lysates,

**Table 2.1** Common *E. coli* protein expression strains and their features

| Strain* | Derivation | Features |
|---|---|---|
| B834 | B strain | methionine auxotroph; used for $^{35}$S and selenomethionine labelling |
| BL21 | B834 | Lacks *lon* and *ompT* proteases to improve protein stability |
| BLR | BL21 | *recA* mutant BL21 *recA* mutant; stabilizes tandem repeats |
| Origami | K-12 | *trxB/gor* mutant, greatly facilitates cytoplasmic disulphide bond formation |
| OrigamiB | Tuner | BL21 *lacY* deletion, *trxB/gor* mutant, greatly facilitates cytoplasmic disulphide bond formation; allows precise control with IPTG |
| Rosetta | BL21 | Enhances expression of proteins having codons rarely used in *E. coli* (additional tRNAs encoded by chloramphenicol resistant plasmid) |
| Tuner | BL21 | BL21 *lacY* deletion mutant allows precise control with IPTG |
| BL21-AI | BL21 | T7 polymerase gene present under the control of the *araBAD* promoter leads to much tighter arabinose-inducible expression |
| BL21-SI | BL21 | T7 polymerase gene present under the control of the *proU* (salt-inducible) promoter |
| BL21 Star | BL21 | RNAse E deficient strain that reduces mRNA degradation and potentially increases protein yields |
| C41 and C43 | BL21 | BL21 mutants that over-produce membranes; enables expression of membrane-associated and toxic proteins |

* Other common/important derivations available for most of these strains are:

(DE3) in these λDE3 lysogens, the T7 RNA polymerase gene is integrated into the *E. coli* genome under the control of the *lacUV5* (IPTG-inducible) promoter. For use with T7-promoter based plasmids.

LysS/LysE: an additional chloramphenicol resistant plasmid carries the gene for T7 lysozyme under constitutively active promoters. T7 lysozyme inhibits the activity of T7 polymerase thereby reducing basal (uninduced) polymerase activity/ protein expression. LysE express higher levels of T7 lysozyme for tighter control.

pLacI: an additional chloramphenicol resistant plasmid carries the gene for high level production of the *lac* repressor to reduce basal expression.

LysS, LysE and LacI are all available in conjunction with additional rare codon tRNAs on the same chloramphenicol resistant plasmid (see Rosettas strain).

instead magnetic beads (Folkers *et al.*, 2004) or filtration plates (Knaust and Nordlund, 2001) are used to capture soluble product, usually via a His tag on the protein. These methods can be carried out manually but are readily automated using laboratory liquid handling systems. Protocol 2.3 describes the expression screen used in the OPPF, which is largely based on the Qiagen Ni-NTA protocol (www.qiagen.com) (Fig. 2.2).

To a large extent the method of analysis of expression is linked to the number of variables in the screening experiment and hence the number of samples to be assayed. In order to accommodate multifactorial screens incorporating, for example, a variety of different fusion vectors, immunodetection assays have been developed based on dot blots (Vincentelli *et al.*, 2003; Knaust and Nordlund, 2001). These are at best only semiquantitative and alternatives using ELISAs have been reported. For simplicity, we prefer the use of SDS-PAGE and by using precast gels (e.g. 26-well Criterion™ gels from Biorad) and running buffer containing a protein stain (InSite dye from National Diagnostics) screening of the soluble and insoluble fractions of a 96-well expression experiment can be performed in 1.5 h. The information content of a gel is much higher than that of a dot blot, providing both an estimate of yield and integrity of the product and serving as a quality check on whether the molecular weight is as expected. In this respect, probably the most accurate screening method that has been reported used matrix-assisted laser desorption ionization (MALDI)-mass spectrometry to measure the mass of soluble proteins produced in a 96-well format and purified using Ni-NTA ZIP-tips (Huang *et al.*, 2003).

### 2.3.2 Other hosts

The use of other hosts for HTP cloning, expression, and purification has been much more limited than *E. coli*, which remains the system of choice for high-level protein production. However, for many mammalian and viral proteins, problems of

## Protocol 2.3 Ni-NTA Mag bead purification in 96-well microplates

The starting material for this protocol is a deep-well 96 well plate/block containing ~1 ml of *E. coli* expression culture/well. The cells are first harvested by centrifugation at 5000 g for 15 min at 4°C, the media removed from the pellets, and the resulting cell pellet frozen to −80°C for a minimum of 30 min (this freezing step promotes cell lysis) prior to use in this protocol. The protocol is adapted from that used on the QIAGEN BioRobot 8000 with QIAGEN Magnetic Ni-NTA beads but can be easily performed with the aid of a multichannel pipettor (MCP) and either a vigorous orbital microplate shaker (minimum 'throw' 2 mm) or vortex mixer with MTP attachment. Bead volumes may require adjustment if those from another supplier are used.

**1.** Resuspend the cells completely in 230 μl of Lysis Buffer supplemented with 1 mg/ml Lysozyme and either 3 units/ml of Benzonase* (Merck, Germany; purity grade I, ≥25 U/μl, Cat. No. 1.01694.0001) or 400 units/ml of DNAse Type I. This can be done by either repeated aspirate/dispense with a suitable multichannel pipette or on an orbital microtitre plate shaker (~1000 rpm for 30 min). Clear the lysate by centrifuging the deep-well block at 5000 g for 30 min at 4°C.

**2.** Approximately 5 min before the end of the centrifugation run dispense 20 μl of the Ni-NTA magnetic bead suspension (ensure full resuspension before you commence pipetting!) into each well of a flat-bottomed microtitre plate (MTP). Not all microtitre plates are magnet compatible – check before you commence the assay.

**3.** Transfer the supernatant from Step 1 without disturbing the 'insoluble' pellet to each well of the MTP containing the Ni-NTA magnetic beads. Mix for 30 min at room temperature using either a MTP shaker or, alternatively, vortex at 600 rpm using an adapter for MTPs. The pellets may be resuspended in 8 M urea buffered with 100 mM $NaH_2PO_4$ and 10 mM Tris to pH 8.0 for analysis of the 'insoluble' fraction on SDS-PAGE.

**4.** Place the 96-well microplate on the 96-well magnet (we find the QIAgen magnets Type A or B work well but there are many other suitable magnets on the market) for 1 min and remove the supernatant carefully from the beads with a MCP.

**5.** Add 200 μl of Wash Buffer to each well, remove from the magnet, and mix on the microplate shaker (or vortex) for 5 min.

**6.** Place on the 96-well magnet for 1 min, and remove buffer.

**7.** Repeat Steps 5 and 6.

**8.** Add 50 μl of Elution Buffer to each well, mix on the MTP shaker (or vortex) for 1 min, place on the 96-well magnet for 1 min, and transfer the supernatant (eluate) to a fresh MTP for analysis on SDS-PAGE. SDS-PAGE sample buffer may be used instead of elution buffer if you require a more concentrated sample for analysis.

### Notes

The addition of Tween 20 to the buffers is necessary to enable optimal collection of the magnetic beads on the sides of the MTP wells and also to facilitate efficient cell lysis in Step 1. Bead volumes etc. are based on the QIAgen protocol and may require adjustment if reagents from another supplier are used.

*Benzonase is recommended as it is more stable than DNase I. When using Benzonase/DNaseI the crude lysate may be used directly in the binding step but the results will be less informative with regards to the solubility/cellular partitioning of the proteins.

### Buffers

Buffer NPI-10-Tween (Lysis Buffer):
50 mM $NaH_2PO_4$, 300 mM NaCl, 10 mM imidazole, 1% v/v Tween 20, adjust pH to 8.0 using NaOH and filter before use. Store at 4°C. 1% Tween is not required when using Lysozyme-based lysis – 0.05% is sufficient.

Buffer NPI-20-Tween (Wash Buffer):
50 mM $NaH_2PO_4$, 300 mM NaCl, 20 mM imidazole, 0.05% v/v Tween 20, adjust pH to 8.0 using NaOH and filter before use. Store at 4°C.

Buffer NPI-250-Tween, 50 ml: (Elution Buffer):
50 mM $NaH_2PO_4$, 300 mM NaCl, 250 mM Imidazole, 0.05% Tween 20, adjust pH to 8.0 using NaOH and filter before use. Store at 4°C.

DNAse I Stock Solutions (40,000 units/ml):
To a 200,000 unit bottle of DNAse I (Sigma D-4527) add 5 ml of UHQ sterile water. Aliquot into 25 μl aliquots store at −20°C. Make up to 25 ml with Lysis Buffer for use.

Lysozyme:
Weigh out 25 mg of Lysozyme (Sigma L-6876, stored at −20°C), add to the 25 ml of Benzonase/DNAse I working solution described above. Use in assay immediately.

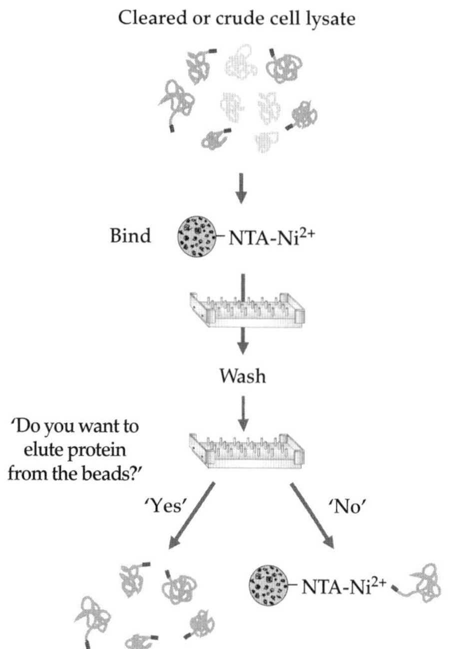


**Figure 2.2** Schematic representation of the Ni-NTA magnetic bead purification protocol in 96-well microplates. His-tagged proteins are purified from a crude cell lysate by specific binding to $Ni^{2+}$-charged magnetic beads. The beads are retained in the plate by the magnet during the multiple washing procedures to remove untagged proteins from the beads. Purified His-tagged proteins may be eluted from the beads either by the addition of a competitor (imidazole) or the addition of SDS-PAGE sample buffer, depending on the assay to be used on the fractions. Reproduced with kind permission from the *Ni-NTA Magnetic Agarose Beads Handbook*, a QIAGEN user protocol.

insolubility limit the number of targets that can be progressed to scale-up and purification.

*2.3.2.1 Yeast*

There are two yeast expression hosts that have an established track record for high-level production of heterologous proteins, namely *Saccharomyces cerevisiae* and *Pichia pastoris*. HTP expression screening using microplate formats has been reported for both these yeasts by Lang and coworkers (Holz *et al.*, 2002, 2003; Boettner *et al.*, 2002). In both cases standard protocols have been miniaturized with cells cultured in either 1.5 ml cultures in 96-deep-well plates for *S. cerevisiae* or 2 ml cultures in 24-deep-well plates for *P. pastoris*. Soluble expression of His-tagged proteins was monitored by SDS-PAGE, in combination with western blotting, following small-scale Ni-NTA purification using filter plates. Levels of expression of around 5 μg/ml were reported and shown to be scaleable in shake flask cultures. No obvious differences between yields from the two yeasts were observed for the same proteins even under controlled fermentation conditions (Prinz *et al.*, 2004).

*2.3.2.2 Higher eukaryotic cells*

Baculovirus infection of insect cells has been widely used, particularly in a pharmaceutical company setting, for producing enzyme targets for inhibitor screening, notably kinases (Chambers *et al.*, 2004). This has led to the development of microplate formats for both baculovirus construction and subsequent expression screening. As with yeast systems, expression of His-tagged proteins is assayed using anti-His antibodies by western blotting (Bahia *et al.*, 2005). Less attention has been paid to the use of mammalian cells for HTP protein expression, though the establishment of a microplate-based system for transient expression of Human Embryonic Kidney 293 (HEK) cells grown in suspension has recently been reported (Davies *et al.*, 2005). With a particular interest in secreted membrane glycoproteins, we have also applied transient expression in HEK cells to carry out the expression of secreted proteins in parallel, as described in Protocol 2.4. Glycoproteins present particular challenges for structural studies since the carbohydrate on the protein introduces potential heterogeneity (Davies *et al.*, 1993). To address this, we have made use of cells mutant for *N*-acetylglucosaminyltransferase 1 ($GnT1^-$) which leads to the addition of high mannose glycans to the expressed glycoprotein (Reeves *et al.*, 2002). These mannose-containing glycans can, in turn, be removed by treatment with endoglycosidase H to produce a protein with a minimal amount of sugar attached.

## 2.4 Purification

In order to obtain the quantities of protein needed for HTP structure determination, scale-up and purification of products identified during expression

## Protocol 2.4 Transient transfection of HEK 293T cells

### Materials

Polyethylenimine (PEI) stock: The transfection reagent (Aldrich, catalogue number 40,872–7 '25 kDa branched PEI'). Stock solution of PEI is prepared as follows: prepare stock solution of 100 mg/ml PEI in water, mix and further dilute to 1 mg/ml, neutralize with HCl and filter sterilize; store 5 ml aliquots frozen.

Preparation of DNA: The plasmid DNA quality is critical for a successful transfection, e.g. prepare using the endotoxin-free mega kit from Qiagen. Acceptable OD ratios should be $>1.9$.

Cells: HEK-293T or 293S GnTI- are grown in DMEM supplemented with l-Gln (Gibco cat. 25030–024) and non-essential amino acids plus 10% FCS. Lower to 2% serum during expression.

### Transfection protocol in 24-well plates

**1.** Seed cells to be 90% confluent on the day of transfection.

**2.** For each well of a 24-well plate dilute 1 μg DNA (typically from 0.5 μg/μl) with 60 μl serum-free DMEM in a 96-well plate then add 2 μl of PEI stock solution (1 mg/ml); mix well by pipetting and incubate for approx. 20 min at RT.

**3.** Add 140 μl DMEM containing 2% FCS to the DNA/PEI cocktail and mix.

**4.** Remove media from the cells in the 24-well plate and add the DNA/PEI cocktail to each well (total vol. 200 μl).

**5.** After 3–4 h at 37°C, top up well with 860 μl of DMEM containing 2% FCS.

**6.** Leave to express for 3–4 days, then collect medium, spin, filter 0.22 μm, then go on to protein purification (method depends on the tags used etc.).

For transfection in roller bottles the protocol is scaled up as follows: For each roller bottle (2125 cm$^2$) 0.5 mg DNA is diluted with 50 ml serum-free DMEM and 1 ml PEI added. Following incubation at room temperature for 20 min, 100 ml DMEM containing 2% FCS is added to the transfection mixture and this added to the cells. After 4 h the media in the roller bottle is topped up to 250 ml and the bottle gassed with 5% $CO_2$ in air and incubated for 3–4 days at 37°C with continuous rolling.

screening is necessary. The larger volumes required present a challenge for high-throughput facilities as plate-based formats are no longer viable. In this section, the technologies available for scaling-up of cultures, parallel purification of proteins, and quality assessment of the protein produced are reviewed.

### 2.4.1 Culture systems

*2.4.1.1 E. coli*

The most commonly used culture system for scale-up of *E. coli* is the shake flask. This conventional methodology has been shown to produce the quantity and quality of protein required with good scalability from plate-based expression screening. In general, cell-line, medium, growth conditions, and induction method are determined by the small-scale expression screen. Typically, the shake-flask is filled to one-quarter of its capacity to allow for adequate aeration of the culture. The generic OPPF protocols for *E. coli* culture and induction using both IPTG induction and autoinduction are given in Protocol 2.5.

Another simple culture system for recombinant protein production from *E. coli* is the use of 2-L polyethylene terephthalate (PET) beverage bottles which are cheap and disposable avoiding the cost and time for sterilization (Millard *et al.*, 2003). These bottles are readily available, inexpensive and can be used in conventional flask holders. Up to 1-L cultures can be grown per bottle, although lower volumes (100–250 ml) are more common for high-density cultures. The system has proved successful for growth of *E. coli* using Luria broth (LB), terrific broth (TB), and M9 media (Millard *et al.*, 2003). In addition, Sreenath *et al.* (2005) have reported the use of PET bottles for production of proteins using autoinduction medium. Protocols for growth in conventional shake-flasks can be readily converted to the PET bottle system.

*2.4.1.2 Insect and mammalian cells*

Insect cells are conveniently grown to multilitre scale in suspension using Erhlemeyer flasks (either glass or plastic). Alternatively, disposable plastic Wave™ Bioreactors (Wave Biotech) offer an affordable and easy-to-use solution for growing 5–10-L volumes of cell culture (Weber *et al.*, 2002). For HEK 293T cells we have developed a scale-up process based on growth in attached culture in roller bottles (see Protocol 2.4). Recently, there have been several reports of the use of suspension cultured HEK cells for large-scale transient expression of proteins (Pham *et al.*, 2003; Durocher *et al.*, 2002; Baldi *et al.*, 2005).

### 2.4.2 Selenomethionine labelling

Labelling of proteins with selenomethionine (SeMet) represents a standard approach to determining phases during protein structure solution by X-ray crystallography using multiwavelength anomalous dispersion methods (Hendrickson *et al.*, 1990). Therefore, HTP structural genomics not only requires a high yield of pure protein, but also a high percentage incorporation of selenium. Substitution of methionine for selenomethionine during *in vivo* protein expression can present many problems. Many *E. coli* cell lines have their own methionine sources, therefore a methionine auxotroph is required to achieve high levels of incorporation. One such auxotroph is the BL21-derived strain, B834. In addition, the biomass of cultures and the yield of protein are often depressed when compared to non-SeMet containing medium due to the toxic side-effects of SeMet labelling of all the proteins within the cell. SeMet labelling within the OPPF is performed using the methods described in Protocol 2.6 for both IPTG induction and autoinduction of protein expression. These methods routinely produce greater than 99% selenomethinone incorporation as determined by mass spectrometry.

If a methionine auxotroph is not used, the methionine within the culture can be replaced by SeMet using a methionine biosynthesis inhibition or 'poisoning' method such as that outlined in Protocol 2.7. This method has the advantage of using any *E. coli* strain available and thus the initial growth of the culture is not reduced, however the method is more laborious than using the methionine auxotroph.

Stols *et al.* have reported the successful SeMet labelling of proteins (over 95% incorporation) with IPTG induction and a 'poisoning' pathway using the 2-L PET bottle methodology (Stols *et al.*, 2004). This was taken a stage further by Sreenath *et al.* who utilized the bottles with autoinduction of the methionine auxotroph B834 to give over 90% SeMet

#### Protocol 2.5 Scale-up of *E. coli* culture

**IPTG induction protocol**

**1.** A single colony is inoculated into 20 ml of GS96 medium with 1% w/v glucose and antibiotics in a 50 ml falcon tube. The culture is incubated at 37°C and 225 rpm overnight (~16 hours).

**2.** The overnight culture is diluted 1 in 100 into 1 L of GS96 medium supplemented with 1% w/v glucose and antibiotics. The cultures are then incubated at 37°C and 225 rpm until $OD_{595} \approx 0.6$.

**3.** The temperature is reduced to 20°C. After 30 min at 20°C, the cultures are induced with 0.5 mM IPTG and are incubated for a further 20 h at 20°C and 225 rpm.

**4.** Product is harvested by centrifugation at 6000 g for 15 min and then frozen.

**Auto-induction protocol using OnEx Solutions from Merck Biosciences**

**1.** As for IPTG induction, a single colony is inoculated into 20 ml of GS96 medium with 1% w/v glucose and antibiotics in a 50 ml tube. The culture is incubated at 37°C and 225 rpm overnight (~16 hours).

**2.** The overnight culture is diluted 1 in 100 into 1 L of GS96 medium supplemented with 20 ml OnEx Solution 1, 50 ml OnEx Solution 2 and 1 ml OnEx Solution 3 along with the relevant antibiotics. The cultures are incubated at 37°C and 225 rpm for 4 h.

**3.** After this 4-h growth period, the temperature is reduced to 25°C and the cultures incubated for a further 20 h.

**4.** The cultures are then centrifuged at 6000 g for 15 min to obtain the pellet containing the protein, which can be frozen.

incorporation (Sreenath *et al.*, 2005). In contrast to *E. coli* expression systems, there are only a few reports of biosynthetic labelling of proteins with selenomethionine in eukaryotic cells (insect, mammalian, or yeast) (Laurila *et al.*, 2005; McWhirter *et al.*, 1999; Davis *et al.*, 2001).

### 2.4.3 Cell lysis

At scale-up, cells are generally lysed by mechanical means using either a French press or cell disruptor or by sonication (Protocol 2.8). Before breakage, the cell pellet is resuspended in buffer to give a uniform cell suspension. The lysis buffer often contains a small percentage of detergent which assists with the lysis and aids solubilization of the protein after it has been removed from the cell. In liquid homogenization using a French press or cell disruptor, cells are sheared by forcing the cell suspension through a narrow aperture under high pressure. Sonication is widely used in both HTP and non-HTP laboratories for cell breakage and makes use of high frequency sound waves to agitate and lyse the cells. The sound waves are emitted through a probe which is immersed in the cell suspension. As sonication causes heating in the sample, the sound waves are applied in short bursts and the sample is immersed in an ice bath. The sample volume and acoustic power applied are critical in the yield of protein recovered and in minimizing heating of the sample, which can lead to protein degradation (Feliu *et al.*, 1998).

### 2.4.4 Chromatography

The main challenge in protein purification for HTP facilities is to combine generic multidimensional chromatography for all target proteins with the parallelization of purification runs. This requirement has been addressed by the commercial development of semiautomated chromatography systems, notably the Äkta Explorer 3D and Äkta Xpress workstations, both available from GE Healthcare (www.chromatography.amershambiosciences.com).

#### Protocol 2.6 Expression of SeMet-labelled proteins I

**IPTG induction protocol with the methionine auxotroph B834**

**1.** A single colony is inoculated into 100 ml of SeMet media containing antibiotics (Molecular Dimensions). The SeMet media consists of 100 ml SeMet Base, 5 ml nutrient mix, and 0.4 ml 10 mg/ml selenomethionine in a 250 ml shake-flask. The culture is incubated at 37°C and 225 rpm overnight (~16 h).
**2.** The culture is centrifuged at 5000 g for 10 min and the supernatant discarded.
**3.** The pellet is then resuspended in 50 ml of PBS and recentrifuged at 5000 g for 10 min.
**4.** The supernatant is removed and the pellet is inoculated into 1 L of SeMet media (1 L SeMet Base, 50 ml nutrient mix, and 4 ml 10 mg/ml selenomethionine) which contains antibiotics.
**5.** The 500 ml cultures in 2 L shake-flasks are incubated at 37°C and 225 rpm until an $OD_{595}$ of ~0.6 is reached.
**6.** The temperature is reduced to 20°C. After 30 min at 20°C, the cultures are induced with 0.5 mM IPTG and are incubated for a further 20 h at 20°C and 200 rpm.
**7.** Product is harvested by centrifugation at 6000 g for 15 min and frozen for later processing.

**Autoinduction protocol with the methionine auxotroph B834**

**1.** A single colony is inoculated into 100 ml of glucose-free SeMet media (Molecular Dimensions) containing antibiotics and incubated at 37°C and 225 rpm overnight (~16 h).
**2.** The culture is centrifuged at 4000 g for 5 min and the supernatant discarded.
**3.** The pellet is then resuspended in 25 ml of PBS and recentrifuged at 4000 g for 5 min.
**4.** The supernatant is removed and the pellet resuspended in 2 ml PBS before inoculation into 1 L of autoinduction SeMet media (1 L SeMet Base, 20 ml OnEx Solution 1, 50 ml OnEx Solution 2, 1 ml OnEx Solution 3, 50 ml SeMet glucose-free nutrient mix and 4 ml 10 mg/ml selenomethionine) which contains antibiotics.
**5.** The 500 ml cultures in 2 L shake-flasks are incubated at 37°C and 225 rpm for 6 h.
**6.** The temperature is reduced to 25°C and the cultures incubated for a further 18 h.
**7.** The culture is then centrifuged at 6000 g for 15 min to obtain the pellet.

**Protocol 2.7 Expression of SeMet labelled proteins II**

**IPTG induction protocol using the 'poisoning' method**

This method follows the same protocol as for production of SeMet labelled protein using the auxotroph, however on addition of IPTG, the following chemicals are also added to each 500 ml culture:

2.5 ml 10 mg/ml SeMet
5 ml 10 mg/ml lysine
5 ml 10 mg/ml threonine
5 ml 10 mg/ml phenylalanine
5 ml 5 mg/ml leucine
5 ml 5 mg/ml isoleucine
5 ml 5 mg/ml valine.

**Autoinduction protocol using the 'poisoning' method**

The method follows that for autoinduction using the auxotroph B834, however at Stage 4, 9 ml 10 mg/ml selenomethionine is added along with the following amino acids:

5 ml 10 mg/ml lysine
5 ml 10 mg/ml threonine
5 ml 10 mg/ml phenylalanine
5 ml 5 mg/ml leucine
5 ml 5 mg/ml isoleucine
5 ml 5 mg/ml valine.

**Protocol 2.8 Lysis using cell disruption or sonication**

**1.** Cell pellets are thawed for 15 min at room temperature prior to resuspension.
**2.** The cell pellets are resuspended in ~30 ml per 10 g pellet of Cell Lysis Buffer (50 mM Tris pH 7.5, 500 mM NaCl, 20 mM imidazole, and 0.2% Tween) with protease inhibitors and DNaseI as required.
**3.** The lysate is passed through the basic Z cell disruptor at 30 kpsi. At this stage the lysate should no longer be viscous and the cells fully lysed.
or
The lysate is sonicated on ice at 20–25% power, 9.9 sec pulse on, 9.9 sec pulse off (500 W) for ~15 min. At this stage the lysate should no longer be viscous and the cells fully lysed.
**4.** The lysate is spun at 4°C and 30,000 g for 30 min to remove cell debris.
**5.** The cleared lysate is then filtered through a 45 μm membrane before purification.

In this section the use of these systems for HTP protein purification is described.

*2.4.4.1 Chromatography strategy*

The chromatographic strategy is determined by the construct chosen during cloning and the number of chromatography steps desired (Fig. 2.3). In general, constructs designed for HTP projects contain an affinity tag to aid in purification, as discussed in Section 2.2.2.2. Therefore the first step in purification of a recombinant protein is often affinity chromatography as this is highly specific for the target protein. The most common tag used is hexahistidine, which binds to immobilized metal affinity chromatography (IMAC) beads, although other tags such as GST and maltose binding protein are also used. As the affinity step usually follows a simple bind–elute protocol, this method both purifies and concentrates the protein. For example the target protein is eluted from a 1-ml IMAC column within 2–4 ml of buffer containing 0.5 M imidazole. Protocol 2.9 outlines a simple method for IMAC purification in which 20 mM imidazole is used in the binding and wash buffers to reduce levels of proteins that bind non-specifically.

If the sample is of the purity required, only buffer exchange is needed to give the final product. If further purification is required, a size exclusion chromatography step (SEC) is usually carried out. This step both removes contaminants from the sample, typically giving protein products of >90% purity, and gives information regarding the

### Protocol 2.9 IMAC-SEC purification

Charge a column using 1.5 × column volume 0.2 M nickel sulphate.

1. Wash with 1.5 × column volume water.
2. Equilibrate with 2 × column volume Binding/Wash Buffer (50m M Tris pH 7.5, 500 mM NaCl, 20 mM imidazole).
3. Load the filtered lysate containing the protein of interest.
4. Wash the column with 5 × column volume Binding/Wash Buffer or until the absorbance at 280 nm becomes stable.
5. Elute the protein with 5 × column volume of Elution Buffer (50 mM Tris pH 7.5, 500 mM NaCl, 500 mM imidazole) collecting the peak detected at 280 nm.
6. Inject this peak directly onto an equilibrated gel filtration column (S200 or S75).
7. Elute the protein with 1.2 × gel filtration column volume of Gel Filtration Buffer (20 mM Tris pH 7.5, 200 mM NaCl, reducing agents as desired) collecting fractions.
8. Run an SDS-PAGE gel on the fractions and pool those containing the protein of interest.

After affinity purification, the protein usually has high purity; however some non-specific binding to the resin can occur, along with binding of any truncations of the target protein.

oligomeric state of the protein when the retention volume is compared to that of standard proteins. In the event that a third chromatography step is necessary to achieve acceptable purity, then an ion exchange step is normally inserted between the affinity and SEC columns. This usually consists of affinity chromatography followed by ion exchange and then size exclusion chromatographies. As the affinity elution buffer is usually not compatible with the ion exchange chromatography step, a buffer exchange is normally included between these stages.

In many cases the recombinant protein expression construct has been designed to include a protease cleavage site (Section 2.2.2.2). Therefore tag cleavage is included in the purification strategy. There are two basic tag removal strategies: either cleavage of the protein in solution or when bound to an affinity column. In general, tagged-proteases are used in order to facilitate their removal from the protein sample. In the first case, the strategy may include an automated affinity step followed by buffer exchange into the conditions compatible with the protease. Conveniently, both 3C and Tev proteases cleave efficiently in the SEC buffer (Protocol 2.9). The protease is then added off-line and incubated in solution; for Tev and 3C this is typically overnight at 4°C. Postcleavage, a second automated affinity chromatography step followed by size exclusion chromatography can be performed in order to gain pure cleaved protein (Kim *et al.*, 2004).

This process can be fully automated by performing the protease cleavage step within a Superloop™ connected to the purification instrument. For on-column cleavage, either an untagged protease or a protease with a different tag to the protein of interest is used. Here, the target protein is bound to the affinity column, one column volume of protease solution applied, and the column incubated for the amount of time required for cleavage to occur. The column is then washed to remove both the protein of interest and the protease. Further purification is then necessary to remove the protease from the sample. On-column cleavage strategies are more readily automated than an in-solution strategy; however, in some cases cleavage is less likely to go to completion. This is most likely due to the protein being tightly bound to the resin and therefore the protease is sterically hindered from approaching the cleavage site.

#### 2.4.4.2 *Instrumentation*

Automation of the strategies described above can be achieved using instruments supplied by GE Healthcare. Many of the procedures described above are preprogrammed to allow ease of implementation of HTP automated purification techniques.

The Äkta Explorer 3D is equipped with positions for up to seven column and has two loop positions. This instrument allows sequential automation of up to six affinity purifications followed by a further chromatography step (e.g. size exclusion or desalting) in series, or up to four affinity purifications followed by two further chromatography steps (e.g. desalt and ion exchange) in series. The eluates from the columns are followed by a UV detector measuring protein absorbance at 280 nm. Such procedures have been implemented in many structural genomic laboratories and have been reported on by Kim *et al.* and Sigrell *et al.* (Kim *et al.*, 2004; Sigrell *et al.*, 2003). The instrument can be used to purify up to six affinity-tagged proteins per day, resulting in 1–50 mg of protein per run at over 90% purity (Sigrell *et al.*, 2003). Standard protocols supplied with the Äkta Explorer 3D are listed in Table 2.2.

The Äkta Xpress consists of multiple modules (systems with either two or four modules are currently available) each with five column positions and five loop positions (Fig. 2.3). Each unit functions in a similar way to the Äkta Explorer 3D system, allowing parallelization of up to four affinity purifications with one further chromatographic step or up to three affinity purifications with two further chromatography steps. This system has many advantages over the Äkta Explorer 3D, as each unit can be run independently, allowing different purification strategies to be run in parallel. In addition, the increased number of inlets, outlets, and loops allow for greater flexibility in protocols. In the OPPF, an Äkta Xpress consisting of four modules has been used to purify 16 His-tagged proteins using an IMAC–gel filtration strategy in one overnight run (~11 h purification time).

#### 2.4.4.3 *Secreted proteins*

Secreted proteins from eukaryotic cell lines present problems in purification due to both the volume of media involved and components of the media which are incompatible with affinity chromatography. If the media are compatible with an affinity chromatography step, then simple programmes similar to those described in Section 2.4.4.1 can be used. However, it is recommended that wash steps are interspersed with sample loading to maximize binding and reduce any risk of the column becoming blocked. A standard protocol for affinity purification of secreted proteins is given in Protocol 2.10. This protocol has been automated and combined with further chromatographic steps (desalt or size exclusion) using the Äkta Xpress system within the OPPF.

Some media are incompatible with affinity chromatography, for example many media used for baculovirus growth contain EDTA which strips the metals from IMAC columns. To avoid this problem, the media can be buffer exchanged using dialysis, stirred-cell, or cross-flow filtration. Alternatively, an additional chromatography step can be introduced in which interfering agents are removed, for example using lentil lectin affinity column to select glycosylated proteins.

**Table 2.2** Preprogrammed Äkta protocols

| Protocol | ÄKTA 3D | ÄKTA Xpress |
|---|---|---|
| 1. Affinity – desalt | ✓ | ✓ |
| 2. Affinity – gel filtration | ✓ | ✓ |
| 3. Affinity – desalt – ion exchange | ✓ | ✓ |
| 4. Affinity – desalt – ion exchange – desalt | | ✓ |
| 5. Affinity – desalt – ion exchange – gel filtration | | ✓ |
| 6. Affinity – on-column cleavage – desalt | | ✓ |

### 2.4.5 Quality assessment

For high-throughput purification pipelines, quality assessment (QA) procedures are required which can be carried out in parallel in relatively high throughput. The QA steps routinely carried out in the OPPF are summarized in Table 2.3. Other methods that have been used in HTP projects include one and two-dimensional NMR screening and differential scanning calorimetry.

#### 2.4.5.1 *Mass spectrometry*

Mass spectrometry (MS) is widely used to ascertain the purity, total mass of the protein produced, and detect any covalent modifications (Cohen and Chait, 2001). Both electrospray ionization (ESI) and MALDI may be used although for intact proteins; ESI has the advantage of being accurate to ~1 Da. Using the simple protocol described in Protocol 2.11, the MS of whole protein samples can be readily automated without the need for sample preparation. This method has proved successful for the

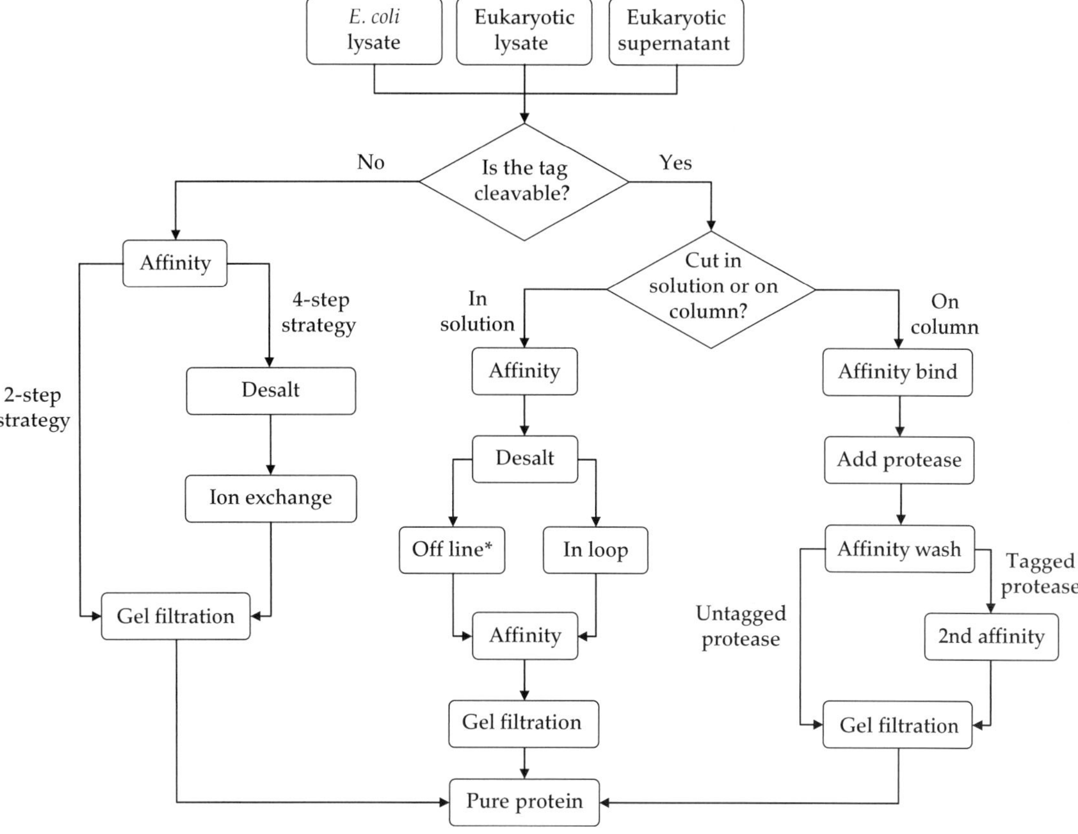


**Figure 2.3** Schematic representation of the workflow for automated protein purification strategies. (* offline tag cleavage involves a manual step.)

### Protocol 2.10 Secreted protein purification

1. Load 100 ml of media containing the protein of interest through an affinity chromatography column.
2. Wash the column with 2 × column volume of wash buffer.
3. Repeat Steps 1 and 2 until all the media has been loaded.
4. Wash the column with 10 × column volume of wash buffer to removed proteins which have bound non-specifically to the resin.
5. Elute the protein of interest with up to 5 × column volume of elution buffer.

For IMAC, the column needs to be charged and equilibrated in Wash Buffer before starting the procedure. Wash Buffer consists of 50 mM Tris pH 7.5, 500 mM NaCl, 20 mM imidazole and Elution Buffer 50 mM Tris pH 7.5, 500 mM NaCl, 500 mM imidazole.

measurement of protein masses from around 8 kDa to 100 kDa.

MS is also used routinely to evaluate the level of selenomethionine incorporation in a protein sample as the theoretical mass is increased by 47 Da per methionine in the sequence. MS can be used to further analyse protein samples by use of tryptic digest followed by MS/MS analysis. This allows *de novo*

**Protocol 2.11 Automated desalting protocol for ESI-MS analysis of intact proteins**

1. Samples are diluted to $\sim$10 μM in a 96-well PCR plate.
2. The protein is injected onto a 1 cm × 5 μm C4 precolumn (Anachem) using an autosampler.
3. The column is washed to waste with water containing 2% MeOH, 0.5% formic acid, and 0.2% trifluoroacetic acid, thus removing salts from the sample.
4. The protein is then eluted to ESI-MS with 20% water/ 80% acetonitrile containing 0.1% formic acid.
5. Simple MS measurements are taken and the raw data deconvoluted to give an accurate mass of the intact protein sample.

**Table 2.3** Routine methods for quality assessment undertaken in the Oxford Protein Production Facility

| Quality assurance technique | Protein characteristic |
|---|---|
| SDS-PAGE | Purity and denatured molecular weight |
| UV spectrometry | Protein concentration |
| Protein assay (BioRad) | Protein concentration |
| Size exclusion chromatography | Purity and oligomeric state |
| Dynamic light scattering | Polydispersity and oligomeric state |
| LC-ESI-MS | Accurate denatured molecular weight, purity, and bound ligands |

sequencing of regions of the protein, although full sequence coverage is rare in this type of experiment (Cohen and Chait, 2001). In addition, information regarding post-translational modifications such as phosphorylation and glycosylation can be attained using MS.

#### 2.4.5.2 *Light scattering*

Dynamic light scattering (DLS) is used to determine the oligomeric state and monodispersity of a protein. A coherent, monochromatic beam of light is passed through a protein sample in solution and the particles scatter the light in all directions. As the particles move according to Brownian motion, they cause time-dependent fluctuations in the scattering of the light. By measuring the time-dependence of the fluctuations, the diffusion coefficient of the particles can be calculated. From the Stokes–Einstein equation, the viscosity of the medium and this diffusion coefficient, the diameter of the particles can be calculated. Using a globular model, the particle diameter can be used to estimate the total molecular weight, and thus the oligomeric state of the protein present. If more than one oligomeric state exists within a sample, such heterogeneity can reduce the crystallizability of the sample (Habel *et al.*, 2001). DLS is therefore a simple, fast measure of the multimeric state and monodispersity of a protein sample. DLS techniques have been taken further with the introduction of 96- and 384-well format equipment such as the DynaPro™ plate reader from Wyatt Technology.

## 2.5 Summary and future perspectives

A major driving force for the development of high-throughput cloning, expression, and protein purification has been the arrival of the post-genomic era, where the emphasis has changed from DNA sequencing to structural (and functional) proteomics. A wide variety of HTP methodologies have been successfully implemented in major world-wide efforts to generate protein for crystallographic and NMR structure determination. For the development of HTP protein production pipelines, many of the initial projects have focused on bacterial genomes, where high-throughput production of proteins expressed solubly in *E. coli* is readily achieved. In the future, the emphasis will shift increasingly to eukaryotic and viral genomes where production of soluble proteins remains a key issue. To what extent will current methodologies be adequate for tackling structural proteomics in the future? Clearly, ligation-independent cloning methods have been a major success in providing completely generic protocols that are readily automated. The inherent difficulties of expressing eukaryotic proteins solubly means a greater degree of oversampling is required, in terms of the number of constructs made. Whilst 96-well plates have been robust in current work, higher density 384- and even 1536-well plates would have significant advantages for generating and analysing these greater numbers of constructs. In addition, a significant number of proteins may only become fully ordered when binding to partner proteins of a hetero-oligomeric macromolecular assembly. In such cases, coexpression strategies using either dual or multiple expression vectors will be required. It is also expected that there will be increased use of higher-throughput baculovirus and mammalian expression systems involving the development of process automation.

The advent of parallelized protein purification systems and high performance chromatography media represent major steps forward in HTP projects. Further developments that would be of benefit include a greater integration of purification and QA, such as on-line mass spectrometry. Refolding of insoluble proteins as a rescue strategy has so far met with limited success; further attempts at developing more generic protocols would appear worthwhile. Underpinning all aspects of HTP protein production is the need to record and track data at every stage of the process using a laboratory information management system (LIMS). Although a number of projects have developed such systems, there remains the need for a widely available solution to support both large and small-scale laboratories. A European project to address this issue has recently been initiated (www.mole.ac.uk/lims/project/).

Although much has been achieved by HTP protein production projects to date, there is clearly scope for continued technology development. As we move into the next phase, there is a real prospect of making exciting additions to basic knowledge of protein structure and function.

## Acknowledgements

We are grateful to Robin Aplin, Nahid Rahman, Dave Alderton, Weixian Lu, and Andrew Farmer (BD Clontech) for their help in developing the protocols described in this chapter. The OPPF is supported by a grant from the UK Medical Research Council and is part of the Structural Proteomics IN Europe (SPINE) consortium (European Commission Grant No.QLG2-CT-2002–00988).

## References

Abergel, C., Coutard, B., Byrne, D., Chenivesse, S., Claude, J. B., Deregnaucourt, C., *et al.* (2003). Structural genomics of highly conserved microbial genes of unknown function in search of new antibacterial targets. *J. Struct. Funct. Genomics* **4**, 141–157.

Alexandrov, A., Dutta, K. and Pascal, S. M. (2001). MBP fusion protein with a viral protease cleavage site: one-step cleavage/purification of insoluble proteins. *Biotechniques* **30**, 1194–1198.

Aslandis, C. D. and de Jong, P. J. (1990). Ligation-independent cloning of PCR products (LIC-PCR). *Nucleic Acids Res.* **18,** 6069–6074.

Bahia, D., Cheung, R., Buchs, M., Geisse, S. and Hunt, I. (2005). Optimisation of insect cell growth in deep-well blocks: development of a high-throughput insect cell expression screen. *Protein Expr. Purif.* **39**, 61–70.

Baker, R. T. (1996). Protein expression using ubiquitin fusion and cleavage. *Curr. Opin. Biotechnol.* **7**, 541–546.

Baldi, L., Muller, N., Picasso, S., Jacquet, R., Girard, P., Thanh, H. P., *et al.* (2005). Transient gene expression in suspension HEK-293 cells: application to large-scale protein production. *Biotechnol. Prog.* **21**, 148–153.

Berrow, N. S., Alderton, D., Sainsbury, S., Nettleship, J., Assenberg, R., Rahman, N., Stuart, D. I. and Owens, R. J. (2007). A versatile ligation-independent cloning method suitable for high-throughput expression screening applications. *Nucleic Acids Res.* **35**, E45, 1–12.

Boettner, M., Prinz, B., Holz, C., Stahl, U. and Lang, C. (2002). High-throughput screening for expression of heterologous proteins in the yeast *Pichia pastoris*. *J. Biotechnol* **99**, 51–62.

Braun, P., Hu, Y., Shen, B., Halleck, A., Koundinya, M., Harlow, E. and LaBaer, J. (2002). Proteome-scale purification of human proteins from bacteria. *Proc. Natl. Acad. Sci. USA* **99**, 2654–2659.

Butt, T. R., Jonnalagadda, S., Monia, B. P., Sternberg, E. J., Marsh, J. A., Stadel, J. M., Ecker, D. J. and Crooke, S. T. (1989). Ubiquitin fusion augments the yield of cloned gene products in *Escherichia coli*. *Proc. Natl. Acad. Sci. USA* **86**, 2540–2544.

Catanzariti, A. M., Soboleva, T. A., Jans, D. A., Board, P. G. and Baker, R. T. (2004). An efficient system for high-level expression and easy purification of authentic recombinant proteins. *Protein Sci.* **13**, 1331–1339.

Chambers, S. P., Austen, D. A., Fulghum, J. R. and Kim, W. M. (2004). High-throughput screening for soluble recombinant expressed kinases in *Escherichia coli* and insect cells. *Protein Expr. Purif.* **36**, 40–47.

Cohen, S. L. and Chait, B. T. (2001). Mass spectrometry as a tool for protein crystallography. *Annu. Rev. Biophys. Biomol. Struct.* **30**, 67–85.

Cordingley, M. G., Register, R. B., Callahan, P. L., Garsky, V. M. and Colonno, R. J. (1989). Cleavage of small peptides in vitro by human rhinovirus 14 3C protease expressed in *Escherichia coli*. *J. Virol.* **63**, 5037–5045.

Davies, A., *et al.* (2005). Optimisation and evaluation of a high-throughput mammalian protein expression system. *Protein Expr. Purif.* **42**, 111–121.

Davis, S. J., Greene, A., Lullau, E. and Abbott, W. M. (1993). Expression of soluble recombinant glycoproteins with predefined glycosylation: application to the crystallization of the T-cell glycoprotein CD2. *Protein Eng.* **6**, 229–232.

Davis, S. J., Ikemizu, S., Collins, A. V., Fennelly, J. A., Harlos, K., Jones, E. Y. and Stuart, D. I. (2001). Crystallization and functional analysis of a soluble deglycosylated form of the human costimulatory molecule B7-1. *Acta Crystallogr. D* **57**, 605–608.

Durocher, Y., Perret, S. and Kamen, A. (2002). High-level and high-throughput recombinant protein production by transient transfection of suspension-growing human 293-EBNA1 cells. *Nucleic Acids Res.* **30**, E9.

Dyson, M. R., Shadbolt, S. P., Vincent, K. J., Perera, R. L. and McCafferty, J. (2004). Production of soluble mammalian proteins in *Escherichia coli*: identification of protein features that correlate with successful expression. *BMC Biotechnol.* **4**, 32.

Feliu, J. X., Cubarsi, R. and Villaverde, A. (1998). Optimized release of recombinant proteins by ultrasonication of *E. coli* cells. *Biotechnol. Bioeng.* **58**, 536–540.

Folkers, G. E., van Buuren, B. N. and Kaptein, R. (2004). Expression screening, protein purification and NMR analysis of human protein domains for structural genomics. *J. Struct. Funct. Genomics* **5**, 119–131.

Gia-Fen, T. C. and Inouye, M. (1990). Suppression of the negative effect of minor arginine codons on gene expression; preferential usage of minor codons within the first 25 codons of the *Escherichia coli* genes. *Nucleic Acids Res.* **18**, 1465–1473.

Habel, J. E., Ohren, J. F. and Borgstahl, G. E. (2001). Dynamic light-scattering analysis of full-length human RPA14/32 dimer: purification, crystallization and self-association. *Acta Crystallogr. D* **57**, 254–259.

Hammarstrom, M., Hellgren, N., van Den Berg, S., Berglund, H. and Hard, T. (2002). Rapid screening for improved solubility of small human proteins produced as fusion proteins in *Escherichia coli*. *Protein Sci.* **11**, 313–321.

Hartley, J. L., Temple, G. F. and Brasch, M. A. (2000). DNA cloning using *in vitro* site-specific recombination. *Genome Res.* **10**, 1788–1795.

Haun, R. S., Serventi, I. M. and Moss, J. (1992). Rapid, reliable ligation-independent cloning of PCR products using modified plasmid vectors. *Biotechniques* **13**, 515–518.

Hendrickson, W. A., Horton, J. R. and LeMaster, D. M. (1990). Selenomethionyl proteins produced for analysis by multiwavelength anomalous diffraction (MAD): a vehicle for direct determination of three-dimensional structure. *EMBO J.* **9**, 1665–1672.

Holz, C., Hesse, O., Bolotina, N., Stahl, U. and Lang, C. (2002). A micro-scale process for high-throughput expression of cDNAs in the yeast *Saccharomyces cerevisiae*. *Protein Expr. Purif.* **25**, 372–378.

Holz, C., Prinz, B., Bolotina, N., Sievert, V., Bussow, K., Simon, B., Stahl, U. and Lang, C. (2003). Establishing the yeast *Saccharomyces cerevisiae* as a system for expression of human proteins on a proteome-scale. *J. Struct. Funct. Genomics* **4**, 97–108.

Huang, R. Y., Boulton, S. J., Vidal, M., Almo, S. C., Bresnick, A. R. and Chance, M. R. (2003). High-throughput expression, purification, and characterization of recombinant *Caenorhabditis elegans* proteins. *Biochem. Biophys. Res. Commun.* **307**, 928–934.

Kim, Y., Dementieva, I., Zhou, M., Wu, R., Lezondra, L., Quartey, P., *et al.* (2004). Automation of protein purification for structural genomics. *J. Struct. Funct. Genomics* **5**, 111–118.

Knaust, R. K. and Nordlund, P. (2001). Screening for soluble expression of recombinant proteins in a 96-well format. *Anal. Biochem.* **297**, 79–85.

Laurila, M. R., Salgado, P. S., Makeyev, E. V., Nettleship, J., Stuart, D. I., Grimes, J. M. and Bamford, D. H. (2005). Gene silencing pathway RNA-dependent RNA polymerase of Neurospora crassa: yeast expression and crystallization of selenomethionated QDE-1 protein. *J. Struct. Biol.* **149**, 111–115.

Lesley, S. A., Kuhn, P., Godzik, A., Deacon, A. M., Mathews, I., Kreusch, A., *et al.* (2002). Structural genomics of the *Thermotoga maritima* proteome implemented in a high-throughput structure determination pipeline. *Proc. Natl. Acad. Sci. USA* **99**, 11664–11669.

Luan, C. H., Qiu, S., Finley, J. B., Carson, M., Gray, R. J., Huang, W., *et al.* (2004). High-throughput expression of *C. elegans* proteins. *Genome Res.* **14**, 2102–2110.

Malakhov, M. P., Mattern, M. R., Malakhova, O. A., Drinker, M., Weeks, S. D. and Butt, T. R. (2004). SUMO fusions and SUMO-specific protease for efficient expression and purification of proteins. *J. Struct. Funct. Genomics* **5**, 75–86.

Marsischky, G. and LaBaer, J. (2004). Many paths to many clones; a comparative look at high-throughput cloning methods. *Genome Research* **14**, 2020–2028.

McWhirter, S. M., Pullen, S. S., Holton, J. M., Crute, J. J., Kehry, M. R. and Alber, T. (1999). Crystallographic analysis of CD40 recognition and signaling by human TRAF2. Proc. Natl. Acad. Sci. USA **96**, 8408–8413.

Millard, C. S., Stols, L., Quartey, P., Kim, Y., Dementieva, I. and Donnelly, M. I. (2003). A less laborious approach to the high-throughput production of recombinant proteins in *Escherichia coli* using 2-liter plastic bottles. *Protein Expr. Purif.* **29**, 311–320.

Parks, T. D., Leuther, K. K., Howard, E. D., Johnston, S. A. and Dougherty, W. G. (1994). Release of proteins and peptides from fusion proteins using a recombinant plant virus proteinase. *Anal. Biochem.* **216**, 413–417.

Pham, P. L., Perret, S., Doan, H. C., Cass, B., St-Laurent, G., Kamen, A. and Durocher, Y. (2003). Large-scale transient transfection of serum-free suspension-growing HEK293 EBNA1 cells: peptone additives improve cell growth and transfection efficiency. *Biotechnol. Bioeng.* **84**, 332–342.

Porath, J. (1992). Immobilized metal ion affinity chromatography. *Protein Expr. Purif.* **3**, 263–281.

Prinz, B., Schultchen, J., Rydzewski, R., Holz, C., Boettner, M., Stahl, U. and Lang, C. (2004). Establishing a versatile fermentation and purification procedure for human proteins expressed in the yeasts *Saccharomyces cerevisiae* and *Pichia pastoris* for structural genomics. *J. Struct. Funct. Genomics* **5**, 29–44.

Reeves, P. J., Callewaert, N., Contreras, R. and Khorana, H. G. (2002). Structure and function in rhodopsin: high-level expression of rhodopsin with restricted and homogeneous N-glycosylation by a tetracycline-inducible N-acetylglucosaminyltransferase I-negative HEK293S stable mammalian cell line. *Proc. Natl. Acad. Sci. USA* **99**, 13419–13424.

Sigrell, J. A., Eklund, P., Galin, M., Hedkvist, L., Liljedahl, P., Johansson, C. M., Pless, T. and Torstenson, K. (2003). Automated multi-dimensional purification of tagged proteins. *J. Struct. Funct. Genomics* **4**, 109–114.

Smith, D. B. and Johnson, K. S. (1988). Single-step purification of polypeptides expressed in *Escherichia coli* as fusions with glutathione S-transferase. *Gene* **67**, 31–40.

Sreenath, H. K., Bingman, C. A., Buchan, B. W., Seder, K. D., Burns, B. T., Geetha, H. V., *et al.* (2005). Protocols for production of selenomethionine-labeled proteins in 2-L polyethylene terephthalate bottles using auto-induction medium. *Protein Expr. Purif.* **40**, 256–267.

Stols, L., Gu, M., Dieckman, L., Raffen, R., Collart, F. R. and Donnelly, M. I. (2002). A new vector for high-throughput, ligation-independent cloning encoding a tobacco etch virus protease cleavage site. *Protein Expr. Purif.* **25**, 8–15.

Stols, L., Millard, C. S., Dementieva, I. and Donnelly, M. I. (2004). Production of selenomethionine-labeled proteins in two-liter plastic bottles for structure determination. *J. Struct. Funct. Genomics* **5**, 95–102.

Studier, F. W. (2005). Protein production by auto-induction in high density shaking cultures. *Protein Expr. Purif.* **41**, 207–234.

Studier, F. W., Rosenberg, A. H., Dunn, J. J. and Dubendorff, J. W. (1990). Use of T7 RNA polymerase to direct expression of cloned genes. *Method Enzymol.* **185**, 60–89.

Vincentelli, R., Bignon, C., Gruez, A., Canaan, S., Sulzenbacher, G., Tegoni, M., *et al.* (2003). Medium-scale structural genomics: strategies for protein expression and crystallization. *Acc. Chem. Res.* **36**, 165–172.

Walhout, A. J., Temple, G. F., Brasch, M. A., Hartley, J. L., Lorson, M. A., van den Heuvel, S. and Vidal, M. (2000). GATEWAY recombinational cloning: application to the cloning of large numbers of open reading frames or ORFeomes. *Method Enzymol.* **328**, 575–92.

Weber, W., Weber, E., Geisse, S. and Memmert, K. (2002). Optimisation of protein expression and establishment of the Wave Bioreactor system for Baculovirus/insect cell culture. *Cytotechnology* **38**, 77–85.

Xu, X. and Jones, I. M. (2004). Rapid parallel expression in *E. coli* and insect cells: analysis of five lef gene products of the *Autographa californica* multiple nuclear polyhedrosis virus (AcMNPV). *Virus Genes* **29**, 191–197.

CHAPTER 3

# Automation of non-conventional crystallization techniques for screening and optimization

Naomi E. Chayen

## 3.1 Introduction

The availability of high-quality crystals is crucial to the structure determination of proteins by X-ray diffraction. It is still not understood why some proteins crystallize with ease while others stubbornly refuse to produce suitable crystals. Producing high-quality crystals has always been the bottleneck to structure determination and with the advent of structural genomics this problem is becoming increasingly acute. In spite of impressive advances in throughput, the crystallization problem has not been solved and better crystallization techniques need to be designed in order to overcome this hurdle (Service, 2002; Chayen, 2004). Finding favourable conditions for crystallization is usually achieved by screening of the protein solution with numerous crystallizing agents in order to find 'hits' that indicate which conditions may be suitable for crystal growth. Optimization of the crystallization conditions is done either by fine tuning of the parameters (precipitant, pH, temperature, additives, etc.) involved, or by manipulation of the crystallization phase diagram with the aim of guiding the experiment in the direction that will produce the desired results. This chapter highlights a variety of non-standard experimental methods of screening and optimization techniques with a focus on those that have been automated and can be adapted to high-throughput trials.

## 3.2 The basis of crystallization

Crystallization is a phase transition phenomenon. Crystals grow from an aqueous protein solution when the solution is brought into supersaturation (Ataka, 1993). Supersaturation is achieved by varying the concentrations of precipitant, protein and additives, pH, temperature, and other parameters (McPherson, 1999; Ducruix and Giegé, 1992; Ducruix and Giegé, 1999).

### 3.2.1 The crystallization phase diagram

The crystallization process can be illustrated by a phase diagram that shows which state (liquid, crystalline, or amorphous solid [precipitate]) is stable under a variety of crystallization parameters. It provides a means of quantifying the influence of the parameters such as the concentrations of protein, precipitant(s), additive(s), pH, and temperature on the production of crystals. Hence phase diagrams form the basis for the design of crystal growth conditions (McPherson, 1999; Ducruix and Giegé, 1992; Ducruix and Giegé, 1999; Chayen *et al.*, 1996; and references therein).

Crystallization proceeds in two phases, nucleation and growth. Nucleation, which is a prerequisite for crystallization, requires different conditions from those of growth. Once the nucleus has formed, growth follows spontaneously (Ataka, 1993;

McPherson, 1999; Ducruix and Giegé, 1992; Ducruix and Giegé, 1999; Chayen *et al.*, 1996).

Figure 3.1 shows an example of a typical crystallization phase diagram. The figure schematically illustrates four areas: (i) an area of very high supersaturation where the protein will precipitate; (ii) an area of moderate supersaturation where spontaneous nucleation will take place; (iii) an area of lower supersaturation just below the nucleation zone where crystals are stable and may grow but no further nucleation will take place (this area is referred to as the metastable zone which is thought to contain the best conditions for growth of large well ordered crystals); and (iv) an undersaturated area where the protein is fully dissolved and will never crystallize (Chayen *et al.*, 1996; Chayen, 2005).

The four common methods of crystallization 'travel' through the phase diagram, each via a different route (dashed lines). Batch crystallization involves mixing of protein and the crystallizing agents at conditions that aim to achieve supersaturation immediately upon mixing. This is in contrast to all other crystallization methods (based on diffusion) in which the protein solution is undersaturated at the outset of the experiment and gradually reaches supersaturation by equilibration with a reservoir solution which contains the crystallizing agents. The dynamic nature of the diffusion methods enables a self-screening process to take place (dashed lines Fig. 3.1) as the trials make their way to the nucleation zone and thereafter to the metastable region (Chayen, 2004; Chayen, 2005).

In an ideal experiment, once nuclei have formed, the concentration of protein in the solute will drop, thereby naturally leading the system into the metastable zone (Fig. 3.1) where growth should occur, without the formation of further nuclei (McPherson, 1999; Ducruix and Giegé, 1992;

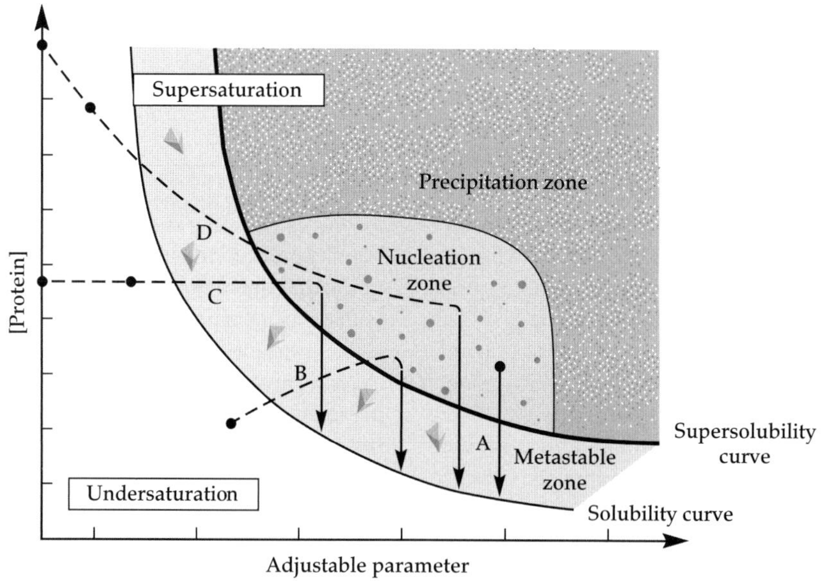


**Figure 3.1** Schematic illustration of a protein crystallization phase diagram. The adjustable parameter can be precipitant or additive concentration, pH, temperature etc. The four major crystallization methods are represented, showing their different routes of reaching the nucleation and metastable zones, assuming the adjustable parameter is precipitant concentration. The black circles represent the starting conditions. Two alternative starting points are shown for free interface diffusion and dialysis because the undersaturated protein solution can contain either protein alone, or protein mixed with a low concentration of the precipitating agents. (A) Batch (B) Vapour diffusion (C) Dialysis (D) Free interface diffusion. The solubility is defined as the concentration of protein in the solute that is in equilibrium with crystals. The supersolubility curve is defined as the line separating conditions where spontaneous nucleation (or phase separation, precipitation) occurs from those where the crystallization solution remains clear if left undisturbed. Reprinted from *Current Opinion in Structural Biology*, **14**, Chayen, pp. 577–583, Copyright (2004), with permission from Elsevier.

Ducruix and Giegé, 1999; Chayen, 2005; Bergfors, 1999). However, the ideal experiment does not often happen and more often than not, excess nucleation occurs, resulting in the formation of numerous, low-quality crystals. The aim is therefore to devise methods that will enable the experimenter to lead the experiment from nucleation to growth, in order to ensure the desired results.

## 3.3 Practical considerations

Although phase diagrams offer one of the basic and most important pieces of knowledge necessary for growing protein crystals in a rational way, they are not often employed because accurate quantitative phase diagrams require numerous solubility measurements. (The solubility is defined as the concentration of protein in the solute that is in equilibrium with crystals.) Reaching equilibrium can take several months because proteins diffuse slowly. An additional limiting factor is that solubility measurements require large amounts of sample (Ataka, 1993; Chayen *et al.*, 1996).

The area of conditions called the 'metastable zone' is situated between the solubility and supersolubility curves on the crystallization phase diagram (Fig. 3.1). The supersolubility curve is defined as the line that separates the conditions where spontaneous nucleation (or phase separation or precipitation) occurs, from those where the crystallization solution remains clear if left undisturbed (Ducruix and Giegé, 1992; Ducruix and Giegé, 1999).

The supersolubility curve is less well defined than the solubility curve but, experimentally, it is found to a reasonable approximation much more easily. It has been reported that for practical purposes, it is sufficient to obtain the supersolubility curve. To construct it, one must set up crystallization trials, varying at least two conditions (one of which is typically the protein concentration) and plot their outcomes on a two-dimensional parameter grid. The supersolubility curve can be obtained rapidly using robots and can aid in the separation of nucleation and growth using seeding and other means. A diagram containing the supersolubility curve (and not the solubility curve) is called a 'working phase diagram' (Chayen, 2005; Saridakis and Chayen, 2003). Uses of such phase diagrams are described in Sections 3.5.1 to 3.7.

## 3.4 Automation and miniaturization of screening procedures

There have never been any set rules or recipes to determine where to start when attempting to crystallize a new protein, hence the most common means to get started is by using multifactorial screens, in other words by exposing different concentrations of the protein to be crystallized to numerous different crystallization agents, buffers, temperatures, etc., usually in combinations that have been successful with other proteins. Automation of screening procedures (e.g. Chayen *et al.*, 1990; Weber, 1990; Soriano and Fontecilla-Camps, 1993) began in the 1990s but only became wide spread with the appearance of structural genomics, for which automation has become vital. Robotics are now available for most methods of crystallization (e.g. Luft *et al.*, 2003; Mueller *et al.*, 2001; Walter *et al.*, 2003; Santarsiero *et al.*, 2002; Chayen *et al.*, 1992).

### 3.4.1 The microbatch method

The first semi-high-throughput automated system to dispense crystallization trials of less than 1 μl was designed in 1990 to deliver batch trials under oil (Chayen *et al.*, 1990). The method was named microbatch to define a microscale batch experiment. It was designed to obtain maximum information on the molecule to be crystallized while using minimal amounts of sample. In order to prevent the evaporation of such small volumes, the trials are dispensed and incubated under low density (0.87 $g/cm^3$) paraffin oil (Fig. 3.2). The crystallization drops remain under the oil since the aqueous drops are denser than the paraffin oil.

Microbatch can be performed either manually or automatically (Chayen *et al.*, 1992). It is the simplest crystallization method and therefore can be easily performed in high-throughput trials. Current robots can dispense microbatch trials down to 1 nl volumes. Depending on the type of oils used to cover the trials, this technique can be harnessed for both screening and optimization experiments.

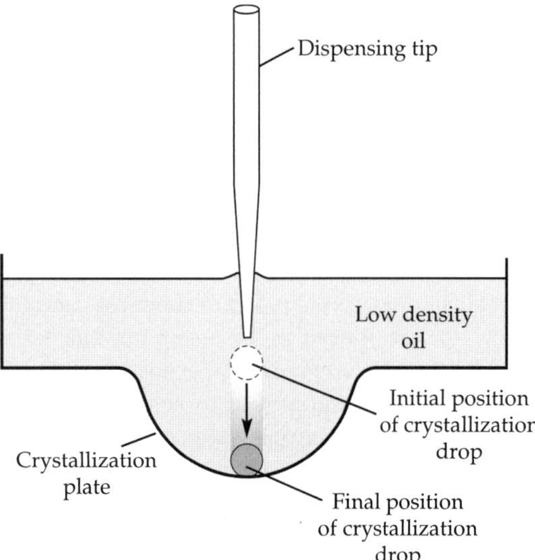


**Figure 3.2** Dispensing of a microbatch trial under oil. The dashed circle represents the initial position of the crystallization drop at the time of dispensing. The grey circle represents the final position of the drop after it had made its way to the bottom of the well (due to its higher density compared to that of the oil). Modified from Chayen (1997). The role of oil in macromolecular crystallization. *Structure* **5**, 1269–1274, Copyright Elsevier.

### 3.4.2 The effect of different oils

The microbatch method is essentially a batch experiment in which the macromolecule and the crystallizing agents are mixed at their final concentrations at the start of the experiment; thus, in order to obtain crystals, supersaturation must be attained upon mixing. Consequently, the volume and the composition of a trial remain constant. This is in contrast to all other crystallization methods which undergo a self-screening process until equilibrium is reached (Fig. 3.1). The stability of the batch is an important benefit for conducting diagnostic studies on the process of crystal growth since the history of the sample can be followed reliably. However, this benefit may become a handicap in the case of screening for crystallization conditions since it is likely that the gradual change of conditions (*en route* to equilibrium), which takes place by the other methods, may be the crucial factor for the formation of crystals (Chayen *et al.*, 1996; Chayen, 1997a; D'Arcy *et al.*, 1996).

A modification of the original microbatch method provides a means of simultaneously retaining the benefits of a microbatch experiment while gaining the inherent advantage of the self-screening process of a diffusion trial (D'Arcy *et al.*, 2003). This modification is based on the following rationale: water can evaporate through different oils at different rates. Paraffin oil acts as a good sealant allowing only a negligible amount of water evaporation through the paraffin oil during the average time required for a crystallization experiment. In contrast, water can diffuse freely through silicone oils. A mixture of paraffin and silicone oils permits partial diffusion, depending on the ratio at which they are mixed (D'Arcy *et al.*, 1996).

It has been shown that for screening purposes it is preferable to use silicone oil or a mixture of paraffin and silicone oils (D'Arcy *et al.*, 1996, 2003). This allows some evaporation of the drops, leading to a higher number of 'hits' and faster formation of crystals compared to trials which are set under paraffin oil. In the case of optimization, where the conditions need to be known and stable, the trials must be covered by paraffin oil, or allowed to evaporate for a set time and then covered (as described in Section 3.5.5 and Protocol 3.7).

Microbatch can be used for almost all the known precipitants, buffers, and additives, including detergents. The oils do not interfere with the common precipitants such as salts, polyethylene glycols (PEG), jeffamine MPD, and even glycerol and ethanol. Microbatch, though, can not be used for crystallization trials containing small volatile organic molecules such as dioxane, phenol, or thymol since these molecules dissolve into the oil (Chayen, 1998). Such substances are only applied in microbatch if the experimenter aims to explore the effect of progressive disappearance of these additives from the solution.

### 3.4.3 Fine tuning of the conditions around the screen

Once a 'hit' has been found, the next step is to conduct a finer screen around these conditions by varying the concentrations of protein, precipitant, addition of additives, etc. In most cases, more than three different ingredients may be needed in each trial, making it difficult to manually pipette all the ingredients into one drop directly under the oil. Thus the procedure in Protocol 3.2 should be followed.

### 3.4.4 Crystallization of membrane proteins in microbatch

An increasing number of membrane proteins, in a variety of different detergents, have been crystallized in microbatch under oil. Some of these had failed to crystallize by all methods other than microbatch. Dispensing is quick and simple, even when performed manually, and the drops in oil do not spread out as they do in vapour diffusion over the siliconized coverslips (Chayen, 2006). Using robots thousands of microbatch trials can be dispensed in high-throughput mode in nanolitre volumes.

The protocol for setting up microbatch experiments containing membrane proteins is identical to that described in Protocols 3.1 and 3.2.

### 3.4.5 Harvesting and mounting crystals from microbatch

Harvesting crystals from microbatch is slightly more difficult than harvesting from coverslips or from standard sitting drops (Protocol 3.3). However, after some practice it can be achieved easily (Chayen, 1998, 2006).

### 3.4.6 Diffusion techniques

Although microbatch is the simplest method of crystallization, it is a relatively new technique and many experimenters still prefer to use vapour diffusion which has been around and has worked well for over 40 years. Hence, there has also been major development in automating and scaling down the quantities of sample using the popular vapour diffusion method (both sitting and hanging drops). An increasing variety of robots are available commercially.

The liquid–liquid free interface diffusion (FID) method, in which protein and precipitant solutions are carefully superimposed and left to slowly mix diffusively, was least used in the past due to handling difficulties. However, in the last 4 years the free interface technique has experienced a revival for both screening and optimization procedures. The

#### Protocol 3.1 Setting up a screen in a microbatch experiment manually

**Equipment and reagents**

Screening solutions from commercial or home made kits
Protein solution
Pipette of 1–2 μl
Optional: automated hand held pipette
Light microscope
Paraffin oil (Hampton Research or Molecular Dimensions)
Silicone oil (Hampton Research or Molecular Dimensions)
Microbatch plates (e.g. Nunc, Terazaki, Douglas Instruments)

**Method**

1. Pipette or pour 6 ml of paraffin oil into a microbatch plate. The oil will spread over the plate and cover the wells.
2. Withdraw 1 μl of the screen solution from its container using a pipette.
3. Insert the pipette tip into the well under the surface of the oil and dispense the 1 μl drop onto the floor of the plate. As you withdraw the tip from the oil, the drop will detach from it and fall to the bottom of the well within a few seconds (Fig. 3.2). Pipettes usually have two stops when pressing them. Dispense the drop into the oil while holding the pipette on the first stop; otherwise you will introduce air bubbles into the drop.
4. Add 1 μl of protein solution to the same well in the same way. The two (separate) 1-μl drops join and become a 2-μl drop. If the drops don't coalesce, mix them gently with the pipette tip.
5. Incubate at the temperature of your choice.
6. Observe trials regularly under a light microscope.

**Method for setting up and using a robot**

There are several robots for setting up screening experiments in microbatch (e.g. Luft *et al.*, 2003; Chayen *et al.*, 1992; DeLucas *et al.*, 2003). The precipitant solutions are transferred simultaneously from stock screening solutions to crystallization plates by any number of syringes, depending on the robotic system. The drops are dispensed under oil and the protein is added to the precipitant drops using a dedicated syringe for the protein solution, either simultaneously with the screening solutions or at a later stage. Some of the robots have a routine of mixing the drops.

most suitable vessel for this method is an X-ray capillary, the geometry of which reduces undesired convective effects (García-Ruíz *et al.*, 2002). A crystallization cassette has also been designed for holding many capillaries that can be put into contact with various precipitant solutions for the use of this method within a high-throughput environment (Ng *et al.*, 2003).

A microfluidic chip has been developed for rapid screening of protein crystallization conditions (Hansen *et al.*, 2002) using the free interface diffusion method. The chip is comprised of a multilayer, silicon elastomer and has 480 valves operated by pressure. The valves are formed at the intersection of two channels separated by a thin membrane. When pressure is applied to the top channel it collapses

### Protocol 3.2 Fine tuning of conditions in microbatch

#### Equipment and reagents

0.5-ml Eppendorf tubes
Coverslips
Pipette
Microbatch plates
Protein solution
Crystallization reagents
Paraffin oil

#### Method

**1.** Mix the protein solution and the crystallizing agents in an Eppendorf tube. (If quantities are very small, mix on a coverslip.) When mixed, draw the drop with a pipette tip and dispense under paraffin oil as described in Protocol 3.1.
**2.** Incubate at the temperature of your choice.
**3.** Observe trials under a light microscope.

When performed by a robot, the different ingredients are placed in different channels/syringes of a dispensing system and dispensed simultaneously under the oil by the action of motorized syringes. All robots have routines whereby they pick up chosen stock solutions and dispense them into a well to which protein is added simultaneously or later on (Chayen, 2006).

### Protocol 3.3 Two alternative ways of harvesting crystals from microbatch

#### Equipment and reagents

Cryoprotectant solution
Precipitate solution at ~5% higher concentration than that in the drops
Micro tools (Hampton Research)
Standard pipette
Scalpel
Loops
Depression plates

#### Method

**1.** Add a few microlitres of cryoprotectant solution to the drop containing the crystals.
**2.** After several minutes check that the crystals are not cracked or dissolved by looking at them under the microscope. If they crack or dissolve, adjust the concentration of cryoprotectant or change cryoprotectant.
**3.** Take the crystals directly out of the oil using a loop and freeze.

If the above protocol proves tricky, harvest in the following way:

**1.** Add harvest solution (of ~5% higher concentration of precipitant than that in the drop) into the well containing the crystals. If you have a 1-μl drop, add 5–10 μl of harvest solution.
**2.** Wait a short while (up to 15 min) to allow the crystals to equilibrate.
**3.** Withdraw the enlarged drop using a standard 10–100 μl pipette which had its tip cut off with a scalpel in order to widen its bore.
**4.** If the crystals stick to the vessel, loosen them gently inside the drop using micro tools or very thin strips of filter paper (the edge of the strip that will touch the crystal is best torn rather than 'cleanly' cut with scissors).
**5.** Transfer the drop into a depression well containing more harvest solution.
**6.** From this stage onwards, handle the crystals as you would from a standard diffusion trial.

the membrane into the lower channel; when the pressure is removed the membrane springs open. Each trial uses 10 nl of protein sample and 144 trials are performed at a time. This device has produced diffracting crystals in volumes of 5 to 20 nl, including new crystals that were not detected by other crystallization methods. A chip for performing larger numbers of trials has recently been devised.

## 3.5 Automation of optimization experiments

Screening procedures can readily be automated and adapted to high throughput. Optimization techniques that are based on control of the crystallization environment (as opposed to just fine tuning the conditions of the initial hit) are more difficult to automate and to adapt to high throughput. However, more effort is currently being invested in automation and miniaturization of such techniques and some examples are given in the following sections.

### 3.5.1 Use of the crystallization phase diagram for optimization

A very common occurrence in crystallization is the formation of clusters of non-diffracting crystals or crystalline precipitate that can not be improved by merely fine tuning the crystallization parameters. In such cases a working phase diagram can be constructed based on the conditions that give the low quality crystals (Protocol 3.4). Such a diagram determines at which conditions the protein precipitates, at which the solution remains clear, and where crystals form (Fig. 3.3). This information can then be used to determine the appropriate conditions for separating the phases of nucleation and growth. Working phase diagrams can be generated manually,

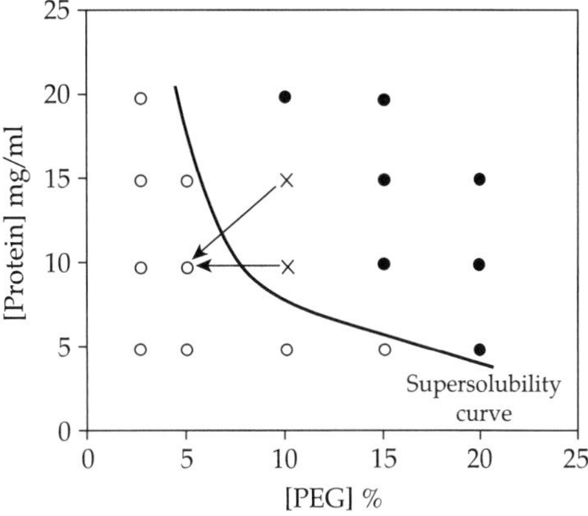


**Figure 3.3** A working phase diagram of a protein. X represent conditions obtained from screening 'hits', giving low quality crystals or crystalline precipitate. Open circles represent clear drops; full circles represent precipitate. The working phase diagram has been plotted from results of trials set up at conditions ranging from above and below the conditions of the 'hits'. Arrows represent two options of transfer/dilution from conditions that would give poor crystals to clear conditions just below the supersolubility curve. The diagonal arrow represents dilution with buffer, where both the concentrations of protein and the precipitant are reduced. The horizontal arrow represents transfer of a drop to lower precipitant concentration or dilution with the protein solution. In these cases the concentration of the protein in the drop in not reduced.

### Protocol 3.4 Constructing a working phase diagram

**Method**

**1.** Note the conditions in a 'hit' of a screen that have yielded crystals of some sort or a crystalline precipitate.

**2.** Dispense 12–24 trials using a crystallization method of your choice, varying the protein concentration and precipitant concentration in steps on a matrix grid. The dispensing can be done either manually or using a robot.

**3.** The concentrations should be above and below the concentration that gave the conditions of the hit. Alternatively the pH or another parameter can be varied.

**4.** Plot the results and you will obtain the supersolubility curve. Figure 3.3 shows a practical example: if a screen has produced 'hits' at conditions of 10–15 mg/ml of protein and 5–10% PEG, the experiments are set at concentrations ranging from 5–20 mg/ml protein versus 2.5–15 or 20% PEG, thus covering a range of conditions above and below the 'hits'.

but of course can be done far quicker if performed using robots.

### 3.5.2 Separation of nucleation and growth

The most commonly used technique for separation of nucleation and growth is seeding (Stura, 1999; Bergfors, 2003; and references therein). Although very successful, seeding often involves handling of fragile crystals and, other than streaking, seeding manoeuvres are not very simple to perform.

An alternative means of conducting trials with a similar outcome to seeding, but without the need to handle crystals, is by dilution. An additional benefit of dilution methods is that they are more amenable to automation compared with seeding techniques (Chayen, 2005). The aim of dilution is to start the trial at nucleation conditions and after a given time

#### Protocol 3.5 Dilution in microbatch and vapour diffusion

**Equipment and reagents**

Microbatch plates
Paraffin oil
Pipette or robot
Plates for hanging or sitting drops
Coverslips
Sealing tape
Protein solution
Buffer

**Method**

**1.** Set up trials under conditions that would give you the low-quality crystals.
**2.** Dilute the trials at given times after set up by adding a volume of buffer or of protein in buffer at a volume which is 5–10% of the total drop volume.

In microbatch, the crystallization drops are diluted directly. In the case of vapour diffusion, either the drops themselves or the reservoirs can be diluted.

If performing manually, use a pipette of your choice. If performed by a robot, the robotic system is programmed to revisit the drops and/or the reservoirs at given times in order to add the diluent (Chayen, 2006).

#### Protocol 3.6 'Backing off' in hanging drops

**Equipment and reagents**

Linbro type plates
Any oil of viscosity of ~5
Pipettes
Coverslips
Pasteur pipette with rubber squeezer
Protein solution
Precipitating agents

**Method**

**1.** Prepare 6–10 reservoirs with a solutions containing precipitant concentration that would result in producing a clear drop if crystallization drops were set up and left to incubate under these conditions. Determine these concentrations from the region in the phase diagram that is just under the supersolubility curve (Fig. 3.3).
**2.** Grease the rim of the plate with oil using a Pasteur pipette and cover the reservoirs with cover slips.
**3.** Set up 6–10 trials under conditions that would give you the low quality crystals.
**4.** At a given time after set up (based on when the first crystals were seen as explained in Section 3.5.3) transfer (either manually or by a robot) one of the cover slips onto a reservoir containing the lower precipitant concentration (Figs 3.3 and 3.4).
**5.** After a further time interval transfer a second cover slip, then the third, fourth, etc. (Figs 3.3 and 3.4).
**6.** Leave one or two drops without transferring them and one or two drops under the low precipitant concentration to act as controls.
**7.** Incubate and wait at least 1 week. The time for the formation of crystals will be longer in the transferred drop compared to the control experiment that had not been transferred, however the crystals at some of the transfer times are likely to be fewer and better ordered.

to 'back off' to conditions of growth (Fig. 3.3). Dilution can be achieved using all crystallization methods. Protocols for 'backing off' experiments in the most common techniques, namely microbatch and vapour diffusion, are given in Protocol 3.5.

### 3.5.3 Determination of the timing of dilution

Dilution must be performed before crystals are visible. The time of dilution is selected by reference to the time which it took to see the first crystals in the initial screens (Saridakis *et al.*, 1994). For example if crystals appeared within 24 hours, nucleation would have occurred anytime between setting up the experiments to several hours before the crystals appear. Hence the dilution should be done at intervals of 1–2 hours after set up. If crystals appear after 4 days, dilution should be performed at intervals of 8–12 hours. Trials that are diluted too soon will produce clear drops while those that are too late will yield low-quality crystals.

### 3.5.4 'Backing off' experiments in hanging drops

Dilution experiments involve revisiting the crystallization drops which may cause disruption to the trial. An alternative way of backing off without touching the trial drop can be achieved in hanging drops in the following way : the coverslips holding the drops are incubated for some time over reservoir solutions that normally give many small crystals (Protocol 3.6). After a given time (before crystals are visible), the coverslips are transferred over reservoirs containing lower precipitant concentrations that would normally yield clear drops (Fig. 3.4). As in dilution, the time of transfer is selected by reference to the time which it took to see the first crystals in the initial screens. The transfer lasts 1–2 seconds. This technique has produced significant improvement in crystal order of a number of proteins (e.g. Saridakis and Chayen, 2000, 2003; Krengel *et al.*, 2006).

### 3.5.5 Control of evaporation in microbatch

An alternative to dilution in microbatch is to approach the optimization strategy from the opposite

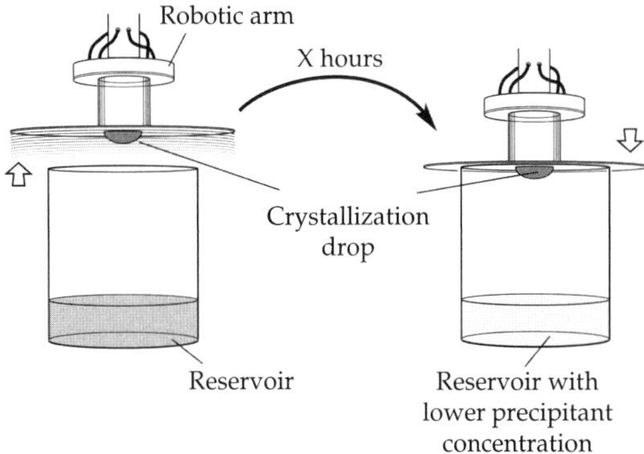


**Figure 3.4** Transfer of a hanging drop from nucleation to optimal growth conditions. Modified from Chayen and Saridakis (2002), *Acta Cryst. D* **58**, 921.

#### Protocol 3.7 Induction and subsequent arrest of nucleation in microbatch

**Equipment and reagents**

Microbatch plates
Paraffin oil
Protein solutions
Precipitant solutions
Pipette or robot
Crystallization reagents

**Method**

1. Set up several crystallization plates containing microbatch trials under a layer of paraffin oil so that the oil just covers the trials (Fig. 3.5a).
2. Allow to incubate for a given time.
3. Top up the oil (at different times for the different trays) so that it fills the dish (Fig. 3.5b).
4. Continue to incubate as standard microbatch trials.
5. Observe daily.

A robot will first dispense the trials under the lower volume of oil. A robotic arm is programmed to add oil at various time intervals after setting up the trials.

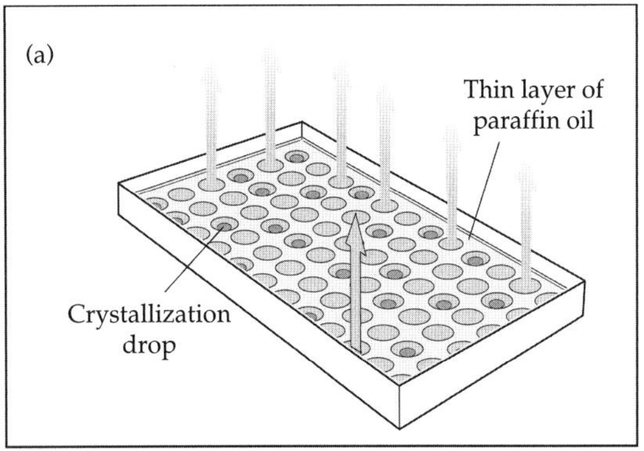


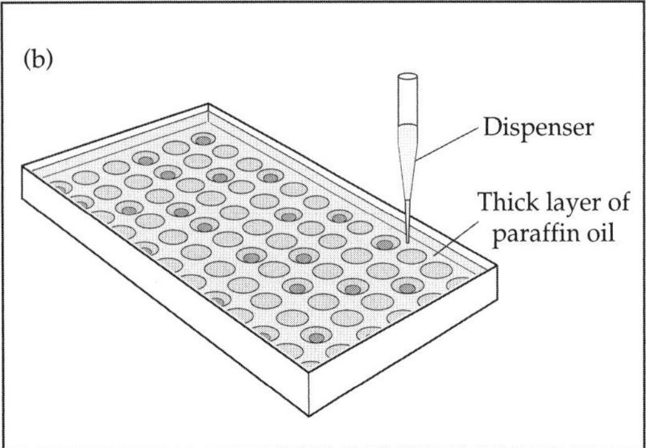


**Figure 3.5** Controlled evaporation of crystallization drops in microbatch. (a) Trials incubated under a thin layer of oil that allows concentration via evaporation, thus leading to nucleation. (b) Arrest of evaporation/concentration by addition of oil to produce a thicker layer above the trials. Modified from Chayen and Saridakis (2002), *Acta Cryst. D* **58**, 921–927, with permission from the IUCr.

side of the phase diagram. Contrary to dilution, where nucleation is the starting point, the starting conditions can be at metastable or unsaturated conditions where the solution is clear. Nucleation can be induced by allowing the crystallization trials to concentrate, and can then be arrested before the nucleation becomes excessive (Protocol 3.7). This is achieved by controlled concentration of the drops through a thin layer of paraffin oil and arrest of the evaporation by increasing the thickness of the oil layer after a given time (Chayen and Saridakis, 2002). Arrest of evaporation at the early stages of nucleation results in the formation of fewer crystals of better quality. This method can also be used when useful crystals take an excessively long time to grow in conventional trials.

## 3.6 Control of the evaporation kinetics of vapour diffusion trials

The formation of numerous, small crystals instead of the desired large single ones can occur due to the crystallization process taking place too rapidly. A means to slow down the equilibration rate in vapour diffusion and thereby approach supersaturation more slowly in order to avoid the formation of 'showers', twinned crystals, or precipitate, is by placing a mixture of paraffin and silicone oils as a barrier over the reservoirs of hanging or sitting drop trials (Chayen, 1997b). Paraffin oil totally blocks all vapour diffusion, resulting in drying out of the drops, while silicon oil allows diffusion to take place fully. The equilibration rate is therefore dictated by the ratio of the two oils and by the thickness of the oil–mixture layer (Protocol 3.8). Equal volumes of paraffin and silicon, in a layer of 250–500 μl, placed over 0.6–1 ml reservoirs in Linbro type plates were found to be optimal. This method has been successful in crystallizing problematic target proteins (e.g. Mandelman *et al.*, 2002; Isupov *et al.*, 2004). The advantage of this technique is that no change is required to the crystallization conditions nor the vessel used. It has recently been automated for adaptation to structural genomics projects by adding one extra step to the procedure used by existing crystallization robots (Chayen and Saridakis, 2002).

## 3.7 Influencing the crystallization environment

There are numerous parameters that can influence crystallization. The crystallization environment, such as growth of crystals in the presence of magnetic fields (Ataka and Wakayama, 2002), electric fields (Charron *et al.*, 2003), high pressure (Lorber *et al.*, 1996), centrifugation (Lenhoff *et al.*, 1997), levitation (Rhim and Chung, 1990), microgravity (Chayen and Helliwell, 1999), and many others can have a significant effect on the quality of the crystals obtained. Most of these though, require expensive, dedicated apparatus and have not yet been widely applied. An environment that is applied more often in comparison to others is gelled media.

## Protocol 3.8 Setting up an oil barrier over vapour diffusion trials

### Equipment and reagents

Paraffin oil
Silicone oil
Crystallization plates for hanging and sitting drops
Coverslips
Sealing tape
Protein solution
Crystallization reagents

### Method for preparation of the oils

**1.** Mix paraffin and silicone oils in equal volumes.
**2.** Shake well and allow to stand for several minutes. The oils are totally miscible once the bubbles have disappeared.

### Method for setting up hanging drops

**1.** Use a Linbro type plate for hanging drops.
**2.** Coat the lips of the reservoirs with grease or oil (unless your plates are pregreased).
**3.** Pipette 0.6–1 ml of the reservoir solution which gives you the showers of crystals into each well.
**4.** Pipette volumes ranging from 0.1– 0.5 ml of a mixture of paraffin and silicone over all the reservoirs except for one reservoir. The oil will form a layer above the reservoirs (Fig. 3.6b).
**5.** Dispense the hanging drops on the coverslips as usual by mixing the protein solution with the reservoir solution. Use the reservoir without oil as your source of precipitant for all the drops.
**6.** Invert the cover slips and place over the wells containing the oil layer.
**7.** Place the last drop over the reservoir without the oil. This drop will act as your control.
**8.** Incubate at the temperature of your choice.
**9.** Wait patiently for the results because in trials containing an oil barrier, crystals require longer periods (e.g. 8–10 days compared to 12–24 hours) to grow to full size but their quality is improved.
**10.** If the quality of the crystals is not sufficiently improved, repeat the protocol using different ratios of paraffin and silicon.

### Method for sitting and sandwich drops

In the case of sitting and sandwich drops, set up the trials as you would normally do and place the layer of oil above the reservoir before sealing the plates with tape (Chayen, 2006).

*Warning:* This technique does not work with PEG or MPD concentration above 13% but is very effective at concentrations below 13% and at all concentrations of all salts.

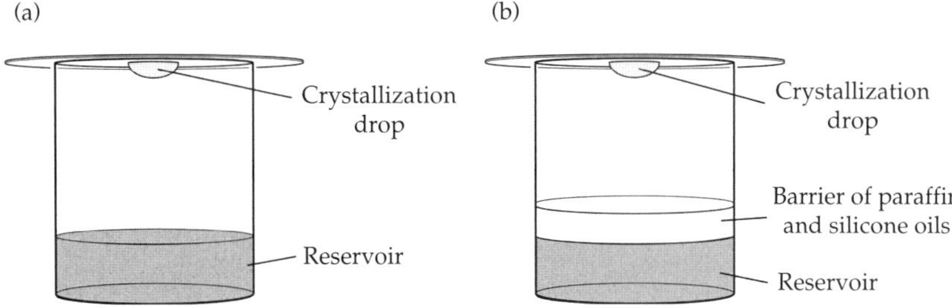


**Figure 3.6** Application of an oil barrier to improve crystals in hanging drops. (a) A standard hanging drop trial; (b) a trial containing an oil barrier. Modified from Chayen (1997b). *J. Appl. Cryst.* **30**, 198–202, with permission from the IUCr.

### 3.7.1 Crystallization in gelled media

The quality of crystals can be significantly improved when grown in gel media. Growth in gel reduces both convection of the protein molecules, thus favouring their slow diffusion to the crystal faces, and sedimentation of the growing crystals, in some way mimicking a microgravity environment (Robert *et al.*, 1999; and references therein).

Gelled trials are performed in vapour diffusion, counter diffusion, and in batch. To date, the most simple and speedy procedure for performing trials

## Protocol 3.9 Preparation of the stock solutions of gels

### Equipment and reagents

Glass beaker
Stirrer
Electric device for supporting the beaker and stirrer
Tetramethyl orthosilane (TMOS)
Deionized water

### Method

Prepare a 4 ml stock solution of TMOS at 5% (v/v) as follows:

**1.** Add 0.2 ml TMOS solution to 2 ml deionized water in a glass beaker.
**2.** Stir the solution vigorously using a stirrer at high speed or by hand. At first, a phase separation occurs which looks like oil drops in the solution. When stirring vigorously these drops disperse.
**3.** Top up the solution to 4 ml with deionized water.
**4.** Stir vigorously for an additional 10–15 min keeping the beaker covered. The mixture should by now have become a clear solution.

## Protocol 3.10 Dispensing microbatch crystallization trials in gels

### Equipment and reagents

Freshly made 5% (v/v) gel solution
Paraffin oil
Microbatch plates
Protein solution
Crystallizing reagents
Eppendorf tubes
Coverslips
Pipette

### Method for manual dispensing

**1.** Mix the protein, crystallizing agents, buffer, etc., and the freshly made gel solution in an Eppendorf tube or on a coverslip. The gel solution should be at a final concentration of at least 0.2% (v/v).
**2.** When mixed, draw a drop of a volume of your choice with a pipette tip and dispense under the oil as described in Protocol 3.1.

### Method for dispensing by robot

**1.** Choose a robotic dispensing system consisting of several channels/syringes in which precipitant, buffer, protein, and additives can be put into different syringes and dispensed simultaneously or in rapid succession.
**2.** Fill a crystallization tray with paraffin oil as done for standard microbatch trials (by hand or robot).
**3.** Place the gel solution while it is still in liquid form into one of the channels/syringes of the dispensing system in the same way as for the other components of the crystallization mixture.
**4.** Dispense the gel solution under the oil simultaneously with all the other components to form one drop.
**5.** Incubate the trials at the selected temperature(s).
**6.** After a given time (12–16 hours for the gel type and concentrations described here) polymerization occurs and the drop gels.
**7.** Harvesting crystals from the gelled drops is done in the same way as from standard microbatch trials since the gel is quite tenuous.
**8.** Test a variety of final concentrations of the gel ranging from 0.2–0.5%.

Gels have also been used as a slowing down mechanism in various guises such as a permeable 'plug' between the two solutions or, in the method of gel acupuncture, as a barrier for the precipitant solution on its way into the protein-containing capillary (Robert *et al.*, 1999).

in gel media is in microbatch (Protocols 3.9 and 3.10). The process is automated and is as simple as setting the trials in standard microbatch. Tetramethyl orthosilane (TMOS) has been found to be the most suitable gel for such microbatch experiments (Moreno *et al.*, 2002).

## 3.8 Concluding remarks

Obtaining high-quality crystals is becoming increasingly crucial to progress in the postgenomic era. Whether in the individual laboratory, or as part of structural genomics projects, it is always vital to have a portfolio of crystallization techniques that can be applied, especially in the cases of proteins that are proving difficult to crystallize. In order to be useful to structural genomics projects, it is also important to miniaturize and automate as many techniques as possible.

This chapter has described several alternative methods to those that are most commonly used. The use of phase diagrams for optimization of the crystallization conditions, control of crystallization kinetics, dynamic separation of nucleation and growth, and crystallization in gels can now be performed automatically while using nanolitre or microlitre volumes of sample. These methods provide a variety of avenues to be explored, either in parallel to conventional methods or when screening and subsequent fine tuning of the initial screening conditions have failed to produce high-quality crystals.

## References


Ataka, M. (1993). Protein crystal growth: An approach based on phase diagram determination. *Phase Transitions* **45**, 205–219.

Ataka, M. and Wakayama, N. I. (2002). Effects of a magnetic field and magnetization force on protein crystal growth. Why does a magnet improve the quality of some crystals? *Acta Crystallogr. D* **58**, 1708–1710.

Bergfors, T. (2003). Seeds to crystals. *J. Struct. Biol.* **142**, 66–76.

Bergfors, T. M., ed. (1999). *Protein Crystallization: Techniques, Strategies, and Tips*. International University Line, La Jolla, USA.

Charron, C., Didierjean, C., Mangeot, J. P. and Aubry, A. (2003). The 'Octopus' plate for protein crystallization under an electric field. *J. Appl. Cryst.* **36**, 1482–1483.

Chayen, N. E. (1997a). The role of oil in macromolecular crystallization. *Structure* **5**, 1269–1274.

Chayen, N. E. (1997b). A novel technique to control the rate of vapour diffusion, giving larger protein crystals. *J. Appl. Cryst.* **30**, 198–202.

Chayen, N. E. (1998). Comparative studies of protein crystallization by vapour-diffusion and microbatch techniques. *Acta Crystallogr. D* **54**, 8–15.

Chayen, N. E. (2003). Crystallisation of membrane proteins in oils. In: *Methods and Results in Crystallization of Membrane Proteins*, Iwata, S., ed. International University Line, USA, pp. 131–139.

Chayen, N. E. (2004). Turning protein crystallisation from an art into a science. *Curr. Opinion Structural Biol.* **14**, 577–583.

Chayen, N. E. (2005). Methods for separating nucleation and growth in protein crystallisation. *Prog. Biophys. Molec. Biol.* **88**, 329–337.

Chayen, N. E. (2006). Optimization techniques for automation and high throughput. In *Methods in Molecular Biology, Macromolecular Crystallography Protocols*, Vol. 1, Doublie, S., ed. Humana Press, New Jersey, pp. 175–190.

Chayen, N. E. and Helliwell, J. R. (1999). Space-grown crystals may prove their worth. *Nature*, **398**, 20.

Chayen, N. E. and Saridakis, E. (2002). Protein crystallization for genomics: towards high-throughput optimization techniques. *Acta Crystallogr. D* **58**, 921–927.

Chayen, N. E., Stewart, P. D. S. and Blow, D. M. (1992). Microbatch crystallization under oil – a new technique allowing many small-volume crystallization trials. *J. Crystal Growth* **122**, 176–180.

Chayen, N. E., Stewart, P. D. S., Maeder, D. and Blow, D. M. (1990). An automated-system for microbatch protein crystallization and screening. *J. Appl. Cryst.* **23**, 297–302.

Chayen, N. E., Boggon, T. J., Casetta, A., Deacon, A., Gleichmann, T., Habash, J., *et al.* (1996). Trends and challenges in experimental macromolecular crystallography. *Quart. Rev. Biophys.* **29**, 227–278.

D'Arcy, A., Elmore, C., Stihle, M. and Johnston, J. E. (1996). A novel approach to crystallising proteins under oil. *J. Cryst. Growth* **168**, 175–180.

D'Arcy, A., MacSweeny, A., Stihle, M. and Haber, A. (2003). Using natural seeding material to generate nucleation in protein crystallization experiments. *Acta Crystallogr. D* **59**, 1343–1346.

DeLucas, L. J., Bray, T. L., Nagy, L., McCombs, D., Chernov, N., Hamrick, D., *et al.* (2003). Efficient protein crystallization. *J. Struct. Biol.* **142**, 188–206.

Ducruix, A. and Giegé, R. (1992). *Crystallization of Nucleic Acids and Proteins, A Practical Approach*. Oxford University Press, Oxford.

Ducruix, A. and Giegé, R. (1999). *Crystallization of Nucleic Acids and Proteins, A Practical Approach*, 2nd edn. Oxford University Press, Oxford.

García-Ruíz, J. M., Gonzalez-Ramirez, L. A., Gavira, J. A. and Otálora, F. (2002). Granada Crystallisation Box: a new device for protein crystallisation by counter-diffusion techniques. *Acta Crystallogr. D* **58**, 1638–1642.

Hansen, C. L., Skordalakes, E., Berger, J. M. and Quake S. R. (2002). A robust and scalable microfluidic metering method that allows protein crystal growth by free interface diffusion. *Proc. Natl. Acad. Sci. USA*, **99**, 16531–16536.

Isupov, M. N., Brindley, A. A., Hollingsworth, E. J., Murshudov, G. N., Vagin, A. A. and Littlechild, J. A. (2004). Crystallization and preliminary X-ray diffraction studies of a fungal hydrolase from Ophiostoma novo-ulmi. *Acta Crystallogr. D* **60**, 1879–1882.

Krengel, U., Dey, R., Sasso, S., Okvist, M., Ramakrishnan, C. and Kast, P. (2006). Preliminary X-ray crystallographic analysis of the secreted chorismate mutase from Mycobacterium tuberculosis: a tricky crystallization problem solved. *Acta Crystallogr. F* **62**, 441–445.

Lenhoff, A. M., Pjura, P. E., Dilmore, J. G. and Godlewski, T. S. Jr. (1997). Ultracentrifugal crystallization of proteins: transport-kinetic modelling, and experimental behavior of catalase. *J. Cryst. Growth* **180**, 113–126.

Lorber, B., Jenner, G. and Giege, R. (1996). Effect of high hydrostatic pressure on nucleation and growth of protein crystals. *J. Cryst. Growth* **158**, 103–117.

Luft, J. R., Collins, R. J., Fehrman, N. A., Lauricella, A. M., Veatch, C. K. and DeTitta, G. T. (2003). A deliberate approach to screening for initial crystallization conditions of biological macromolecules. *J. Struct. Biol.* **232**, 170–179.

Mandelman, D., Gonzalo, P., Lavergne, J.-P., Corbier, C., Reboud, J.-P. and Haser, R. (2002). Crystallization and preliminary X-ray study of an N-terminal fragment of rat liver ribosomal P2 protein. *Acta Crystallogr. D* **58**, 668–671.

McPherson, A. (1999). *Crystallization of Biological Macromolecules.* Cold Spring Harbor Laboratory Press, Cold Spring Harbor.

Moreno, A., Saridakis, E. and Chayen, N. E. (2002). Combination of oils and gels for enhancing the growth of protein crystals. *J Appl. Cryst.* **35**, 140–142.

Mueller, U., Nyarsik, L., Horn, M., Rauth, H., Przewieslik, T., Saenger, W., Lehrach, H. and Eickhoff, H. (2001). Development of a technology for automation and miniaturization of protein crystallization. *J. Biotechnol.* **85,** 7–14.

Ng, J. D., Gavira, J. A. and García-Ruíz, J. M. (2003). Protein crystallization by capillary counterdiffusion for applied crystallographic structure determination. *J. Struct. Biol.* **142**, 218–231.

Rhim, W.-K. and Chung, S. K. (1990). Isolation of crystallizing droplets by electrostatic levitation. *Methods* **1**, 118–127.

Robert, M.-C., Vidal, O., Garcia-Ruiz, J.-M. and Otálora, F. (1999). Crystallization in gels. In: *Crystallization of Nucleic Acids and Proteins*, Ducruix, A. and Giegé, R., eds. Oxford University Press, Oxford, pp.149–175.

Santarsiero, B. D., Yegian, D. T., Lee, C. C., Spraggon, G., Gu, J., Scheibe, D., *et al.* (2002). An approach to rapid protein crystallization using nanodroplets. *J. Appl. Cryst.* **35**, 278–281.

Saridakis, E. and Chayen, N. E. (2000). Improving protein crystal quality by decoupling nucleation and growth in vapor diffusion. *Protein Sci.* **9**, 755–757.

Saridakis, E. and Chayen, N. E. (2003). Systematic improvement of protein crystals by determining the supersolubility curves of phase diagrams. *Biophys J.* **84**, 1218–1222.

Saridakis, E., Shaw Stewart, P. D., Lloyd, L. F. and Blow, D. M. (1994). Phase-Diagram and Dilution Experiments in the Crystallization of Carboxypeptidase-G(2). *Acta Crystallogr. D* **50**, 293–297.

Service, R. (2002). Structural genomics. Big biology without the big commotion. *Science* **298**, 948–950.

Soriano, T. M. and Fontecilla-Camps, J. C. (1993). Astec – an automated-system for sitting-drop protein crystallization. *J. Appl. Cryst.* **26**, 558–562.

Stura, E. A. (1999). Seeding. In: *Crystallization of Nucleic Acids and Proteins*, Ducruix, A. and Giegé, R., eds. Oxford University Press, Oxford, pp.177–208.

Walter, T. S., Brown, D. J., Pickford, M., Owens, R. J., Stuart, D. I. and Harlos, K. (2003). A procedure for setting up high-throughput nanolitre crystallization experiments. I. Protocol design and validation. *J. Appl. Cryst.* **36**, 308–314.

Weber, P. C. (1990). A protein crystallization strategy using automated grid searches on successively finer grids. *Methods* **1**, 31–37.

CHAPTER 4

# First analysis of macromolecular crystals

**Sherin S. Abdel-Meguid, David Jeruzalmi, and Mark R. Sanderson**

## 4.1 Macromolecular cryocrystallography

The past decade and a half have witnessed an almost complete revolution in the way that macromolecular diffraction data are recorded. The promise of diffraction data measurements essentially free of the effects of radiation damage has driven a change from older methods requiring crystalline samples to be mounted in thin glass capillaries for measurements at ambient temperatures (or so) to newer, experimental schemes that enable measurements at cryogenic temperatures from crystals mounted in free-standing films. Preparation of macromolecular crystalline samples for measurements at cryogenic temperatures can be considered to have three separable stages. These are: cryoprotection, shock-cooling, and cryogenic transfer to the X-ray diffraction camera. Each of these will be described in detail below. However, as in any experimental technique, numerous variations to solve specific problems have been, and continue to be, reported and the interested reader is urged to consult issues of the *Journal of Applied Crystallography* and *Acta Crystallographica*, section D. The theoretical and experimental aspects of macromolecular cryocrystallography have been reviewed in Garman and Schneider, 1997 and Rodgers, 1997.

## 4.2 Cryoprotection of macromolecular crystals

Cryoprotection of crystalline samples involves either the introduction of antifreeze agents into and around the crystal (which contain 40–90% solvent) or a removal of water from its exterior. The objective of this procedure is to enable the water in and around a crystal to form stabilizing, vitreous ice upon shock-cooling, rather than crystalline ice that would damage the lattice. Cryoprotection is commonly accomplished by one of three methods: (1) slow equilibration of crystals in a cryostabilization buffer; (2) quick passage of the crystal through a cryostabilization buffer ('quick dip' method); or (3) transfer of the sample into water-immiscible hydrocarbons. These are discussed in detail below. The slow equilibration approach allows the cryostabilization buffer, containing the antifreeze agent, to thoroughly permeate the crystal. Formation of crystalline ice in and around the crystal is thus suppressed during shock cooling. However, some samples do not tolerate exposure to antifreeze agents for the time required for a full equilibration. In such cases, the quick-dip method or a transfer to a water immiscible hydrocarbon is indicated (Hope, 1988; Kwong and Liu, 1999; Riboldi-Tunnicliffe and Hilgenfeld, 1999). These methods suppress formation of damaging crystalline ice around the crystal by either coating the surface with antifreeze-containing buffer or an immiscible hydrocarbon. Formation of crystalline ice around the crystal is suppressed, while the water within the crystal (thought to be kinetically less prone to form crystalline ice) remains in a native state. The goal of all of these approaches is to define a reproducible handling procedure that enables shock cooling of a macromolecular crystal with its X-ray diffractive properties (as observed at ambient or subzero temperatures) preserved intact (or in some cases improved). The appropriate experimental protocol will vary with the crystalline sample.

### 4.2.1 Identification of a cryostabilization buffer

Preparation of a crystal cryostabilization buffer for a newly prepared macromolecular crystal is an extension of the efforts to devise a crystal stabilization buffer; in some cases the buffers may be one and the same. The major objective is to introduce the crystal – in a controlled manner – into solutions containing synthetic mother liquor supplemented with an antifreeze agent, which include various polyhydric alcohols, low molecular weight polyethylene glycols, sugars, organic solvents (Garman and Schneider, 1997), and salts (Rubinson *et al.*, 2000) (Table 4.1). Making the assumption that the introduction of any non-native component into a crystal has the potential to cause damage, the investigator should aim to use the lowest concentration that is necessary to cryopreserve the crystal for data collection (Mitchell and Garman, 1994). The nature and concentration of the appropriate cryosolvent will vary with the crystalline sample.

### 4.2.2 Transfer of macromolecular crystals into cryostabilization buffer

The three common methods in use for performing crystal cryoprotection prior to shock-cooling are discussed in the following sections. The first two methods involve replacement of solvent in/around the crystal with a synthetic buffer, while the third approach involves replacement of solvent on the surface of the crystal with various types of hydrocarbons. The method of cryoprotection will usually vary with the crystalline sample.

**Table 4.1** Antifreeze agents used in the shock cooling of macromolecular crystals

| Agent | Concentration range (%) |
|---|---|
| Glycerol* | 10–40 |
| Ethylene glycol | 10–40 |
| Propylene glycol | 10–40 |
| 2,4 Methyl pentane diol | 10–40 |
| Sucrose | up to 30 |
| Alcohols (methanol, ethanol, isopropanol) | up to 10 |
| Low molecular weight polyethylene glycols | up to 25 |
| Mineral oil | 100 |
| Salts (Lithium, sodium, and magnesium salts) | varies |
| Mixtures of the above | varies |

#### *4.2.2.1 Slow equilibration in cryoprotection buffer*

Slow equilibration into a cryostabilization buffer is carried out by serially transferring the macromolecular crystal into a progressively higher concentration of cryosolvent. This procedure is initiated by transferring crystals into synthetic mother liquor supplemented with 5% antifreeze agent. Crystal transfers can performed using a fibre loop, as described below. The crystal is then serially transferred into a progressively (e.g. 5% steps) higher concentration of antifreeze agent until the concentration of agent determined to be non-damaging is reached. Allow the crystal to equilibrate in the cryostabilizing buffer for 15 min to several hours or even over night; the equilibration time will vary with the crystal. The time required for diffusion into a crystal is dependent on many factors (size of small molecule, temperature, size of crystal, nature of channels in the crystal, etc.). The little data that exists suggests that this time could be on the order of minutes to hours (Wyckoff *et al.*, 1967).

Variations to the slow-equilibration method include cross-linking of crystals prior to transfer into cryobuffers (Lusty, 1999), transfer of crystals into cryobuffer by dialysis, or the introduction of the cryobuffer during crystallogenesis.

#### *4.2.2.2 Rapid passage through cryoprotection buffer*

Rapid passage, also known as the 'quick-dip' method, involves mounting a crystalline sample in a fibre loop and rapidly (1–5 s) dragging the sample through a cryostabilization buffer. This method is especially useful for crystalline samples that are damaged by prolonged exposure to antifreeze-containing buffers. The simplicity and speed offered by the 'quick-dip' method makes it attractive when rapidly preparing shock-cooled samples.

*4.2.2.3 Transfer into an immiscible hydrocarbon*

A third method for cryoprotecting macromolecular crystals involves replacing the solvent around the crystal with a water immiscible hydrocarbon prior to shock cooling. In a manner similar to the 'quick-dip' method, a loop-mounted crystal is passed through a small drop of a hydrocarbon (e.g. Para tone-N, dried paraffin oil, etc.). During this time, the aqueous mother liquor is sloughed-off as a result of being moved through the hydrocarbon. The crystal will often remain in the loop during this protocol. If necessary, a paper wick can be used to aid removal of the aqueous buffer. One deficit of this method is that crystals that are mechanically sensitive do not respond well to being passed through highly viscous hydrocarbons.

## 4.3 Shock-cooling of macromolecular crystals

Preparation of a shock-cooled macromolecular crystal involves the rapid introduction of a loop-mounted sample into a cryogen. Introduction into the cryogen must be rapid in order to ensure that aqueous solvent within the crystal cools as a vitreous and not crystalline ice, which would damage the sample. Three types of cryogens are currently in widespread use and these are freshly thawed liquid propane (or ethane), liquid nitrogen, or cold gaseous nitrogen (Garman and Schneider, 1997; Rodgers, 1997; Garman, 2003).

### 4.3.1 Mounting a crystal in a fibre loop

Figure 4.1 demonstrates the procedure for mounting a macromolecular crystal in a free-standing film, which is supported by a thin fibre loop (Fig. 4.2).

### 4.3.2 Shock-cooling into freshly thawed liquid propane

The use of liquid propane as a cryogen to prepare shock-cooled macromolecular crystals has several advantages. Shock cooling by plunging into a cold liquid enables an efficient transfer of heat away from the sample in comparison to the use of a gaseous cryogen, such as cold nitrogen gas. Propane also displays a high heat capacity and a large differential between its melting and boiling temperatures, which minimizes formation of an insulating layer of cryogen near the crystal due to boiling (Kriminski

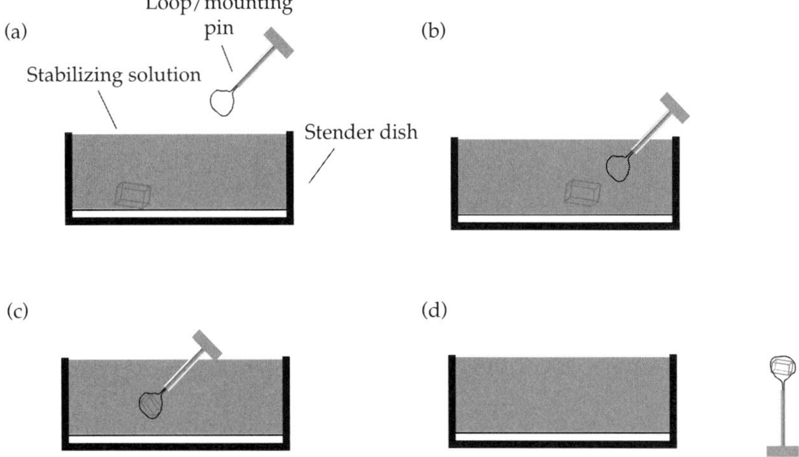


**Figure 4.1** Mounting a macromolecular crystal into a fibre loop. (a) A crystal and fibre loop/mounting pin is selected. The mounting procedure is performed under a low-power dissecting microscope. (b) The loop is used to gently nudge the crystal off the bottom of the dish. (c) As the crystal falls through the solution, the loop is used to 'catch' the crystal. The dislodged crystal will be more buoyant in a viscous solution; this makes the mounting procedure easier to perform successfully. (d) The loop with crystal in tow is passed through the air–water interface. Frequently, at this final step, the crystal will slip off the loop frustrating the mounting process. In such a case, repeat the procedure above.

*et al.*, 2003). However, this method has some important disadvantages, which include additional hardware requirements and the necessity to work with a flammable/explosive material. Finally, commercial shipping of propane requires special paperwork and handling due the associated hazards.

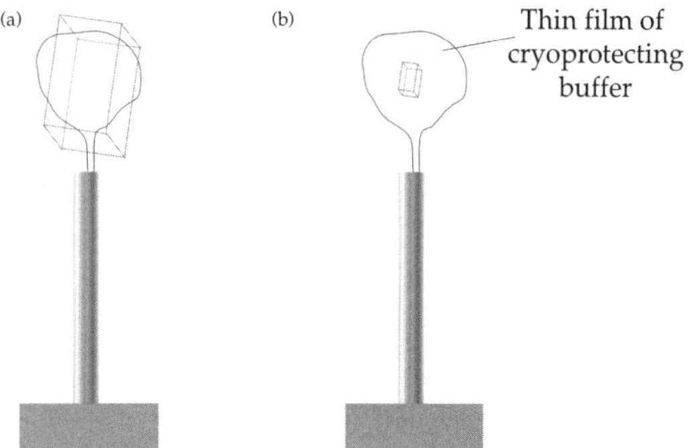


**Figure 4.2** Examples of loop-mounted macromolecular crystals. (a) In this type of mount, the fibre loop and the crystal are of similar sizes. This produces a mount that is nearly dry, which enables faster shock cooling (due to lack of an insulating layer of cryoprotecting buffer). (b) A second-type of fibre-loop mount in which the diameter of the loop is much larger than the crystal. The crystal is entirely suspended in a thin film of cryoprotecting buffer. This type of mount is better suited to crystals that would be damaged by the type of mount depicted in Fig. 6a. The extra buffer in the plane of the loop presents an insulating layer during shock cooling and leads to a highly anisotropic specimen, which may introduce absorption errors into the final data set. The mount depicted in Fig. 6b should be used only when necessary.

### 4.3.3 Shock-cooling into liquid nitrogen

The use of liquid nitrogen as a cryogen exploits its liquid state to enable efficient heat transfer away from the crystalline sample (as compared to gaseous nitrogen), but without the specialized hardware and safety concerns associated with use of liquid propane. An additional advantage is that commercial shipping of samples prepared in this way is much less cumbersome since transport companies do not generally consider them to be hazardous. However, liquid nitrogen is not as favourable a cryogen as liquid propane due to a lower heat capacity and a smaller differential between its melting and boiling temperatures, which can result in less efficient cooling due to the formation of an insulating layer of vapour around the crystal during the initial plunge. There may be instances where the differences in the cryogens (liquid propane vs. liquid nitrogen) might be crucial to the preparation of

### Protocol 4.1 Shock-cooling into freshly thawed liquid propane

**Materials**

Stainless steel Dewar (1000 ml, to hold the sample stage)
Tall Dewar (to hold cryocanes containing processed crystalline samples)
Cryovials (one per sample)
5–10 liters of liquid nitrogen (varies with the number of samples)
Sample stage (wire rack, CrystalCap holder, available from Hampton Research)
Solidified propane (prepared by completely submerging a 50-ml tube containing liquid propane in liquid nitrogen, each sample to be shock-cooled might consume up to 1.5 ml of liquid propane. Liquid propane is prepared by condensing gaseous propane into a tube that has been cooled in a bath of liquid nitrogen. Left submerged in the bath, liquid propane will eventually solidify
Vial canes and sleeves to hold the processed samples
Storage Dewar

**Procedure (Fig. 4.3)**

**1.** Immediately prior to a session of shock cooling of crystalline samples, a tube of solid propane is allowed to thaw; the resulting liquid is used to fill 1–4 cryovials at a time. If the loop-mounting process (see above) becomes prolonged (>10–15 min), the liquid propane might begin to solidify. In such a case, the mounting process should be halted and the solidified propane in the cryovials should be thawed. The stock of thawed liquid propane should be maintained cold by partial immersion in a bath of liquid nitrogen. Liquid propane should always be handled with appropriate hand and eye protection. Prolonged skin contact with liquid propane at any temperature can cause severe burns due to evaporative freezing (Hicks *et al.*, 1979).

**2.** Several propane-filled cryovials are placed on the sample stage. The sample stage should be placed in an appropriately sized stainless steel Dewar. The Dewar should be filled with liquid nitrogen such that the lip of the cryovials is slightly above the level of the liquid nitrogen. Liquid propane should be allowed to thaw in the cryovials and achieve temperature equilibration with the liquid nitrogen bath (1–3 min).

**3.** A cryoprotected and loop-mounted crystal is shock-cooled by rapid plunging it into a cryovial containing liquid propane. The time that the mounted crystal spends exposed to air should be kept to a minimum to avoid drying of the sample. Failure to plunge the loop-mounted sample into propane rapidly enough will damage the sample due to the formation of crystalline ice. During this procedure, hand motions should be fluid and smooth and might require some practice. Shock cooling by this method can be performed with crystalline samples equilibrated at any temperature (e.g. 25°C, 4°C, etc.).

**4.** Propane in the cryovial should be permitted to solidify (5–10 min). The loop-mounted crystal will thus become entombed in solid propane. The cryovial containing the shock-cooled sample may be transferred to a cryocane for storage in a long-term storage Dewar vessel. Shock-cooled crystals prepared in this way may be stored in a storage Dewar indefinitely.

## Protocol 4.2 Shock-cooling into liquid nitrogen

### Materials

Items from Protocol 4.1, with the exception of propane
Crystal wand (available from Hampton Research, Inc.)
Cryotongs (available from Hampton Research, Inc.)
Vial clamp (available from Hampton Research, Inc.)

### Procedure (Fig. 4.3)

**1.** A loop-mounted crystal is attached to the magnetic crystal wand, which is used to rapidly plunge the crystal into liquid nitrogen. The time that the mounted crystal spends exposed to air should be kept to a minimum to avoid drying of the sample. Shock cooling of crystalline samples into liquid nitrogen ($t = -180°C$) can be performed with crystalline samples that are equilibrated at any temperature (e.g. 25°C, 4°C, etc.).

**2.** While in the liquid nitrogen bath, the sample (held by the magnetic wand) is manoeuvred into a cryovial, which is held submerged in the liquid with a vial clamp.

**3.** Once secured in a cryovial, the processed samples are placed on cryocanes, which are placed into a liquid nitrogen Dewar vessel for storage.

## Protocol 4.3 Shock-cooling into gaseous nitrogen

### Materials

Gaseous nitrogen stream ($t = -180°$)
X-ray diffraction camera

### Procedure (Fig. 4.4)

**1.** The X-ray camera is prepared to accept a loop-mounted sample. The X-ray camera should be properly aligned with the X-ray beam and the cold nitrogen cold stream should be aimed at the eucentric point of the X-ray camera.

**2.** The gaseous nitrogen stream (which is aimed at the presumptive location of the cooled crystal) is temporarily diverted with an obstruction such as an index card. Some cold-streams include hardware to enable the temporary diversion of the stream from the crystal.

**3.** With the cold nitrogen stream diverted, the loop-mounted crystal is positioned on the goniometer head.

**4.** Rapid cooling of the sample is achieved by quickly removing the obstruction, which re-establishes the original flow of the stream. The time that the mounted crystal spends exposed to air should be kept to a minimum to avoid drying of the sample.

**5.** Samples prepared in this way are ready for X-ray diffraction or can be recovered (see below) for storage and later use.

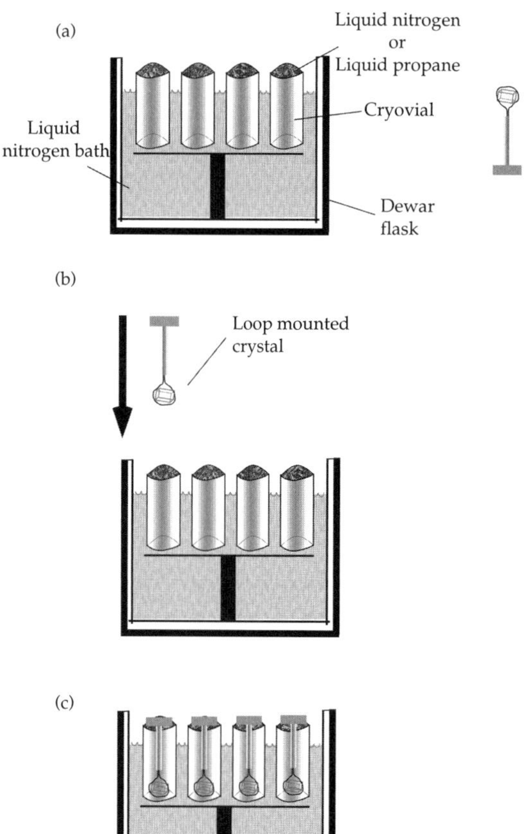


**Figure 4.3** Procedure for shock-cooling a crystal using a liquid cryogen. (a) The cryogens (liquid nitrogen or propane) are maintained in a bath of liquid nitrogen. Cryovials containing cryogens are placed in a bath of liquid nitrogen. (b) A loop-mounted crystal (see Fig. 4.3a and b) is quickly and with a smooth motion plunged into the cold cryogenic liquid. (c) Prepared samples are transferred to a storage Dewar as described in the text (Section 4.3.2 and 4.3.3).

cooled samples. In practice, however, these differences are not very significant when working with small to moderate size crystals (up to 200 microns$^3$) (Kriminski *et al.*, 2003).

### 4.3.4 Shock-cooling into gaseous nitrogen

Using a gaseous nitrogen stream as a cryogen is the simplest of the three methods of preparing shock-cooled crystals, since it requires no special equipment other than that to be used for the measurement of X-ray diffraction. Of special note is that gaseous nitrogen exhibits the lowest heat capacity in comparison to liquid propane and liquid nitrogen and thus gives rise to a slower rate of cooling (Kriminski *et al.*, 2003). In some cases, this may prove to be an advantage. The efficiency and reproducibility of preparing samples using this method can be highly dependent on the type of cold stream being used, how it is being operated, and the dexterity of the operator. Finally, the requirement for a cold-stream to prepare samples precludes the flexibility of sample preparation (e.g. the samples need to stable at room temperature, which is how most X-ray cameras are arranged) found with the use of other cryogens.

## 4.4 Assembly of the mounted crystal onto the X-ray diffraction camera

## 4.5 Storage and transport of macromolecular crystals

The availability of intense synchrotron X-radiation at dedicated centres far from the investigator's home laboratory requires a reliable means of transporting cryocooled macromolecular crystals. Commercial shipping companies will accept for transport properly prepared 'dry shipping' Dewar containers. These containers have all the elements of a conventional storage Dewar in addition to a material that absorbs liquid nitrogen and can maintain cryogenic temperatures in the absence of any liquid. Such containers are, therefore, not at risk of spilling liquid if tipped over. Cryocooled samples can be kept cold in this way for up to 2 weeks.

## 4.6 Crystal lattices and symmetry

A crystal is made up of repeating units called unit cells. Each unit cell in the crystal has the same number of atoms or molecules arranged in a pattern that repeats regularly in three-dimensions (Fig. 4.6). It is the regularity or periodicity that makes a crystal diffract X-rays, while it is the content of the unit cell that gives a crystal its unique diffraction pattern. Furthermore, the degree to which all unit cells and their content have the same orientation in a crystal is directly proportional to its diffraction resolution.

A unit cell is a parallelepiped (Fig. 4.7) that often contains more than one molecule. The molecules

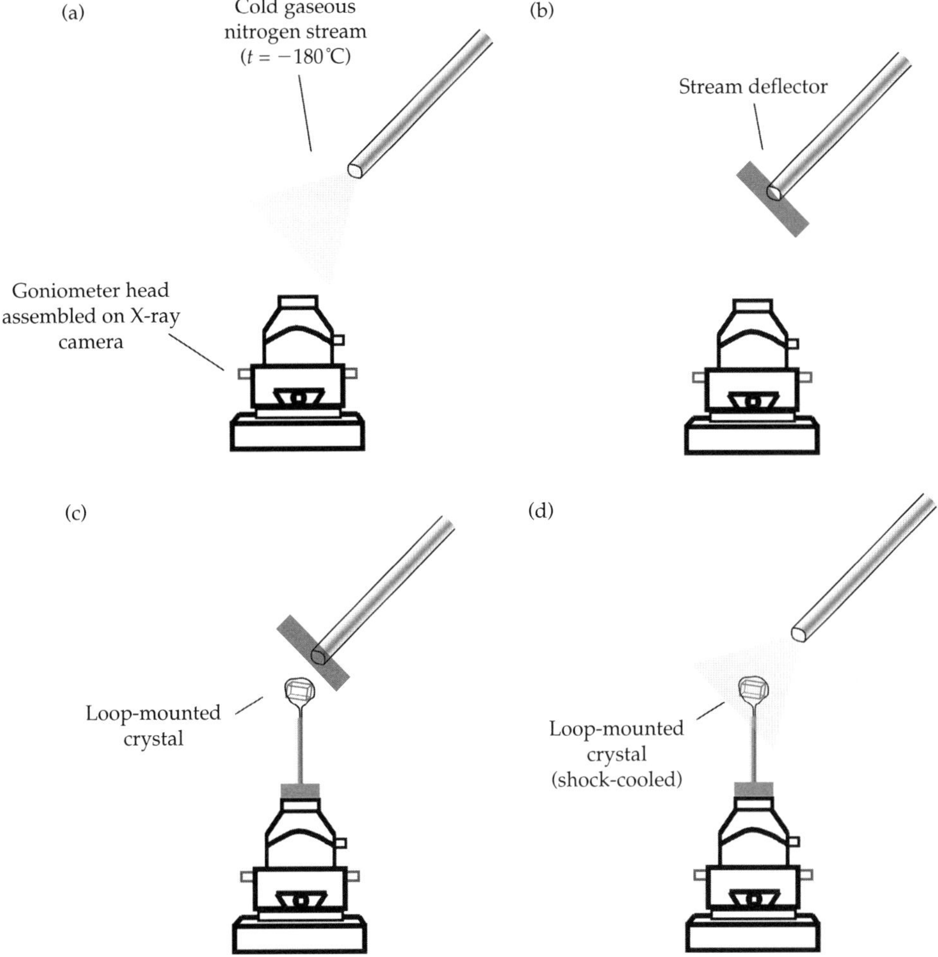


**Figure 4.4** Procedure for shock-cooling a crystal using a gaseous cryogen. (a) The goniometer head is assembled on the X-ray camera and the cold gaseous nitrogen stream is centred on the eucentric point of the camera. (b) The stream is deflected with some appropriate obstruction. (c) The loop-mounted crystal is placed on the goniometer head. (d) The flow of cold nitrogen is re-established, thus shock-cooling the crystal.

in a unit cell are related to each other by crystallographic symmetry. The size and shape of the unit cell are described by a set of lengths ($a$, $b$, and $c$), representing the edges of the cell, and three angles ($\alpha$, $\beta$, and $\gamma$) between these edges (Fig. 4.7). The angle $\alpha$ is between $b$ and $c$, $\beta$ is between $a$ and $c$, while $\gamma$ is between $a$ and $b$. These six parameters are known as the lattice constants. The lengths are usually chosen to have the shortest possible dimensions and the angles closest to 90°. The unique portion of the unit cell is called the asymmetric unit; it is the portion that is not related to other portions by crystal symmetry. The asymmetric unit can contain more that one molecule, often related by non-crystallographic symmetry. The content of the unit cell can be generated by employing crystal symmetry on the asymmetric unit. Elements of crystal symmetry include rotation and screw axes, mirror and glide planes, and an inversion centre (Table 4.2). Glide planes are not relevant in macromolecular crystallography due to the chirality of the biological building blocks.

Crystals are grouped into seven crystal systems based on their symmetry. Table 4.3 describes the

## Protocol 4.4 Assembly of samples prepared with propane

### Materials

Aligned X-ray camera
Cold gaseous nitrogen stream ($t = -180°$) aimed at the rotation centre of the goniometer
Curved forceps
Gloves or finger cots
Vial clamp
Low-form Dewar flask
Cold crystal wand
Cold cryotongs
Face shield
Lint-free tissue
Goniometer head with magnetic mount

### Procedure (Fig. 4.5)

**1.** The edge of the cold-stream should be positioned approximately 1–2 cm from the presumptive position of the crystal. This is a much larger distance than that used during recording of data and is meant to provide an ample work space.
**2.** A cryovial is selected and placed on a flat surface at ambient temperatures. This allows the external most layer of solid propane to thaw (*c.* 10 s) and will result in release of the loop/mounting pin from the cryovial. At this point, a substantial amount of propane will remain in solid form around the crystal. The slow thawing properties of propane provide a favourable time window to transfer the crystal to nitrogen stream without exposing the crystal to ambient temperatures. Liquid propane should always be handled with appropriate hand and eye protection. Prolonged skin contact with liquid propane at any temperature can cause severe burns due to evaporative freezing (Hicks *et al.*, 1979).
**3.** While the propane immediately around the crystal is still solid, a pair of curved forceps is used to lift the pin (containing the loop-mounted crystal) out of the cryovial. The mounting pin is placed on the magnetic mount that is held on the goniometer head. The solid propane that remains around the crystal will slowly thaw under the cold stream, efficiently transferring the crystal from one cryogen (propane) to another (gaseous nitrogen). Damage to the sample will probably result if propane around the crystal is permitted to thaw while the sample is outside the cold-stream.
**4.** As the propane continues to thaw, the cold-stream should be brought closer to the sample. Optimally, the tip of the nozzle should be brought as close as possible to the sample without casting a shadow on the X-ray detector.
**5.** Any liquid propane that remains on the sample should be wicked away from around the crystal with a lint-free tissue. Thawed liquid propane should be collected and allowed to safely evaporate in a chemical fume hood.

## Protocol 4.5 Assembly of samples prepared with liquid nitrogen

### Materials

Crystal transfer tongs
Curved nose clamp
Shallow Dewar flask
Crystal wand
Insulated gloves
Face shield

### Procedure (Fig. 4.5)

**1.** A vial containing a shock-cooled crystal is placed in a bath of liquid nitrogen, which is housed in a shallow Dewar flask.
**2.** The pin containing the loop-mounted crystal is removed from the cryovial and captured with the magnetic crystal wand.
**3.** The mounting pin (held by the magnetic crystal wand) is manoeuvred into the head of the cryotong. Encased within the head of the cryotong, the loop-mounted crystal will remain at cryogenic temperatures for up to 20 s.
**4.** With the cryotongs in their locked position, the crystal is lifted out of the liquid nitrogen bath and positioned above the magnetic mount, which is held on the goniometer head. The time that the head of the cryotongs spends outside of a cryogen (liquid or gaseous nitrogen) should be kept to a minimum.
**5.** Transfer of the crystal to the cold nitrogen stream is achieved by quickly unlocking the cryotongs.

*Note*: Hand and eye protection should always be used when handling liquid nitrogen.

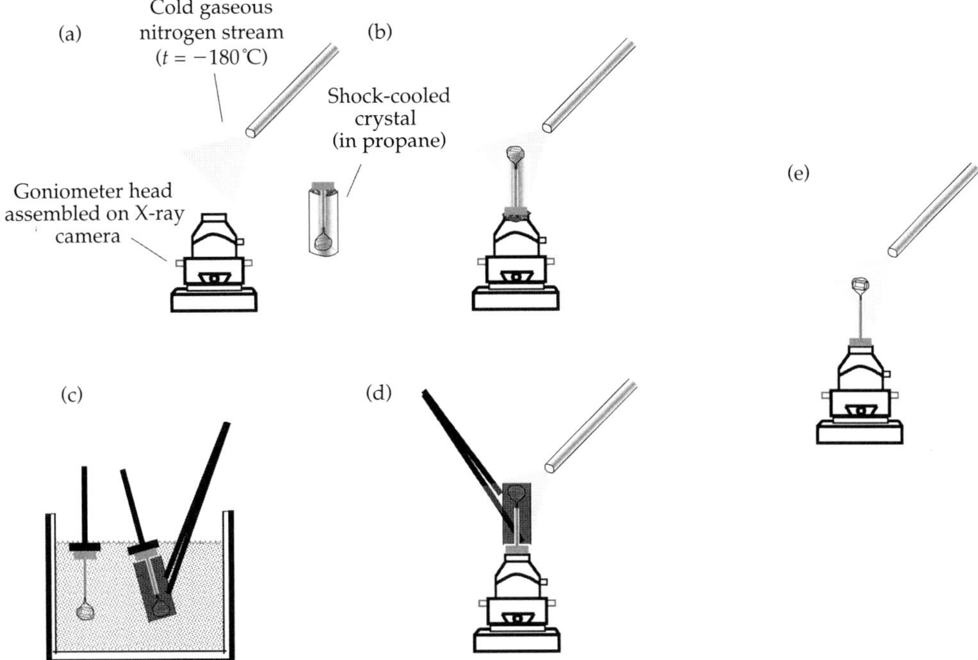


**Figure 4.5** Procedure for transfer of shock-cooled macromolecular crystals to the X-ray camera. (a) The goniometer head is assembled on the X-ray camera and the cold gaseous nitrogen stream is centred on the eucentric point of the camera. (b) For samples prepared with propane or ethane, a cryovial containing a crystal shock-cooled in propane is selected and allowed to partially thaw (5–10 s). Once the external layer of propane has thawed, a pair of forceps is used to lift the loop-mounted crystal, still encased in solid propane, out of the cryovial and placed on the goniometer head. The solid propane is allowed to thaw completely to liquid, which falls away from the crystal. Excess liquid should be collected and disposed of properly. (c) For samples prepared with liquid nitrogen, a cryovial is selected and placed in a bath of liquid nitrogen (held in a shallow Dewar flask). The pin containing the loop-mounted crystal is removed from the cryovial and captured with the magnetic crystal wand as shown above. The mounting pin (held by the magnetic crystal wand) is manoeuvred into the head of the cryotong. (d) With the cryotongs in its locked position, the crystal is lifted out of the liquid nitrogen bath and the mounting pin positioned on the magnetic mount held on the goniometer head. (e) The cryotongs are quickly unlocked to expose the crystal to the nitrogen stream.

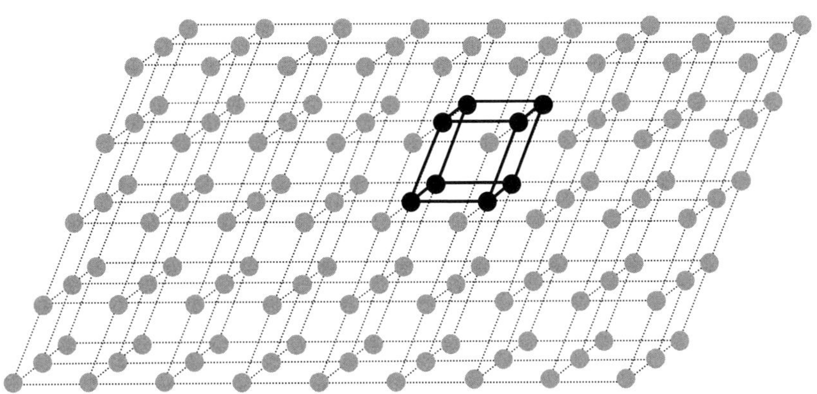

**Figure 4.6** A crystal lattice. Circles represent lattice points; heavy lines represent a unit cell.

### Protocol 4.6 Recovery of shock-cooled crystals from the X-ray diffraction camera

**Materials**

Cryotongs
Shallow Dewar flask
Cane
Cryovial
Vial clamp
Storage Dewar
Liquid nitrogen
Gloves
Face shield

**Procedure**

**1.** Recovery of crystals from the cold stream involves performing the procedure described for the mounting of crystals to X-ray camera but with the steps reversed (Protocols 4.4 and 4.5). Cool the cryotongs, cryovial, and vial clamp immersion in liquid nitrogen.
**2.** The cold cryotongs are used to rapidly capture the mounting pin/loop. This step should be carried out rapidly so that the crystal leaves the cold stream and enters the protected (cold) space within the cryotongs in a minimum amount of time.
**3.** The cryotongs (with enclosed crystal) is rapidly lifted off the goniometer head and plunged into a liquid nitrogen bath.
**4.** The magnetic crystal wand is used to capture the mounting pin.
**5.** While immersed in liquid nitrogen, the mounting pin (attached to crystal wand) is manoeuvred into a cryovial (which is held with the cooled vial-clamp).
**6.** The recovered crystal (in the cryovial) is secured to a cryocane, which is stored in a liquid nitrogen storage Dewar.

### Protocol 4.7 Storage and transport of macromolecular crystals

**Materials**

'Dry Shipping' Dewar (Taylor Wharton: Cryo Express and Cryo Flight)
Hard case
Insulated gloves
Face shield

**Procedure**

**1.** A 'dry shipping' Dewar (at ambient temperature) is charged by filling it slowly with liquid nitrogen. Consult the product literature for specific details.
**2.** Once filling is complete, the absorbent is thoroughly cooled by the liquid nitrogen (approx 7–8 h). This will require topping up of the Dewar several times. A properly charged Dewar should be able to stably maintain liquid nitrogen on standing.
**3.** In preparation for shipping, liquid nitrogen from the 'dry shipping' Dewar is decanted into an appropriate storage Dewar. Allow the 'dry shipping' Dewar to stand for several minutes upside down until it is completely empty of liquid (commercial shipping companies will refuse to ship a container that contains liquid).
**4.** Once dry, cryocanes containing cryocooled crystals are transferred from the liquid nitrogen bath to the shipping Dewar. Samples prepared with liquid propane or liquid nitrogen are handled in the same manner. The temperature inside the shipping Dewar should maintain the propane in the cryovials in solid form.

Many commercial shipping companies will accept a properly prepared shipping Dewar containers for transport around the world. Each investigator should check with the shipping company of choice about requirements, paperwork, and cost.

minimum symmetry, relevant to macromolecules, of the unit cell for each of the seven crystal systems and the restrictions imposed on unit cell lattice constants. For example the triclinic system has no symmetry or restrictions on its lattice constants, while the cubic system has four three-fold axes along the diagonal of the cube, three equal lattice lengths, and three lattice angles equal to 90°. The unit cell in each crystal system is chosen to contain at least one lattice point. A unit cell that has only one lattice point at each of its corners is called primitive (*P*). Note that each corner of a unit cell in a crystal is shared by eight other unit cells; therefore, a primitive cell contains only one lattice point. However, at times it is advantageous to select a unit cell that contains more than one lattice point. These cells are called centred cells; there are three such cells. They are body-centred (*I*; German for Innenzentrierte), where a unit cell contains a second lattice point at the centre of the cell, face-centred (*F*) having an additional lattice point at the centre of each of its six faces and *C*-centred (*C*) having an additional lattice point at only one of the six faces (by definition the *ab* face). Centring expands the volume of the unit cell. Whereas *C*- and *I*-centring double the volume relative to that the *P* cell, *F*-centring quadruples the volume. When centring is included, the number of unique lattices expands to 14 (Table 4.3 and Fig. 4.8). These 14 lattices are known as the Bravais lattices; they were first described by Frankheimer and Bravais in the mid-nineteenth century.

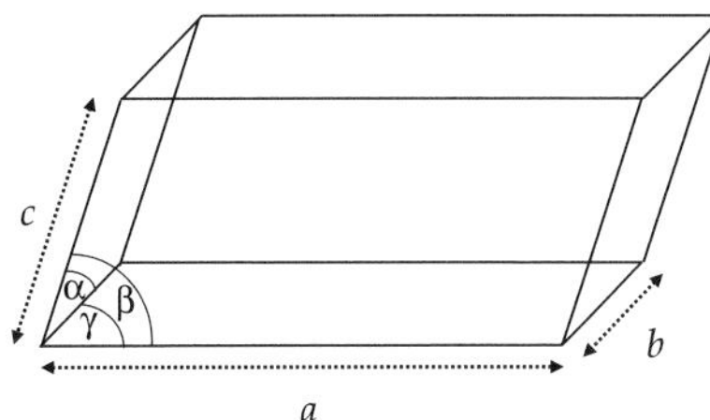


**Figure 4.7** A unit cell, showing the six parameters known as lattice constants.

A unique combination of crystal symmetry elements and centring is called a space group. There are only 230 possible space groups. However, this number is reduced to 65 for biological macromolecules because the chirality of their biological building blocks. The 65 biologically relevant space groups are listed in Table 4.4.

**Table 4.2** Elements of crystal symmetry

| Symmetry element | Description |
|---|---|
| Rotation axes | Counterclockwise rotation of 360°/*n* about an axis, where *n* is 1, 2, 3, 4 or 6 |
| | 2-fold axis is rotation by 180° |
| | 3-fold axis is rotation by 120° |
| | 4-fold axis is rotation by 90° |
| | 6-fold axis is rotation by 60° |
| Screw axes | Same as rotation axis, but followed by a translation of *p*/*n* along the rotation axis, where *p* is an integer <*n* |
| | $2_1$ screw axis is rotation by 180° followed by translation of 1/2 of a unit cell |
| | $3_1$ screw axis is rotation by 120° followed by translation of 1/3 of a unit cell |
| | $3_2$ screw axis is rotation by 120° followed by translation of 2/3 of a unit cell |
| | $4_1$ screw axis is rotation by 90° followed by translation of 1/4 of a unit cell |
| | $4_2$ screw axis is rotation by 90° followed by translation of 1/2 of a unit cell |
| | $4_3$ screw axis is rotation by 90° followed by translation of 3/4 of a unit cell |
| | $6_1$ screw axis is rotation by 60° followed by translation of 1/6 of a unit cell |
| | $6_2$ screw axis is rotation by 60° followed by translation of 1/3 of a unit cell |
| | $6_3$ screw axis is rotation by 60° followed by translation of 1/2 of a unit cell |
| | $6_4$ screw axis is rotation by 60° followed by translation of 2/3 of a unit cell |
| | $6_5$ screw axis is rotation by 60° followed by translation of 5/6 of a unit cell |
| Inversion centre | All points inverted through a centre of symmetry |
| Mirror plane | Reflection through a plane |
| Glide plane | Same as mirror plan, but followed by a translation of half the unit cell parallel to the plane; glide planes are not relevant in macromolecular crystallography due to the chirality of the biological building blocks |

**Table 4.3** The seven crystal systems

| System | Bravais lattices | Minimum symmetry of unit cell | Restriction on lattice constants |
|---|---|---|---|
| Triclinic | *P* | No symmetry | $a \neq b \neq c; \alpha \neq \beta \neq \gamma$ |
| Monoclinic | *P*, *C* | One 2-fold axis, parallel to *b* | $a \neq b \neq c; \alpha = \gamma = 90°; \beta > 90°$ |
| Orthorhombic | *P*, *C*, *I*, *F* | Three mutually perpendicular 2-fold axes | $a \neq b \neq c; \alpha = \beta = \gamma = 90°$ |
| Tetragonal | *P*, *I* | One 4-fold axis, parallel to *c* | $a = b \neq c; \alpha = \beta = \gamma = 90°$ |
| Trigonal/ | *P* | One 3-fold axis, parallel to *c* | $a = b \neq c; \alpha = \beta = 90°; \gamma = 120°$ |
| rhombohedral | (*or R*)[a] | | $a = b = c; \alpha = \beta = \gamma \neq 90°$ |
| Hexagonal | *P* | One 6-fold axis, parallel to *c* | $a = b \neq c; \alpha = \beta = 90°; \gamma = 120°$ |
| Cubic | *P*, *I*, *F* | Four 3-fold axes along the diagonal of the cube | $a = b = c; \alpha = \beta = \gamma = 90°$ |

[a]Rhomobohedral is a subset of the trigonal system in which the unit cell can be chosen on either hexagonal or rhombohedral axes.

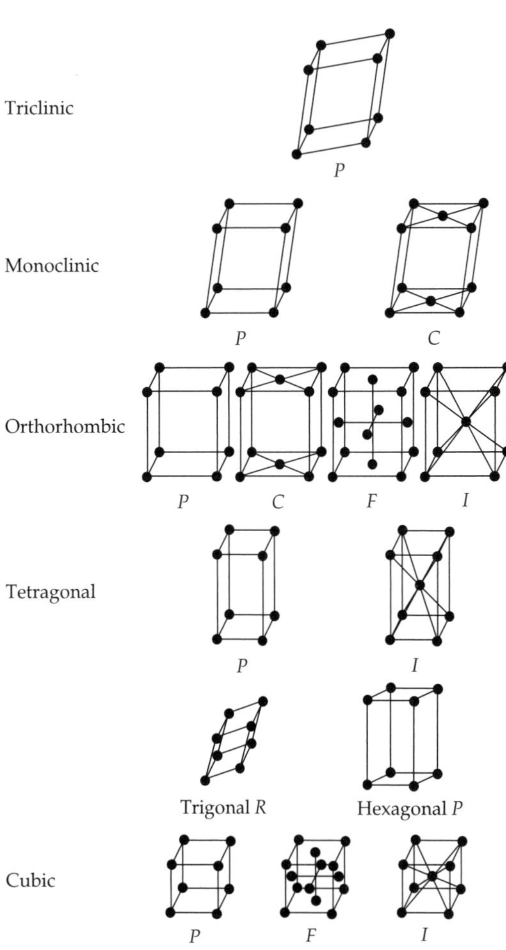


**Figure 4.8** The 14 Bravais lattices. Black circles represent atoms or molecules. *P* cells contain only one lattice point, while *C*- and *I*-centred cells contain two and *F*-centred cells contain four.

## 4.7 Lattice and space-group determination from X-ray data

In the past, the traditional way of determining space-groups and cell dimensions was by analysing X-ray precession photographs of the undistorted lattice for absences and measuring spot separations in order to determine lattice dimensions (Abdel-Meguid *et al.*, 1996). Nowadays, oscillation data collection is usually started on a cryocooled crystal mounted in a loop at an undefined orientation and the space group determined 'on the fly' after 10–20 frames have been collected.

Two processing packages are predominantly used by crystallographers in house and at synchrotrons indexing of oscillation data. HKL2000 (and its predecessor DENZO) written by Zbyszek Otwinowski and Wladek Minor (Otwinowski and Minor, 1997 and 2001) and MOSFLM supported by CCP4 (Leslie, 1993). Both have very powerful autoindexing routines. The code for that within HKL (Otwinowski and Minor, 1997) is yet to be fully disclosed as it is a commercial package and MOSFLM has within it the powerful algorithm DPS (open source) written by Michael Rossman and Cees van Beek which uses a similar method of Fourier indexing (Rossmann and van Beek, 1999; Powell, 1999; Rossmann, 2001) to that of HKL. Both algorithms are extremely powerful but in the author's experience they do not always give the same result with difficult space-group determinations and it is often useful to run both when initially indexing a crystal.

Other good processing packages are d*TREK incorporated in CrystalClear, written by Jim

**Table 4.4** The 65 space groups that are possible for macromolecular crystals

| Crystal system | Diffraction symmetry[a] | Space groups[b] |
|---|---|---|
| Triclinic | $\bar{1}$ | P1 |
| Monoclinic | 2/*m* | P2, P$2_1$, C2 |
| Orthorhombic | *mmm* | P222, P$222_1$, P$2_12_12$, P$2_12_12_1$, C222, C$222_1$, F222, [I222, I$2_12_12_1$] |
| Tetragonal | 4/*m* | P4, (P$4_1$, P$4_3$), P$4_2$, I4, I$4_1$ |
| | 4/*mmm* | P422, (P$4_1$22, P$4_3$22), P$4_2$22, P$42_1$2, (P$4_12_1$2, P$4_32_1$2), P$4_22_1$2, I422, I$4_1$22 |
| Trigonal | $\bar{3}$ | P3, (P$3_1$, P$3_2$), R3 |
| | $\bar{3}$*m* | [P321, P312], [(P$3_1$21, P$3_2$21), (P$3_1$12, P$3_2$12)], R32 |
| Hexagonal | 6/*m* | P6, (P$6_1$, P$6_5$), (P$6_2$, P$6_4$), P$6_3$ |
| | 6/*mmm* | P622, (P$6_1$22, P$6_5$22), (P$6_2$22, P$6_4$22), P$6_3$22 |
| Cubic | *m*3 | P23, P$2_1$3, F23, [I23, I$2_1$3] |
| | *m*3*m* | P432, (P$4_1$32, P$4_3$32), P$4_2$32, F432, F$4_1$32, I432, I$4_1$32 |

[a] The overbar indicates an inversion axis, while *m* represents an mirror plane.
[b] Space groups in brackets and parentheses are indistinguishable from diffraction patterns. Those in parentheses are enantiomorphs.

Pflugrath (Pflugrath, 1997, 1999), which is marketed with MSC X-ray detectors (this program evolved from MADNESS) and XDS written by Wolfgang Kabsch and incorporating the IDXREF autoindexing algorithm (Kabsch, 1988a, 1988b, 1993a, 1993b), which starts by calculating vectors between reflections with low indices and building up to full data indexing. Otwinowski and Minor have written the commercial, macromolecular autoindexing routines within the PROTEUM which supports data collection of Bruker detectors.

ELVES has been developed as an expert system, by James Holton and Tom Alber, to go from data collection frames to structure without human intervention and will obviate the need for intermediate space-group determination described above. Very recently, 12 different European sites have been collaborating to develop a software package known as DNA (automateD collectioN of datA) for the automatic collection and indexing of macromolecular diffraction data. Further information is available at the web site www.dna.ac.uk.

### 4.7.1 Starting out – preliminary data collection and indexing

1. Look at the crystals carefully under a dissecting microscope equipped with polarizers. Often crystals which are hexagonal and tetragonal are easily recognizable from their external morphology and cubic crystals may be identified from their lack of polarization. Other rectangular habits turn out often to be either monoclinic or orthorhombic. Coming at an indexing problem armed with this morphological information is very helpful.

2. Collect a wedge of data (say 10 frames) and also collect a frame at 45° and 90° away from the starting oscillation position (for a crystal say whose habit appears to have faces at 90° to each other and whose space group could be monoclinic, orthorhombic, or cubic). Collecting frames away from the starting oscillation position can save considerable time collecting worthless data if these frames are found to be have pathologies such as very high mosaicity or splitting as they will be probably encountered later in a full data collection when the full oscillation range is swung through.

3. Make sure you have an accurate values for the direct-beam position on the detector you are using and the crystal-film distance. If these have been recorded from a previous successful data collection and processing, time can be saved by having them as starting parameters for indexing. At many synchrotrons, the prerecording of a wax ring will give an accurate crystal to detector distance, sometimes the values displayed on the LED may be fallible.

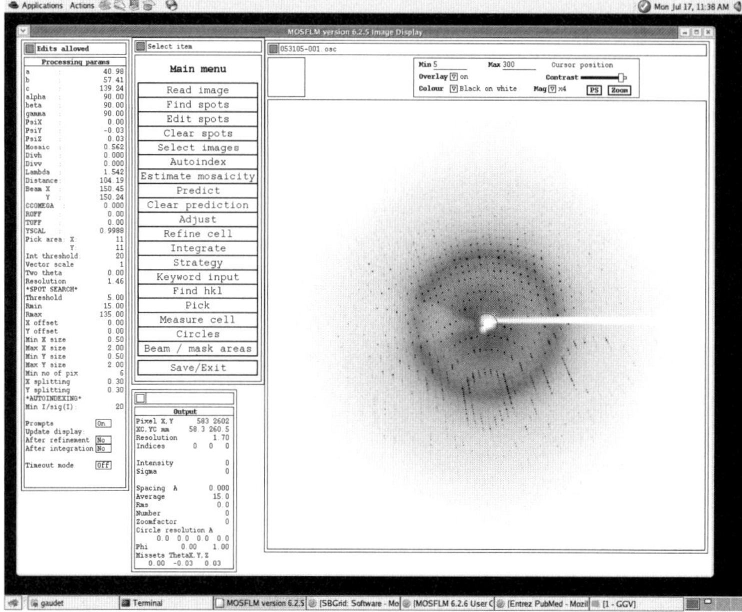


**Figure 4.9** Graphical display of the autoindexing of TRPV ankyrin diffraction data using MOSFLM.

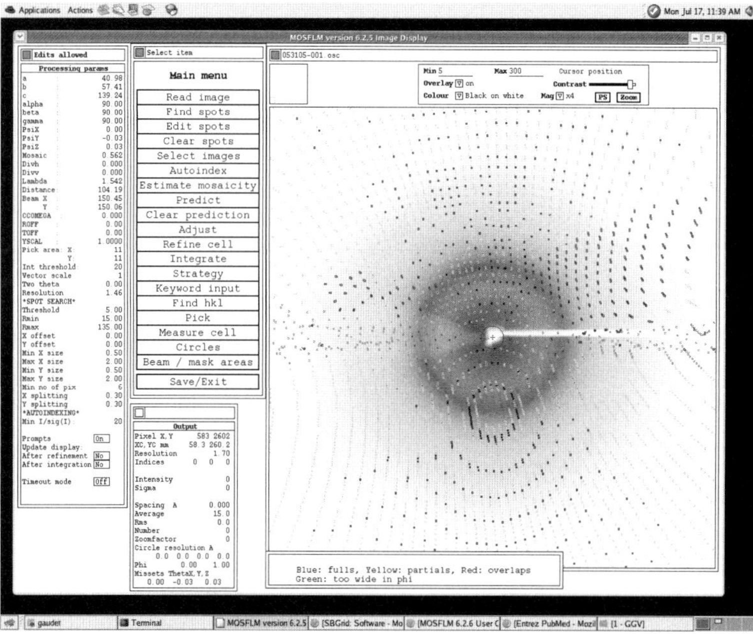


**Figure 4.10** Graphical display of the autoindexing of TRPV ankyrin diffraction data using MOSFLM with the calculated spot predictions superimposed.

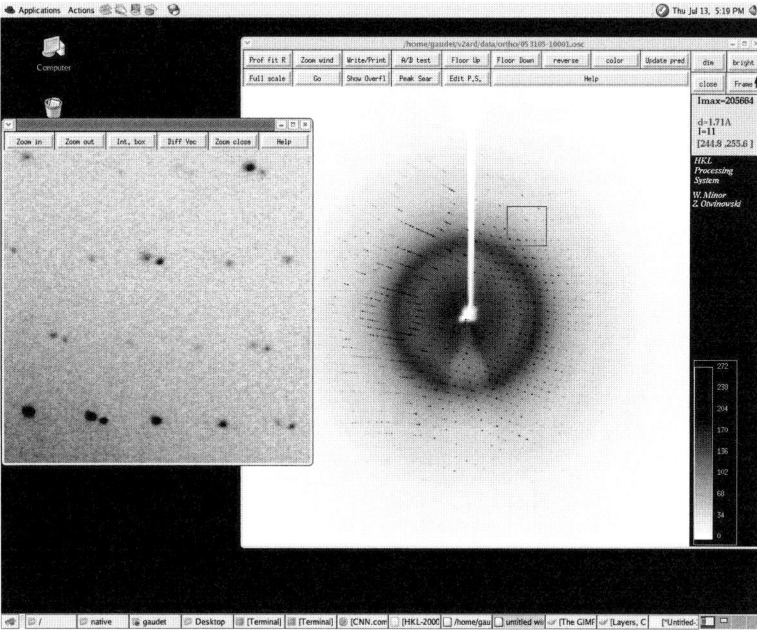


**Figure 4.11** Graphical display of the autoindexing of TRPV ankyrin diffraction data using DENZO.

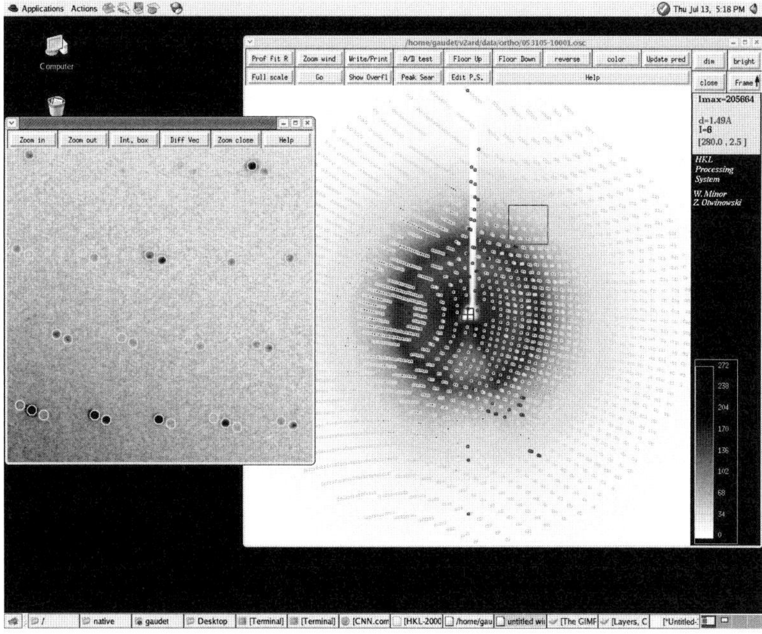


**Figure 4.12** Graphical display of the autoindexing of TRPV ankyrin diffraction data using DENZO with the calculated spot predictions superimposed.

4. Load in a frame 1 of the data and check the resolution to which the diffraction extends. Look for pathologies in the diffraction pattern, such as split spots or ill-defined, blurred reflections. Check also the frames at 45° away and 90° away for similar pathologies. If such pathologies are present, it is worth mounting another crystal as it could be that not all crystals suffer the same problems. Otherwise it could be that the crystals are inherently mosaic (the solution is to check a frame collected on a capillary-mounted crystal as inappropriate cryoprotectant can increase mosaicity), or they could be twinned (the solution is dependent on the nature of the twinning, if the data is indexable this may be sorted out after the event using detwinning software, otherwise effort may have to be spent growing untwinned crystals by changing the conditions or adding additives such as dioxane).

5. Start indexing the frames with your favourite software package, HKL2000, MOSFLM, etc. Two examples are given in Figs 4.9 to 4.12, one for an indexing with MOSFLM (Fig. 4.9 without predictions and Fig. 4.10 showing the predictions) and the other for HKL2000 (Fig. 4.11 without the predictions and Fig. 4.12 with the predictions). The crystal in question is of a rat TRPV2 Ankyrin Repeat Domain protein (Jin *et al.*, 2006). The space group is $P2_12_12_1$ and the cell parameters are $a = 41.0$ Å, $b = 57.5$ Å, $c = 139.5$ Å, $\alpha = \beta = \gamma = 90°$. Once a decision has been made on the space-group it always pays to integrate in the background as the data collection is proceeding. This can be very instructive as integration statistics can indicate whether problems of initial space-group definition have arisen. For example one does not wish to leave a synchrotron and return home to find that the true space-group was of a lower symmetry and insufficient data has been collected or that the data is giving terrible merging statistics as indicated by a worsening R-merge during data collection.

## References


Abdel-Meguid, S. S., Jeruzalmi, D. and Sanderson, M. R. (1996). Preliminary characterization of crystals. *Methods Mol. Biol.* **56,** 55–86.

Garman, E. (2003). 'Cool' crystals: macromolecular cryocrystallography and radiation damage. *Curr. Opin. Struct. Biol.* **13,** 545–551.

Garman, E. F. and Schneider, T. R. (1997). Macromolecular crystallography. *J. Appl. Cryst.* **30,** 211–237.

Hicks, L. M., Hunt J. L. and Baxter C. R. (1979). Liquid propane cold injury: a clinicopathologic and experimental study. *J. Trauma* **19,** 701–703.

Hope, H. (1988). Cryocrystallography of biological macromolecules: a generally applicable method. *Acta Crystallogr. B* **44,** 22–26.

Jin, X., Touhey, J. and Gaudet, R. (2006). Structure of the N-terminal ankyrin repeat domain of the TRPV2 ion channel. *J. Biol. Chem.* **281,** 25006–25010.

Kabsch, W. (1988a). Evaluation of single crystal X-ray diffraction data from a position-sensitive detector. *J. App. Cryst.* **21,** 63–70.

Kabsch, W. (1988b). Automatic indexing of rotation diffraction patterns. *J. App. Cryst.* **21,** 67–71.

Kabsch, W. (1993a). Data collection and processing. In: Sawyer, L., Issacs, N. and Bailey, S., eds. *Proceedings of the CCP4 Study weekend*. Daresbury laboratories, Warrington, UK, pp. 56–62.

Kabsch, W. (1993b). Automatic processing of rotation diffraction data from crystals of initially unknown symmetry and cell constants. *J. App. Cryst.* **26,** 795–800.

Kriminski, S., Kazmierczak, M. and Thorne, R. E. (2003). Heat transfer from protein crystals: implications for flash-cooling and X-ray beam heating. *Acta Crystallogr. D* **59,** 697–708.

Kwong, P. D. and Liu, Y. (1999). Use of cryoprotectants in combination with immiscible oils for flash cooling macromolecular crystals. *J. Appl. Cryst.* **32,** 102–105.

Leslie, A. (1993). Data collection and processing. In: *Proceedings of the CCP4 Study Weekend,* Sawyer, L., Isaacs, N. and Bailey, S. eds. Daresbury Laboratories, Warington, UK, pp. 44–51.

Lusty, C. J. (1999). A gentle vapor-diffusion technique for cross-linking of protein crystals for cryocrystallography. *J. Appl. Cryst.* **32,** 106–112.

Mitchell, E. P. and Garman, E. F. (1994). Flash-freezing of protein crystals: investigation of mosaic spread and diffraction limit with variation of cryprotectant concentration. *J. Applied Crystallogr.* **27,** 1070–1074.

Otwinowski, Z. and Minor, W. (1997). Processing of X-ray diffraction data collected in oscillation mode. *Method Enzymol.* **276,** 286–306.

Otwinowski, Z. and Minor, W. (2001). Denzo and Scalepack 211. In: *International Tables in Crystallography,* vol. F, Rosmann, M. G. and Arnold, E., eds. IUCr Press, pp. 226–235.

Pflugrath, J. W. (1997). Diffraction-data processing for electronic detectors: theory and practise. *Method Enzymol.* **276,** 307–326.

Pflugrath, J. W. (1999). The finer things in X-ray diffraction data collection. *Acta Crystallogr. D* **55,** 1718–1725.

Powell, H. R. (1999). The Rossmann Fourier autoindexing algorithm in MOSFLM. *Acta Crystallogr. D* **55,** 1690–1695.

Riboldi-Tunnicliffe, A. and Hilgenfeld, R. (1999). Cryocrystallography with oil – an old idea revived. *J. Appl. Cryst.* **32,** 1003–1005.

Rodgers, D. (1997). Practical cryocrystallography. In: *Macromolecular Crystallography,* Carter, C. W. and Sweet, R. M., eds. Academic Press, **276,** pp. 183–202.

Rossmann, M. G. and van Beek, C. G. (1999). Data processing. *Acta Crystallogr. D* **55,** 1631–1641.

Rossmann, M. G. (2001). Automatic indexing of oscillation images. In: *International Tables in Crystallography,* vol. F, Rosmann, M. G. and Arnold, E., eds. Kluwer Press, pp. 209–211.

Rubinson, K. A., Ladner, J. E., Tordova, M. and Gilliland, G. L. (2000). Cryosalts: suppression of ice formation in macromolecular crystallography. *Acta Crystallogr. D* **56,** 996–1001.

Wyckoff, H. W., Doscher, M., Tsernoglu, D., Inagami, T., Johnson, L. N., Hardman, K. D., *et al.* (1967). Design of a diffractometer and flow cell system for X-ray analysis of crystalline proteins with application to the crystal chemistry of ribonuclease S. *J. Mol. Biol.* **27,** 563–578.

CHAPTER 5

# In-house macromolecular data collection

**Mark R. Sanderson**

## 5.1 Introduction

In the field of macromolecular crystallography a renaissance occurred in data collection with the advent of two-dimensional area detectors. These rapidly superseded film cameras and diffractometers fitted with proportional counters as the tools of choice for macromolecular data collection, resulting in a huge increase in collection speeds. A typical data collection to high resolution on a diffractometer would take a matter of weeks using, typically, five to ten capillary mounted crystals. For the duration that a crystal lasted in the X-ray beam individual diffraction reflections were then collected in resolution shells, with overlap between shells for scaling. The full dataset was then formed by merging and scaling the resolution shells. The first detectors to be introduced reduced data collection times from weeks to a matter of days, with the possibility of collecting a full data set on a single crystal before it decayed. The advent of cryocooling occurred soon after the introduction of area detectors and this further augmented data quality through collection effectively without decay.

The first detectors to be introduced were two-dimensional Multiwire detectors and the FAST detector. The first multiwire detectors were developed and commercialized by Prof. Nguyen-huu Xuong, Ron Hamlin, and coworkers in San Diego (San Diego multiwire detector SDMW) (Cork *et al.*, 1975; Xuong *et al.*, 1985a; Hamlin, 1981, 1985) and Ron Burns and coworkers at Harvard (Xentronics detector) (Durban *et al.*, 1986). These groups modified the technology used in two-dimensional particle detectors, developed by Charpak to detect high-energy particles in collider experiments (Charpak, 1988; Charpak *et al.*, 1968, 1989), for use in single-crystal X-ray diffraction studies. Xuong, Hamlin, Nielsen, Howard, and coworkers developed a method of collecting screenless images on these detectors (Xuong *et al.*, 1985b, Howard *et al.*, 1985; Edwards *et al.*, 1988). Even though the multiwire detectors heralded a new era of data collection, the older multiwire detectors have now been surpassed by image plate and CCD technology. Having said that, Bruker AXS is developing a new detector based on multiwire technology with a greater sensitivity by incorporating a very narrow wire spacing and hence using the superb feature of these detectors; namely, that they are single photon counters with high detected quantum efficiency (DQE) (Schierbeek, 2006). The multiwire detectors have been excellently reviewed in Garman (1991) and Kahn and Fourme (1997). The FAST system, incorporating a scintillation screen and a TV scanning system, which was developed by Arndt and coworkers at the LMB, Cambridge and commercialized by Enraf-Nonius, Delft, was used in a number of leading crystallographic laboratories (Arndt, 1985, 1982).

In this Chapter I have focused on in-house data collection, as data collection at synchrotrons is covered in this volume by Wasserman *et al.*, Chapter 12 . Naturally, the techniques and strategy for collecting data in-house and at synchrotrons have a great deal in common.

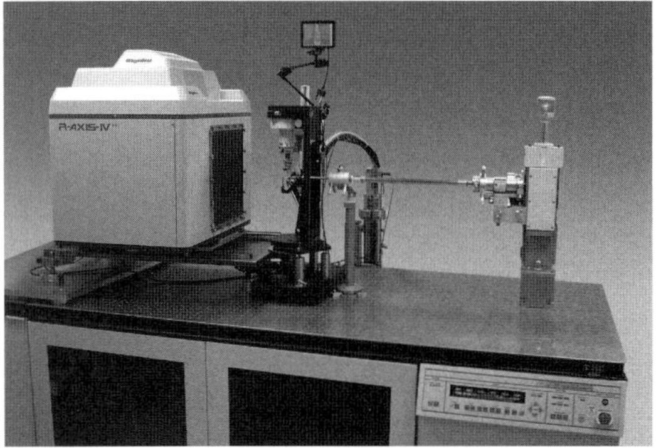

**Figure 5.1** Raxis IV$^{++}$ mounted on a generator. The rotating anode wheel is on the far right and the long optics is clearly visible. The detector is on the left and in front of this is an inverted φ-axis. (Courtesy of Dr Joseph Ferrara, Rigaku Americas Corporation.)

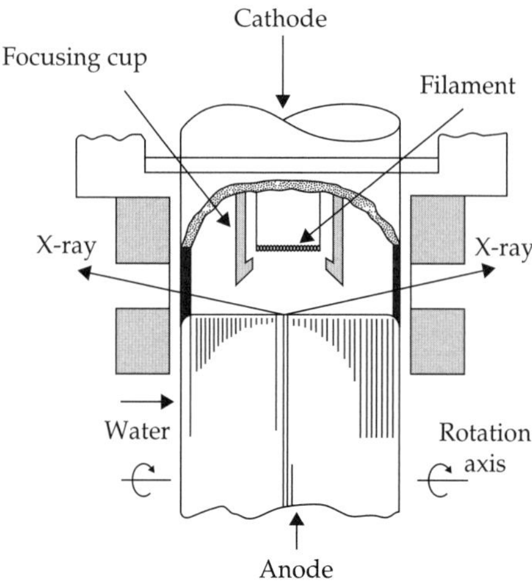


**Figure 5.2** Cut-through diagram of a rotating-anode generator. The anode is water cooled and the whole anode assembly evacuated. The X-rays pass through beryllium windows, which are transparent to X-rays. (Reproduced with permission from Monaco, H. L., Viterbo, D., Scordari, F., Gilli, G., Zanotti, G. and Catti, M. In: *Fundamentals of Crystallography*, Giacovazzo, C., ed. International Union of Crystallography Texts on Crystallography, Oxford Science Publications.)

Data collection systems consist of four major components: a source of X-rays, focusing mirrors (optics), a motor-driven goniostat and crystal viewing assembly, and an area detector for recording diffracted images, as shown in Fig. 5.1. I shall describe below the components that make up this assembly. Before proceeding to discuss X-ray generators, two terms that are widely used in describing the attributes of X-ray generation should be defined. These are flux and brilliance. Flux is defined as the number of photons per second per mrad and brilliance as the number of photons per second per unit phase space volume (with units photons/s/mm$^2$/millirad$^2$). These are important values when comparing X-ray generators with different filament and focal spots sizes. For synchrotrons, these parameters are quoted per 0.1% relative bandwidth.

## 5.2 X-ray generators

The generation of X-rays in the home laboratory for macromolecular purposes is primarily carried out using rotating anode generators. This technology was developed at the Royal Institution in London (Müller, 1929) and at the Laboratory of Molecular Biology, Cambridge (Broad, 1956; Arndt, 2003) and is elegantly simple. The impetus for its development was to attain higher X-ray fluxes, which could be applied to protein and fibre diffraction studies. A rotating anode is shown in Fig. 5.2 and consists of a filament acting as a source of high-energy electrons, which strike a rapidly rotating, water cooled target (usually copper for macromolecular crystallographic purposes). Electrons are ejected from the inner atomic orbitals of the target metal with X-rays being generated when outer orbital electrons fall to refill these vacant inner shells. The clever feature of the rotating anode is that target rotation brings a cool piece of copper into the path of the electron beam allowing much higher voltages to be applied (compared with fixed target sources), hence producing much higher X-ray flux.

Figure 5.2 is a schematic diagram of the rotating anode assembly, showing the anode target, electron gun with the whole assembly evacuated by a turbo-molecular pump backed onto a Ruffing pump. The elements of rotating anode generators have remained essentially the same over the past 40 years, with higher fluxes being generated by improvements in the vacuum system (which increases stability and decreases the arcing), in the bearings, seals, and in the heat dissipation from the target. I shall in this chapter review the most modern generators, although old generators and very old generators are very actively being used as they are mechanically robust and, if well maintained and parts are still available, can see tens of years of service (in the author's own laboratory we have an old Elliot GX-18 big wheel generator with a 70-micron focus, for which we have enhanced the vacuum system by putting in turbo-molecular pumps, which will out-perform all generators on the market except for those of highest brilliance – FR-E, 007 HF, and Microstar, discussed below).

In this section I shall only cover the X-ray generators that are currently being sold commercially. Many laboratories have moved away from collecting a large amount of data in-house to using synchrotrons as their preferred method of acquiring native data and phases for structure solution. These groups consequently only use in-house systems to screen crystals prior to taking them to a synchrotron. In the author's opinion, it really pays dividend to have a very bright X-ray source in-house for three reasons. Firstly, if one has grown a very small crystal of a given macromolecule from initial crystallization trials, being able to characterize it in terms of its space-group, cell dimensions, and cell contents may give a good indication of the direction in which to proceed with further crystallization attempts. For example if one is trying to cocrystallize a protein – DNA complex and there turns out not to be DNA in the lattice as judged by the cell volume and Vm), then clearly conditions have to be altered in order to steer the crystallization attempts in a direction towards obtaining crystals of the complex. The most optimistic chance of performing this operation reliably on small, weakly diffracting crystals is achieved by having an in-house generator of the highest brilliance. Secondly, it may occur that, at a critical time, due to lack of beam time at a synchrotron, one has an urgency to collect and try to phase a structure in-house. Then, with the highest brilliance generator one is going to be able to collect data to the highest possible resolution in the fastest time. Thirdly, if only one crystal is available, and the cryocooling conditions have not been determined, and if the crystal does not 'travel well' to a synchrotron, it pays to have a good source of X-rays close to the where the crystal was grown in order to assess its diffracting power and, if possible, to obtain its space group and cell dimensions.

### 5.2.1 Rigaku MSC X-ray generators

Rigaku FR-E Superbright and Rigaku FR-E Superbright DW are the highest brilliance generators currently on the market and operate with a 70 μm × 70 μm focal spot size and a power of 2 kW, with a maximum operating voltage of 60 kV and a current varying up to 45 mA dependent on the target material. The rotating anode has a diameter of 280 mm and revolves at 9000 r.p.m. The large wheel and the fast rotation allows rapid heat dissipation from the anode and hence permits a higher loading (in terms of voltage and current). This concept, of having a large wheel to dissipate the heat over a large area, came in with the Elliot GX-13 and GX-18 series of generators (which Enraf-Nonius continued to make for a short time after purchasing Elliot from GEC). For the GX-13 and 18, the wheel is a 457 mm (18 inches) in diameter and spins at 3000 r.p.m. Therefore, a larger wheel allows heat dissipation over a larger surface area as compared with its smaller counterpart, and spinning the wheel faster will allow more heat dissipation but will in turn place a greater mechanical demand on the drive motor, bearings, and seals.

The development of better bearings and ferrofluidic seals, as compared to the older lip-seals, has allowed the wheel to be spun faster and cooled more efficiently. The downside of having a very large wheel is that it is more awkward to remove and clean, since handling jigs are required for this operation compared with a standard 99 mm wheel, as found on the RU200 and RU300 generators and on the Rigaku MSC 007HF, which can be easily removed by hand. The FR-E and the FR-E DW (which has the capability of collecting at two wavelengths by having a band of copper and chromium on the same wheel and compatible optics for in-house SAD phasing) are flagship generators but their initial high cost and high running costs, in terms of ferrofluidic seals, has meant that they have only been purchased by more affluent laboratories.

Rigaku MSC has recently introduced a UltraX-18 generator which runs at 18 kW with a wide range of different focal spot sizes ranging from $0.1 \times 12\,mm^2$ through to $0.5 \times 10\,mm^2$ with loadings ranging from 1.2 kW through to 18 kW accordingly for a copper anode target. For this generator the electron gun has been totally redesigned in order to minimize discharge and hence improve stability. The anode is also a new direct drive model which has a low vibration.

Two generators that have become popular are the Rigaku MSC 007 and 007HF, which are very high brilliance X-ray generators incorporating a $70\,\mu m \times 70\,\mu m$ focal spot. The possible loading on the 007HF is higher than the 007 by using a changed bearing design and spinning the wheel at 9000 r.p.m instead of 6000 r.p.m in order to increase heat dissipation.

### 5.2.2 Bruker AXS X-ray generators

Bruker AXS make a high-flux 27 kW generator known as the Microstar. It is belt driven, has a $100\,\mu m$ focal spot, and includes novel features such as variable anode speeds to aid in powering up the generator depending on the state of the vacuum. The output of this generator is $1.7 \times 10^9$ photons/mm$^2$/sec and is configured with Helios multilayer optics and the Platinum 135 CCD (Charged Couple Device) to form the Proteum X8 system. Very recently, they have released the Microstar Ultra whose intensity is $8 \times 10^{10}$ photons/mm$^2$. The increase in intensity has been achieved by increasing the rotation rate of the anode and remodelling it to increase heat dissipation.

## 5.3 X-ray mirrors

Historically, macromolecular data collection on diffractometers embodied a rotating anode generator which was coupled to a graphite monochromator in order to filter off the unwanted $K_\beta$ radiation. After the monochromator, the X-rays then passed through a collimator before striking the crystal. An inherent problem with the use of a graphite monochromator is that it greatly attenuates the beam and only a fraction of the flux arrives at the crystal. One way around this is to use X-ray mirrors, which are coated with nickel in order to filter off the $K_\beta$ radiation, arranged so that the beam strikes the first mirror at grazing incidence, focusing the beam in the horizontal direction, and then strikes the second mirror imparting focusing in the vertical direction. X-ray mirrors of different types (such as toroidal mirrors and Franks optics) had been used extensively in diffraction studies on fibres from biological specimens, but the alignment of these earlier mirrors was difficult and could result in a degree of X-ray scatter whilst aligning. In 1991, Zbyszek Otwinoski and Jerry Johnson produced a set of mirrors which came to be known as the Yale-MSC mirrors, where driving of the motors which bent the mirrors could be controlled remotely, hence protecting the aligner. The mirrors were in turn housed in a X-ray impermeable box, which was flooded with helium in order to the reduce air scatter of the X-rays and reduce air oxidation of the mirror nickel coating. Using these mirrors in combination with an area detector became very popular and meant that the beam could be focused on the crystal and, at the same time, maximizing the flux falling on it.

Subsequently, manufacturers have developed other mirror systems, two of which are in common use. The first are multilayer optics. Here, instead of the X-rays being focused by a bent mirror with the X-ray radiation at grazing incidence, the focusing is performed by a curved sandwich of layers of substrate, which are alternately coated with a light and heavy element. The layers of the 'sandwich' act as a Bragg d-spacing, imparting diffraction from successive layers. In place of a second mirror, the complete composite formed of the multilayer, as described above, is glued a similar sandwich rotated by 90° so that the incident beam is, in effect, focused over the full length of the multilayer mirror. Multilayer mirrors are marketed by OSMIC, with different specifications compatible with a wide range of X-ray sources, and the Helios optics is marketed by Bruker AXS and is configured to their Microstar generator to form the Proteum X8 system. The second is a mirror system that uses an ellipsoidal, monocapillary optic for X-ray focusing; it is marketed by AXCO and is called the S70PX. It fits very neatly into a small space between the source and the crystal. Undesired $K_\beta$ radiation may be filtered off by using nickel filters of a range of thicknesses if it is found that the $K_\beta$ spots are interfering with the indexing of the $K_\alpha$ ones.

## 5.4 Other in-house X-ray sources

Apart from the rotating anode generators described above, there are two other developments in the generation of X-ray radiation in-house. The first is the development of a high-brilliance, sealed tube generator by Ulrich Arndt at the LMB Cambridge and then marketed by Bede Scientific Instruments Ltd. This source consists of an ellipsoidal mirror with the filament at one focus and the crystal at the other. The source uses very low power, 24 W, and has a very high output (Bloomer and Arndt, 1999; Arndt *et al.*, 1998a, b). This source will clearly prove useful where electrical power is restrictive, such as on the space station; therefore protein diffraction data could be collected directly after crystals have been grown in order to answer vexing questions regarding the improvement of crystal quality after growth under zero-gravity conditions. The second development is that of a small synchrotron for the generation of X-rays for in-house protein crystallographic use. This development has been possible by using free electron laser technology to steer the circulating electron beam in a synchrotron with a very small diameter and produce synchrotron radiation in the 1 Ångstrom range. This system is now sold by Lycean and the first one will shortly be installed in a protein crystallography laboratory in the US.

## 5.5 Setting up for data collection

### 5.5.1 Centring crystals

Centring crystals is an activity that is common to the start of all data collection, so I shall begin by describing this and then proceed to discuss the specifics of data collection on two types of image plate detector (the Raxis-IV$^{++}$ and the MAR345) and one CCD detector (the MARCCD) Protocol 5.1 and Protocol 5.2. Centring is a procedure that has to be mastered and the ability to do it quickly comes with practice. Some goniometer heads are supplied with arcs and translation sledges and some with only translation sledges. A very commonly used goniometer head, which is of very high quality, is made by the Huber Company and these are supplied with and without arcs and in a variety of heights. It is best to make sure the arc and translation securing grub-screws are loosened and the arcs and sledges set to zero before commencing. Screw the goniometer onto the goniostat then mount the loop- or capillary-mounted crystal onto the goniometer so that it can be viewed through the telescope or on a viewing screen.

Centring is most easily accomplished by rotating the base plate below the crystal by inserting a thin Allen key (hex-key in America) into a hole in the base plate and turning so as to place a crystal face parallel to one of the translation sledges (for a crystal with oblong or squarish faces) with the $\varphi$ (phi)-axis on the machine set to zero. Then lock the grub-screw securing the goniometer plate so that it will not rotate. It is possible to centre a crystal mounted at a random orientation, but to centre on a crystal face is the easiest way of starting and often will place a crystallographic zone onto the detector. The translation sledge is then adjusted with the goniometer key to bring the crystal into the centre of the cross-hairs. Unlock the $\varphi$ drive on the goniostat and move to $\varphi = 90°$ and repeat the operation to bring the crystal into the centre of the cross-hairs. Crystal centring is then checked at $\varphi$ positions of 180° and 270° and adjustments made so that when the crystal is gently rotated on $\varphi$ there is no side-to-side precession of the crystal relative to the cross-hairs. If there is a small side-to-side precession then very small adjustments are made to the translations to correct this at positions 0° and 180°; 90° and 270°. This manoeuvre takes practice and is very important as at the end of it a perfectly centred crystal is achieved, which both stays in the beam for the full course of the data collection and is centred. As I said above, a crystal face should be aligned parallel with a translation sledge. However, sometimes clear crystal faces may not be distinct and then the centre of mass of the crystal is estimated by eye and placed in the centre of the cross-wires and the translation corrections performed at $\varphi = 0°$ and 90° and then at 0° and 180°; 90° and 270°, as above, so that when the crystal is gently rotated there is no side-to-side precession. If a crystal has been cryocooled with a large amount of liquid around it, this may have the effect of acting like a lens and one has to be careful to correct adjustments on the translations accordingly. With crystal mounting in a capillary, an amount of liquid left around the crystal in the capillary can have the same lensing effect.

### Protocol 5.1 Data collection using the Raxis $IV^{++}$ image plate

The direct beam position is always determined after a change in filament and/or realignment of the mirror system. This consists of placing a thin sheet of attenuating nickel sheet (provided by the manufacturer) on the goniometer so that it is aligned normal to the beam. The main shutter is opened for a very short period of time (the length of time is dependent on the flux of the X-ray generator) but when read should result in a small intense spot at the centre of the plate. It is important, if one does not know the value of the exposure, to err on the side of caution and start with a few tenths of a second and then increase the value rather than over-expose the central region of the film emulsion to excessive radiation. Progressively, increase the value until you are satisfied with the number of counts in this region. The direct beam spot should be easily cursorable and then entered into the software.

ALWAYS check that all the safety shutters are closed before opening the safety enclosure and making adjustments. X-rays are very dangerous and every safety precaution should be taken when using them. A new user should under no circumstances start using X-ray equipment without experienced guidance.

**1.** The telescope on the Raxis $IV^{++}$ is mounted at 45° to the $\varphi$ axis and the goniometer is locked onto the $\varphi$ axis and the crystal centred as described under (see Centring crystals above).
**2.** Crank the detector back to a distance so that 2.8 Ångstrom is collected at the edge. A calibration chart mapping resolution in Ångstroms at the edge of the detector with distance should be readily available to the operator, either in paper form or on the computer terminal. Collect three oscillations of 1°. If the crystal habit looks as if it is either monoclinic, orthorhombic, or tetragonal (i.e. the crystal looks like a small 'shoebox', a rectilinear parallelpiped) then collect frames at 0°, 45°, and 90°. If the crystal appears to have a hexagonal or trigonal habit and the six-fold axis is along the direction of the $\varphi$ rotation axis then collect frames at 0°, 30°, and 60°.
**3.** Once the first frame is recorded, determine the diffraction limit of the crystal from the image. If this limit is less than 2.8 Ångstroms then stop the collection of the remaining two frames and adjust the distance to the diffraction limit so as to fill the whole plate with diffraction intensities. If the crystal diffracts further than 2.8 Ångstroms then stop the collection after the first frame and drive in the detector to, say, 2 Ångstroms and record three more images. Continue to change the detector distance and take frames till the maximum resolution of the diffraction pattern fills the detector. Care must be taken when driving to high resolution not to collide the back stop or the cryocooling equipment with the black paper on the front of the detector.
**4.** Process the frames using CrystalClear, which is the standard software that comes with the instrument, and determine the space group and crystal lattice parameters. (Some crystallographers prefer to export either a single frame or all three to process with other software; personally I prefer to process and set up for collection using CrystalClear and maybe use other software if I have problems easily getting a space group determination or good integration of the data.) Check for spot overlap at the chosen distance (if there is overlap then the detector will have to be moved further out). Also select $\Delta\varphi$ so that there is no spot overlap in the $\varphi$ direction. If there is overlap then $\Delta\varphi$ must be decreased.
**5.** It pays to process the data as it is collected, to observe how far the crystal diffracts in different crystal orientations and how well it merges in terms of R factor in resolution shells and overall cumulative R factor.

## 5.6 Image plate detectors

Image plates are made by coating a flexible, matt plastic backing sheet with a thin layer of emulsion (150 μm) consisting of a europium compound BaFBr(Eu). They are primarily sold by Fuji and are now available for a wide range of imaging purposes, ranging from $^{32}P$ gel scanning in molecular biology, X-ray crystallographic detectors, through to medical X-ray imaging. This emulsion is sensitive to X-rays and when exposed 'colour centres' are raised to an excited state. Once the plate has been exposed for the required period of time the plate is 'read' by scanning across the plate with a finely focused HeNe laser. The 'colour centres' then fall in their energy levels and emit visible light, which is read by a very sensitive photomultiplier and the analogue signal converted into electronic output. For reuse the plate is 'flashed' clean by exposing it to light from the erasing lamps. Effectively, the image plate is like a reusable piece

**Protocol 5.2 Data collection using the MAR345 and the MARCCD using the MARDTB**

**1.** Move the detector back to $\geq$200 mm to give clearance and to avoid damaging the face of the detector.
**2.** Move the back-stop out of the way. This is done using the 'remote control' unit next to the instrument.
**3.** Check that $\varphi$ (PHI) on the drum of the goniostat is set to 0°. Check that $\chi$ (CHI) is set to zero. (Some cryocooling users like to set $\chi = 60$ or 70°; the remote control unit allows toggling CHI between 0 and 60°.)
**4.** (a) Mount the 'crystal and the goniometer' onto the goniostat. Centre the crystal and align it to the centre of the 'cross-wires' using the small viewing screen (see Centring crystals above).

(An option on the MARDTB is to have motorized *x*, *y* translations built into the goniometer; *z*-translation along $\varphi$ is always available. When *x*, *y* motors are present an IUCr, Huber goniometer is not used, but the crystal is mounted directly onto the $\varphi$ axis with a magnetic cap.)

(b) Replace the back-stop.
**5.** Move the detector to 80 nm for the MAR345 or 35 mm for the MARCCD165 (corresponding to a resolution of 1.4 Ångstroms). Unless the crystal being studied is a DNA crystal or an exceptionally well-diffracting protein then the final observed is likely to be less than this.
**6.** Check the crystal diffraction by setting up to take a 0.5° oscillation (for around 2 min) by specifying the required parameters in the file, namely crystal name, distance, and oscillation range. If the diffraction is weak, repeat the exposure with a longer time.
**7.** Index the image and look for tell tale pathologies in the image, such as spot splitting and high spot mosaicity. If the crystal looks ok, proceed.
**8.** Determine the resolution limit of the crystal from this image and drive the detector to the distance such that the edge of the detector corresponds to this diffraction limit, i.e. the diffraction image should fill the detector once recorded at this distance. Confirm that the diffraction spots are well separated at this distance. If not, drive the detector further out and record frames until they are.
**9.** From the oscillation and the newly derived distance, calculate the orientation matrix using the strategy programme within the MAR or by another software package. Determine the start $\varphi$ and the oscillation range in order to collect the highest percentage of the total obtainable data in the shortest time. Confirm that along $\varphi$ there are no overlaps, otherwise $\Delta\varphi$ has to be decreased.
**10.** Optimize the beam using the beam optimization features on the MARDTB.
**11.** Start collecting the data. Try and process the data 'on the fly' during collection and perform data reduction on it in order to see how well it merges. Also inspect the frames closely during data collection, looking for crystal pathologies such as spot splitting and changes in the resolution of crystal diffraction at different crystal orientations.

## Useful points to note

For the MAR345 a distance of 100 mm (50 mm for the MARCCD165) corresponds to a resolution of 1.54 Ångstrom when using Copper K$\alpha$ radiation.

Distance to resolution conversions can be calculated using simple trigonometry by using the expressions (for zero detector swing):

Resolution =
$1.54/(2^* \sin[0.5^* \tan^{-1}(172.5/x)])$ for the MAR345

Resolution =
$1.54/(2^* \sin[0.5^* \tan^{-1}(82.5/x)])$ for the MARCCD

where the wavelength is assumed to be that for Copper K$\alpha$, namely 1.54(178) Ångstroms and x is the crystal to detector distance.

If no image appears initially then this may be because the erase lamp has blown; this may happen especially at synchrotrons, where they see much higher use.

of photographic film without the need for chemical processing and time-consuming scanning, as was the case when film was used to record macromolecular X-ray data. Three image plate systems are now marketed by Rigaku MSC: the Raxis HR (with 350-mm plates for high resolution work); the Raxis-IV$^{++}$; and the Raxis-HTC (with three 300-mm image plates and a very fast readout by the use of a dual read head).

### 5.6.1 Raxis-IV$^{++}$

The detector consists of 30 cm square plates mounted onto a flexible belt. Once exposed the plate is driven round to the back of the detector and sucked by vacuum onto the inside of a metal hemicylinder. The plate is read out by a revolving mirror which spins the length of the axis of the cylinder and, while progressing incrementally, illuminates the plate with

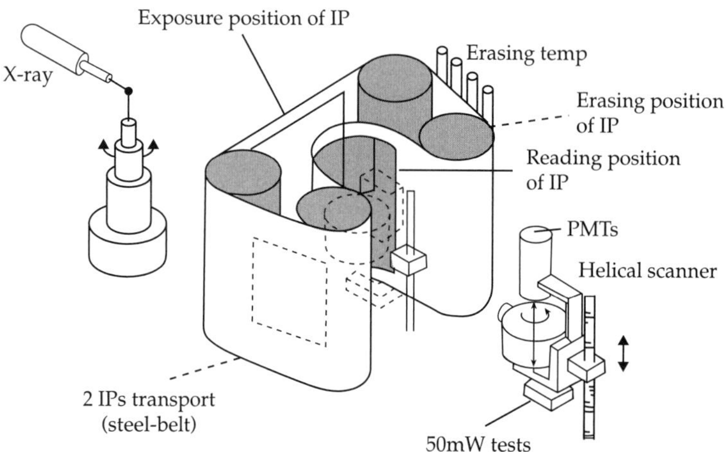


**Figure 5.3** Diagram showing the belt onto which is mounted the image plates and the spinning mirror read-head within the Raxis IV$^{++}$. IP, image plate; PMT. (Courtesy of Dr Joseph Ferrara, Rigaku Americas Corporation.)

HeNe laser light (Fig. 5.3). The emitted visible radiation is fed to a photomultiplier and the image downloaded onto the computer disk. The total plate readout takes 3 min. On moving the exposed plate to the read position, another plate on the belt is driven to the expose position. An earlier image plate detector, the Raxis-II, had a longer read out time of 7 min and employed an illuminated mirror on a carriage moving on a linear motor which scanned across the plate as it was incremented up the plate by a long worm screw. As the mirror on the carriage was illuminated by the laser and focused onto the image plate the emitted visible light returned to a photomultiplier at the end of the track. In contrast, the image plate systems marketed by MAR research employ a spiral readout with the read head spinning from the edge to the centre. The intensities that are read are then transformed into cartesian coordinates and a circular diffraction image downloaded to the disk.

### 5.6.2 CCD detectors

CCD detectors are now used extensively at synchrotrons and increasingly in-house in combination with a high-brilliance X-ray generator (Naday *et al.*, 1998; Westbrook and Naday, 1997). Figure 5.4 shows the principal of the CCD detector, which consists of a scintillation screen attached to the face of the CCD by an array of optical fibres know as an optical taper. When X-rays fall on the scintillation screen emitted photons are fed back to the CCD, which converts the light energy into electrons; these are stored in potential wells which act as registers and are read out to generate the image. CCD detectors are marketed in a variety of active areas depending on the manufacturer of the CCD chip and their size. MAR research makes a CCD, the MARCCD, consisting of a single chip of diameter 165 mm; Rigaku MSC makes the Saturn 944 (for in-house use which has a 94 × 94 mm image area and incorporates a Kodak KAF-4320E CCD chip, giving a 2084 × 2084 pixel area with four port readout) and Jupiter 210 (primarily for synchrotron use with a 210 mm active area) CCD detectors. Bruker AXS manufactures the Platinum 135 CCD (for in-house use) and the Apex II CCD (for high-throughput applications). Oxford Diffraction also market an Xcalibur Nova CCD detector system which incorporates a 165 mmOnyx CCD (with a 2k × 2k Kodak chip) for macromolecular crystallography. The Oxford Nova generator is a high flux sealed tube generator of the Arndt design (see above). Oxford Diffraction also market a range of other CCD detectors; namely the Sapphire, which has a 92 mm (diagonal) active area, and the Ruby with a 135 mm active area, both incorporating a 2k × 2k Kodak chip. The Quantum4 CCD is marketed by the ADSC company (headed

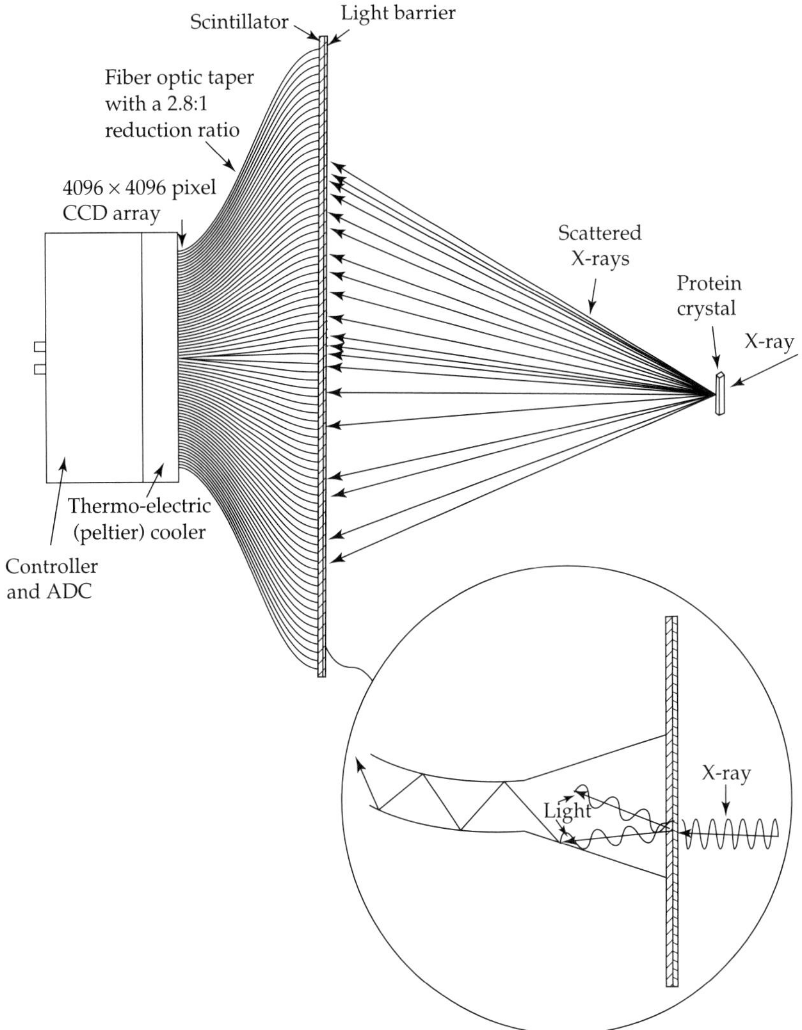


**Figure 5.4** Schematic diagram of a CCD detector. (Reproduced with permission from Nolting, *Methods in Modern Biophysics*, Springer Verlag, 2004.)

by Hamlin, Nielsen, and Xuong) and consists of four chips bonded together to make an active area of 210 $cm^2$ and currently used on many synchrotron beamlines around the world.

### 5.6.3 The MAR345 and MARCCD

Both the MAR345 and the MARCCD are supplied with a MARDTB goniostat (DTB being an abbreviation for Desk Top Beam line). This assembly is supplied with variable apertures ranging from 0 to 5 mm, with 2.5 mm resolution. It also incorporates built-in ion chambers with very wide dynamic ranges, permitting very precise optimization of the beam. Either detector (image plate or CCD) can be tilted in the vertical to a maximum 2θ of 30°. For the MAR345, the detector can be driven from 75 to 425 mm and the CCD detector from 20 to 370 mm using the MARDTB.

## Acknowledgements

I should like to thank Peter Collins and Claudio Klein for their discussion and comments concerning data collection using the MARDTB.

## References

Arndt, U. W. (1982). X-ray television area detectors. *Nucl. Inst. Methods* **201**, 13–20.

Arndt, U. W. (1985). Television area detector diffractomers. *Method Enzymol.* **114**, 472–485.

Arndt, U. W. (2003). Personal X-ray reflections. *Method Enzymol.* **368**, 21–42.

Arndt, U. W., Ducumb, P., Long, J. V. P., Pina, L. and Inneman, A. (1998a). Focusing mirrors for use with microfocus X-ray tubes. *J. App. Cryst.* **31**, 733–741.

Arndt, U. W., Long, J. V. P. and Duncumb, P. (1998b). A microfocus X-ray tube used with focusing collimators. *J. App. Cryst.* **31**, 936–944.

Bloomer, A. C. and Arndt, U. W. (1999). Experiences and expectations of a novel X-ray microsource with focusing mirror. I. *Acta Crystallogr. D* **55**, 1672–1680.

Broad, D. A. G. (1956). Rotating anode X-ray tube. UK Patent Applications Nos. 5172, 5173, 12761, 13376, 38939.

Charpak, G. (1988). Some applications of multiwire chambers to the detection of electromagnetic radiations. *Nucl. Inst. Methods* **269**, 341–345.

Charpak, G., Bouclier, R., Bressani, T., Favier, J. and Zupancic, C. (1968). The use of multiwire proportional counters to select and localize charged particles. *Nucl. Inst. Methods* **62**, 262–268.

Charpak, G., Dominik, W., Santiard, J. C., Sauli, F. and Solomey, N. (1989). Gaseous detectors with parallel electrodes and anode mesh planes. *Nucl. Inst. Methods* **274**, 275–290.

Cork, C., Hamlin, R., Vernon, W. and Xuong, N. H. (1975). Xenon-filled multiwire area detector for X-ray diffraction. *Acta Crystallogr. A* **31**, 702–703.

Durban, R. M., Burns, R., Metcalf, P., Freymann, D., Blum, M., Anderson, J. E., Harrison, S. C. and Wiley, D. C. (1986). Protein, DNA and virus crystallography with a focused imaging proportional counter. *Science* **232**, 1127–1132.

Edwards, S. L., Nielsen, C. and Xuong, N. H. (1988). Screen precession method for area detectors *Acta Crystallogr. B* **44**, 183–187.

Garman, E. F (1991). Modern methods for rapid X-ray diffraction data collection from crystals of macromolecules. In: *Methods in Molecular Biology*, vol. 56, *Crystallographic Methods and Protocols*, Jones, C., Mulloy, B. and Sanderson, M. R., eds. Humana Press.

Hamlin, R. (1985). Multiwire area X-ray diffractometers. *Method Enzymol.* **114**, 416–451.

Hamlin, R., Cork, C., Howard, A., Nielsen, C., Vernon, W., Matthews, D. and Xuong, N. H. (1981). Characteristics of a flat multiwire area detector for protein crystallography *J. Appl. Cryst.* **14**, 85–93.

Howard, A. J., Nielsen, C. and Xuong, N. H. (1985). Software for a diffractometer with multiwire area detector. *Method Enzymol.* **114**, 452–472.

Kahn, R. and Fourme, R. (1997). Gas proportional detectors. *Method Enzymol.* **276**, 268–286.

Müller, A. (1929). A spinning target X-ray generator and its input limit. *Proc. Roy. Soc. London A* **125**, 507–516.

Naday, I., Ross, S., Westbrook, E. M. and Zentai, G. (1998). Charge-coupled device/fiber optic taper array x-ray detector for protein crystallography. *Opt. Eng.* **37**, 1235–1244.

Schierbeek, B., Benning, M., Diawara, Y. and Durst, R. (2006). First results of Åxiom 200, a high-speed, photon-counting X-ray area detector. *Acta Crystallogr. A* **62**, s160.

Westbrook, E. M. and Naday, I. (1997). Charge-couple device-based area detectors. *Method Enzymol.* **276**, 244–268.

Xuong, N. H., Sullivan, D. Nielsen, C. and Hamlin, R. (1985a). Use of multiwire diffractometer as a national resource for protein crystallography. *Acta Crystallogr. B* **41**, 267–269.

Xuong, N. H., Nielsen, C., Hamlin, R. and Anderson, D. (1985b). Strategy for data-collection from protein crystals using a multiwire counter area detector. *J. App. Cryst.* **18**, 342–350.

CHAPTER 6

# Solving the phase problem using isomorphous replacement

Sherin S. Abdel-Meguid

## 6.1 Introduction

From the mid 1950s, when it was first introduced, till the mid 1990s, the method of isomorphous replacement played a central role in the determination of almost all unique macromoleular structures. Isomorphous replacement was the technique used in the first successful high-resolution structure determination of a protein molecule, myoglobin (Kendrew *et al.*, 1958; Kendrew *et al.*, 1960). It was developed by Perutz and coworkers in 1954 (Green *et al.*, 1954) while working on the structure determination of haemoglobin. The technique was introduced to solve the 'Phase Problem', the loss during X-ray diffraction data measurement of the relative phase shifts associated with each diffraction point (maximum). Although the amplitude of a diffraction maximum can be directly measured from diffracting crystals by counting photons or recording intensities, phases are indirectly determined because there are no lenses that can bend and focus X-rays. Thus, the isomorphous replacement method was developed to computationally calculate phases from the intensities of the diffracted waves.

The technique of isomorphous replacement requires the introduction of atoms of high atomic number (heavy atoms; Fig. 6.1) into the macromolecule under study without changing the crystal's unit-cell parameters or orientation of the protein in the cell (Abdel-Meguid, 1996). This is commonly done by soaking native crystals in a solution containing the desired heavy atom. The binding of these atoms to the functional groups of macromolecules is facilitated by the presence of large liquid channels in crystals, in which the functional groups protrude. The addition of one or more heavy atoms to a macromolecule introduces differences in the diffraction pattern of the derivative relative to that of the native. If this addition is truly isomorphous, these differences will represent the contribution from the heavy atoms only; thus the problem of determining atomic positions is initially reduced to locating the position of a few heavy atoms. Once the positions of these atoms are accurately determined, they are used to calculate a set of phases for data measured from the native crystal. Although, theoretically, one needs only two isomorphous derivatives to determine the three-dimensional structure of a biological macromolecule, in practice more than two are needed. This is due to errors in data measurement and scaling and in heavy-atom positions, as well as lack of isomorphism.

The search for isomorphous derivatives is as empirical as searching for crystallization conditions. Numerous heavy atoms must be screened before finding the one or more that binds to the protein without damaging the crystal. The soaking of native crystals in a solution containing heavy atoms gives rise to one of four outcomes. The best outcome would be an isomorphous heavy-atom derivative containing one or a small number of heavy atoms identically attached to each protein molecule in the crystal, resulting in distinct changes between the diffraction patterns obtained from derivative and native crystals. At the other extreme, the soaking process results in no such detectable changes when comparing native and derivative crystals. The two remaining outcomes are either the native crystal gets destroyed during soaking or

| IA | IIA | IIIB | IVB | VB | VIB | VIIB | VIII | | | IB | IIB | IIIA | IVA | VA | VIA | VIIA | VIIIA |
|---|---|---|---|---|---|---|---|---|---|---|---|---|---|---|---|---|---|
| 1 H | | | | | | | | | | | | | | | | 1 H | 2 He |
| 3 Li | 4 Be | | | | | | | | | | | 5 B | 6 C | 7 N | 8 0 | 9 F | 10 Ne |
| 11 Na | 12 Mg | | | | | | | | | | | 13 Al | 14 Si | 15 P | 16 S | 17 Cl | 18 Ar |
| 19 K | 20 Ca | 21 Sc | 22 Ti | 23 V | 24 Cr | 25 Mn | 26 Fe | 27 **Co** | 28 Ni | 29 Cu | 30 Zn | 31 Ga | 32 Ge | 33 As | 34 **Se** | 35 Br | 36 **Kr** |
| 37 Rb | 38 **Sr** | 39 Y | 40 Zr | 41 Nb | 42 **Mo** | 43 Tc | 44 **Ru** | 45 **Rh** | 46 **Pd** | 47 **Ag** | 48 **Cd** | 49 In | 50 **Sn** | 51 Sb | 52 Te | 53 **I** | 54 **Xe** |
| 55 Cs | 56 **Ba** | 57 #**La** | 72 Hf | 73 **Ta** | 74 **W** | 75 **Re** | 76 **Os** | 77 **Ir** | 78 **Pt** | 79 **Au** | 80 **Hg** | 81 **Tl** | 82 **Pb** | 83 Bi | 84 Po | 85 At | 86 Rn |
| 87 Fr | 88 Ra | 89 *Ac | 104 Rf | 105 Db | 106 Sg | 107 Bh | 108 Hs | 109 Mt | 110 Ds | 111 Rg | 112 Uub | 113 Uut | 114 Uuq | 115 Uup | 116 Uuh | 117 Uus | 118 Uuc |

| | | | | | | | | | | | | | | |
|---|---|---|---|---|---|---|---|---|---|---|---|---|---|---|
| #Lanthanides | 58 **Ce** | 59 **Pr** | 60 **Nd** | 61 Pm | 62 **Sm** | 63 **Eu** | 64 **Gd** | 65 **Tb** | 66 **Dy** | 67 **Ho** | 68 **Er** | 69 **Tm** | 70 **Yb** | 71 Lu |
| *Actinides | 90 **Th** | 91 Pa | 92 **U** | 93 Np | 94 Pu | 95 Am | 96 Cm | 97 Bk | 98 Cf | 99 Es | 100 Fm | 101 Md | 102 No | 103 Lr |

**Figure 6.1** The Periodic table showing the elements used successfully as heavy-atom derivatives in bold and underlined. The rest of the elements are shown only for completeness.

the soaking process causes the crystal to become non-isomorphous.

## 6.2 Theoretical basis

Originally, isomorphous replacement phasing of biological macromolecules requires the measurement of at least three X-ray diffraction data sets, a native and two or more derivatives. Therefore, the method was more commonly referred to as the method of multiple isomorphous replacement (MIR). However, the introduction of area detectors and synchrotron radiation allowed for significant improvements in data quality and the ability to use only a single isomorphous derivative if its heavy atom is an anomalous scatterer. The latter is referred to as single isomorphous replacement with anomalous scattering (SIRAS), where the anomalous data is used to break the phase ambiguity.

The theoretical basis of isomorphous replacement can be found in Blundell and Johnson (1976), Drenth (1999), and was recently summarized by Taylor (2003). Here, I will only give a brief overview. As indicated above, an X-ray diffraction experiment only gives us intensities of waves scattered from planes (hkl) in the crystal, but the phase shift associated with each hkl is lost during data measurement. Thus, each measured intensity ($I_{\mathrm{hkl}}$) can be reduced to structure factor amplitude ($F_{\mathrm{hkl}}$) with unknown phases, where $F_{\mathrm{hkl}}$ is proportional to the square root of $I_{\mathrm{hkl}}$. Each structure factor amplitude ($F_{\mathrm{hkl}}$) and its associate phase ($\alpha_{\mathrm{hkl}}$) can be described in terms of a vector quantity, the structure factor ($\mathbf{F}_{\mathrm{hkl}}$). For every hkl, native and derivative structure factors (Fig. 6.2) are related as shown in Eq. 1:

$$\mathbf{F}_{\mathrm{PH}} = \mathbf{F}_{\mathrm{P}} + \mathbf{F}_{\mathrm{H}} \tag{1}$$

where $\mathbf{F}_{\mathrm{PH}}$, $\mathbf{F}_{\mathrm{P}}$, and $\mathbf{F}_{\mathrm{H}}$ are the structure factors of the derivative, the native protein, and the heavy atom, respectively.

Once the heavy-atom position has been determined, its structure factor amplitude $F_{\mathrm{H}}$ and phase $\alpha_{\mathrm{H}}$ can be calculated. Since the structure factor amplitudes for the native ($F_{\mathrm{P}}$) and derivative ($F_{\mathrm{PH}}$) are experimentally measured quantities, it is thus possible to calculate the protein phase angle $\alpha_{\mathrm{P}}$ from the following equations:

$$F_{\mathrm{PH}}{}^2 = F_{\mathrm{P}}{}^2 + F_{\mathrm{H}}{}^2 + 2F_{\mathrm{P}}F_{\mathrm{H}}\cos(\alpha_{\mathrm{P}} - \alpha_{\mathrm{H}}) \tag{2}$$

or

$$\begin{aligned}\alpha_{\mathrm{P}} &= \alpha_{\mathrm{H}} + \cos^{-1}[(F_{\mathrm{PH}}{}^2 - F_{\mathrm{P}}{}^2 - F_{\mathrm{H}}{}^2)/2F_{\mathrm{P}}F_{\mathrm{H}})] \\ &= \alpha_{\mathrm{H}} \pm \alpha' \end{aligned} \tag{3}$$

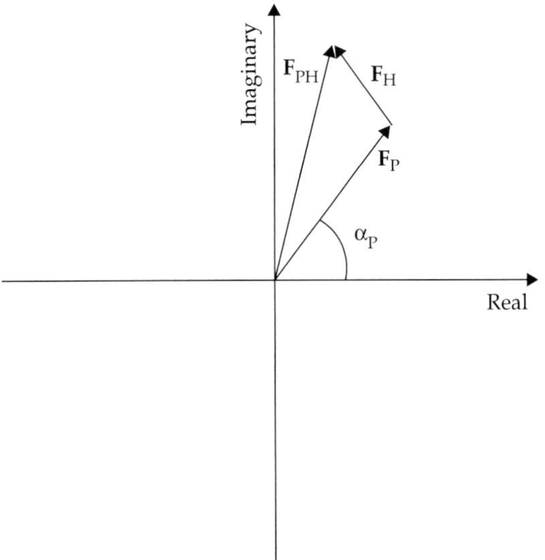


**Figure 6.2** Vector (Argand) diagram showing the relationships between heavy-atom derivative ($\mathbf{F}_{PH}$), native protein ($\mathbf{F}_P$) and heavy atom ($\mathbf{F}_H$); $\alpha_P$ is the phase angle for the native protein. The vectors are plotted in the complex plane.

From Eq. 3 and Fig. 6.3a it is clear that with only one heavy-atom derivative (single isomorphous replacement; SIR) the resultant phase will have two values ($\alpha_{Pa}$ and $\alpha_{Pb}$); one of these phases will represent that of one structure and the other of its mirror image. But, since proteins contain only L-amino acids, this phase ambiguity must be eliminated using a second derivative, the anomalous component of the heavy atom or by solvent levelling (Wang, 1985), as shown diagrammatically in Fig. 6.3b.

Once the phase angle $\alpha_P$ has been determined for every hkl, a Fourier synthesis is used to compute the electron density ($\rho$) at each position (xyz) in the unit cell (the repeating unit forming the crystal lattice) using Eq. 4:

$$\rho(\text{xyz}) = 1/V\,\Sigma_h\Sigma_k\Sigma_l\mathbf{F}_P(\text{hkl})e^{-2\pi i(hx+ky+lz)} \quad (4)$$

where V is the volume of the unit cell, $i$ is the imaginary component $\sqrt{-1}$, and

$$\mathbf{F}_P(\text{hkl}) = F_P(\text{hkl})e^{i\alpha_P} = F_P(\text{hkl})\cos\alpha_P(\text{hkl}) + iF_P(\text{hkl})\sin\alpha_P(\text{hkl}) \quad (5)$$

(a)

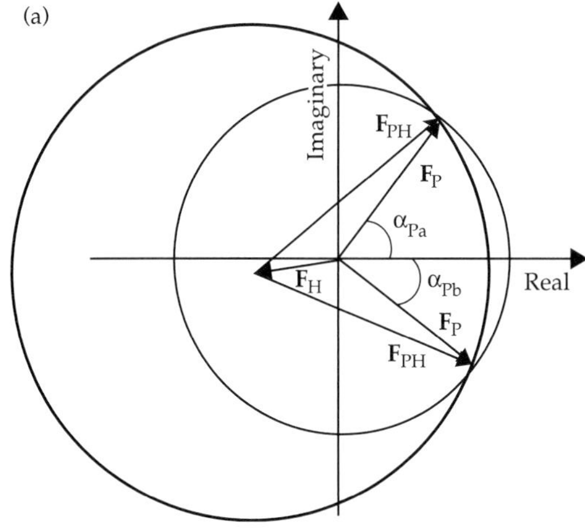


(b)

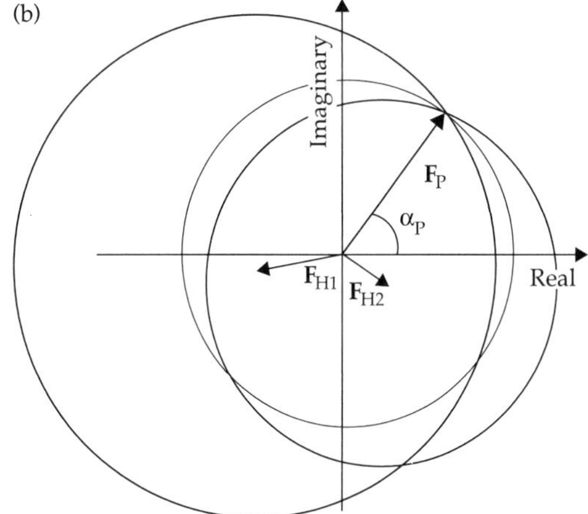


**Figure 6.3** Isomorphous replacement phase determination (Harker construction). (a) Single isomorphous replacement. The circle with radius $\mathbf{F}_{PH}$ represents the heavy-atom derivative, while that with radius $\mathbf{F}_P$ represents the native protein. Note that the circles intersect at two points causing an ambiguity in the phase angle; $\alpha_{Pa}$ and $\alpha_{Pb}$ represent the two possible values. (b) Double isomorphous replacement. The same construction as that in single isomorphous replacement except that an additional circle with radius $\mathbf{F}_{PH2}$ (vector not shown for simplicity) has been added to represent a second heavy-atom derivative. Note that all three circles (in the absence of errors) intersect at one point thus eliminating the ambiguity in the protein phase angle $\alpha_P$. $\mathbf{F}_{H1}$ and $\mathbf{F}_{H2}$ represent the heavy-atom vectors for their respective derivatives.

The electron density map ($\rho$(xyz)) can then be interpreted in terms of a three-dimensional atomic model.

## 6.3 Selection of heavy-atom reagents

Both the size and chemical composition of the molecule under investigation are important criteria to consider when selecting heavy-atom reagents for derivatization. Ones choice must insure that the differences in diffraction amplitudes due to heavy-atom contributions are larger than the errors in data measurement. The size of the heavy atom (atomic number) and the number of sites required for successful phasing are proportional to the size (molecular weight) of the macromolecule. Larger macromolecules may require not only atoms of high atomic number, but also more than one heavy atom per molecule. For example the structure determination of the ribosome required heavy-atom clusters such as $Ta_6Br_{12}{}^{+2}$ (Ban *et al.*, 2000). Therefore, when studying larger macromolecules it may be useful to calculate the magnitude of the change in the diffraction signal before deciding which heavy atom to try. Crick and Magdoff (1956) showed that the average fractional intensity change ($\Delta I/I$) for acentric reflections can be estimated from the following equation:

$$\Delta I/I = (2N_H/N_P)^{1/2}(Z_H/Z_P) \qquad (6)$$

where $N_H$ and $N_P$ are the number of heavy atoms and non-hydrogen protein atoms, respectively; $Z_H$ is the atomic number of the heavy atom and $Z_P$ is 6.7 (the average atomic number of protein atoms).

In the case of small proteins, inspection of the amino acid composition can give valuable insights into which reagents should be tried first. For example if the protein contains no free cysteines or histidines it may be best to start soaking with compounds other than mercurials, or to genetically engineer heavy-atom binding sites as was done with the catalytic domain of $\gamma\delta$ resolvase (Abdel-Meguid *et al.*, 1984; Hatfull *et al.*, 1989). However, assuming a normal distribution of amino acids, one should begin with platinum compounds such as $K_2PtCl_4$ (the most widely successful heavy-atom reagent), which binds mainly to methionine, histidine, and cysteine residues. Petsko *et al.* (1978) have described the chemistry of this reagent in a variety of crystal mother liquors. They also concluded that most other platinum compounds react with proteins in a similar fashion, except for those containing $(Pt(CN)_2)^{2-}$ which bind to positively charged residues. Mercurial compounds are the second most successful group of reagents in derivative preparation; most mercurials either bind to cysteine sulphurs or histidine nitrogens. In addition to platinum and mercury, Fig. 6.1 show the many elements successfully used in isomorphous replacement phasing.

## 6.4 Heavy atoms and their ligands

The preparation of heavy-atom derivatives and selection of reagents have been extensively reviewed (Abdel-Meguid, 1996; Blundell and Johnson, 1976; Petsko, 1985; Kim *et al.*, 1985; Holbrook and Kim, 1985; Garman and Murray, 2003). Historically, heavy atoms have been grouped into class A and class B elements based on their ligand preference. Class A elements prefer 'hard' ligands such as carboxylates and other oxygen containing groups. These ligands are electronegative and form electrostatic interactions with the derivatives; they include the carboxylates of aspartic and glutamic acids and the hydroxyl of serine and threonine. On the other hand, class B elements prefer 'soft' ligands such are those containing sulphur, nitrogen, and halides. These include methionine, cysteine, and histidine. The latter two amino acids are the most reactive amongst all 20 naturally occurring amino acid residues. The cysteine's sulphur is an excellent nucleophile; it will react irreversibly with mercuric ions and organomercurials at wide range of pHs. The thiolate anion also forms stable complexes with class B metals, but since cysteines are almost totally protonated at pH 6 or below, this reaction is more sensitive to pH than that with mercurials. The imidizole side chain of histidine is also highly reactive, particularly above pH 6 where it is unprotonated. It reacts well with reagents containing platinum, mercury, and gold.

In addition to the amino acids described above, several other amino acid residues are also reactive toward compounds containing heavy atoms. These are the side chains of arginine, asparagine, glutamine, lysine, tryptophan, and tyrosine. Those that are not reactive are alanine, glycine, isoleucine,

leucine, phenylalanine, proline, and valine. Arginine forms electrostatic interactions with anionic ligands, while asparagine and glutamine may weakly coordinate to simple metal complexes through their side chain's amide nitrogen. Like arginine, lysine can form electrostatic interactions with anionic ligands at pHs below its pKa. Near or above its pKa of about 9, it can also react with platinum and gold complexes. The indole ring nitrogen of tryptophan can be mercurated, but tryptophan is usually buried in the protein and rarely accessible to solvent. While the phenol hydroxyl oxygen of tyrosine is expected to be a good nucleophile, its pKa is greater than 10. Thus, the tyrosine utility as a ligand for heavy atoms has been in the substitution of the phenolate hydroxyl with iodine (Sigler, 1970). In addition to the side chains of amino acids, the N-terminal amine and the C-terminal carboxylate of a protein are potentially ligands for heavy-atom derivatives.

Dr Bart Hazes (University of Alberta) has further grouped heavy-atom derivatives in six different categories as follow:

*Class A*: This class consists of the alkaline earth metals ($Sr^{2+}$ and $Ba^{2+}$), the lanthanides ($La^{3+}$, $Ce^{3+}$, $Pr^{3+}$, $Nd^{3+}$, $Sm^{3+}$, $Eu^{3+}$, $Gd^{3+}$, $Tb^{3+}$, $Dy^{3+}$, $Ho^{3+}$, $Er^{3+}$, $Tm^{3+}$, and $Yb^{3+}$) and the actinide $(UO_2)^{2+}$. As indicated above, these elements prefer carboxylates and other oxygen containing ligands. They withstand low pH and ammonium sulphate, but they have lower solubility at higher pH and in the presence of phosphates.

*Class B*: As described above, this group contains many of the most popular heavy-atom derivatives containing mercury and platinum, such as *p*-chloromercuribenzoate, $HgC1_2$, mercuric acetate, ethylmercury chloride, $K_2PtC1_4$, $K_2Pt(NO_2)_4$, and $K_2PtC1_6$. In general, these reagents prefer ligands containing sulphur and nitrogen such as cysteine and histidine. This group also consists of many silver, gold, palladium, iridium, osmium, and cadmium containing reagents.

*Anionic*: This group includes heavy-atom reagents that predominantly binds to basic regions on the protein through their overall negative charge, such as iodide, $(HgI_3)^-$, $(Pt(CN)_4)^{2-}$, $(IrC1_6)^{3-}$, and $(Au(CN)_2)^-$.

*Cationic*: This group includes heavy-atom reagents that predominantly bind to acidic regions on the protein through their overall positive charge, such as $(Pt(NH_3)_4)^{2+}$, $(Ir(NH_3)_6)^{3+}$, and $(Hg(NH_3)_2)^{2+}$.

*Hydrophobic*: This group includes the nobel gases krypton and xenon, which bind to hydrophobic pockets in the protein. The main impediment to the use of these gases has been the technical challenge in derivatization under pressure, particularly since pressurized capillaries of glass or quartz are explosion hazards. A special device to make nobel gas derivatives has been described by Schiltz *et al.* (1994), and a commercial one is now being sold by Molecular Structure Corporation for use in cryocrystallography.

*Others*: This includes iodine that can be used to mono- or di-iodinate tyrosine residues.

## 6.5 Preparation of heavy-atom derivatives

Macromolecular crystals grow in an equilibrium state with their mother liquor. Disrupting this equilibrium can often destroy the crystals or their ability to diffract X-rays. This situation can be exacerbated by the transfer of the crystal to a solution containing a heavy atom. Therefore, it is important, once crystals are removed from their sealed environment, to first transfer them to a stabilizing solution and let them re-equilibrate before further transfer to the heavy atom solution. Usually, a stabilizing solution is identical to the mother liquor in which the crystal was grown, but with a higher concentration of precipitant.

The mechanics of derivative preparation is simple; it involves the transfer of one or more native crystals from the stabilizing solution to a solution differing only in the presence of a compound containing the desired heavy atom. However, before attempting to prepare derivatives, it is important to recognize that heavy-atom reagents are very toxic and must be handled with utmost care. These reagents are selected for their strong affinity for biological molecules. Thus, they present real and serious danger to their users. Once crystals have been transferred to the heavy-atom solution, they can be soaked in that solution for a period of time.

Soaking times can be as little as a few hours or as long as several weeks, but usually on the order of 1 to 4 days. Soaking times are dependent on temperature and heavy-atom compound concentration; at lower temperatures and heavy-atom concentrations it may be necessary to soak for longer periods of time. The concentration of the heavy-atom reagent used for derivative preparation will depend on its solubility in the mother liquor. A good starting value is 1 mM, but concentrations as low as 0.05 mM and as high as 100 mM have been reported. The ideal derivative is arrived at by varying soaking time and heavy-atom compound concentration. The latter variable is more useful since mass action can force the formation of a derivative even in the case of weak binding functional groups. Soaking times as short as 1 h combined with concentrations of 0.3 mM were reported to produce good mercury derivatives of iron superoxide dismutase (Ringe *et al.*, 1983). In addition to temperature and heavy-atom compound concentration, the composition of the mother liquor and pH should be considered. Many of the buffers, additives, and precipitants used in mother liquors, such as tris, phosphate, citrate, β-mercaptoethanol, dithiothreitol, and $NH_3$ derived from ammonium sulphate at high pH, may compete with the protein for heavy-atom binding. It may be necessary at times to transfer crystals into more appropriate mother liquor before derivative preparation. For example crystals grown out of ammonium sulphate may be transferred to lithium sulphate to avoid the formation of metal–ammonia complexes, and salts may possibly be replaced by polyethylene glycol (PEG). Such changes in mother liquor are best done incrementally and slowly to avoid shocking the crystals. Also, one should recognize that the solubility of heavy atom reagents in the mother liquor and their binding to functional groups on the protein, are pH dependent. The ideal pH range is 6 to 8; lower pH may result in protonation of glutamic and aspartic acids of proteins, while at higher pH many heavy-atom reagents are labile and form insoluble hydroxides.

As indicated above, the search for suitable heavy-atom derivatives is as empirical as searching for crystallization conditions. To speed up the process and to save time, initial scanning for suitable heavy-atom derivatives can be done visually, using small crystals, by observing deterioration (cracking or dissolution) of the crystals. Concentrations of the heavy-atom reagent and soaking times should be adjusted to insure that the crystals do not show serious cracks; minor surface cracks may not be detrimental to some crystals. However, some crystals are very sensitive to most heavy-atom reagents, they tend to shatter and lose their ability to diffract, even if the reagents are very dilute and soak times are short. This can be overcome by crosslinking the crystals with glutaraldehyde before soaking. A colour change of the crystal during soaking does not always mean that the heavy atom has specifically bound, since non-specific binding may also cause the crystal to change colour. Non-specific binding can be minimized by back-soaking in the stabilizing solution, the solution that does not contain any heavy atom.

Although the soaking method for heavy-atom derivative preparation is by far the simplest and most common, it is not the only method used. One can first derivatize the macromolecule, and then crystallize. This procedure is less frequently used because of drawbacks such as the inability to produce isomorphous crystals due to the disruption of intermolecular contacts by the heavy atoms. Other frequent problems are the introduction of additional heavy-atom sites (a potential complicating factor in phasing) by exposing sites hidden by crystal contacts, and changing the solubility of the derivatized macromolecule.

The above two methods for derivative preparation have been successfully used in the phasing of both nucleic acids and proteins. However, an additional method has been used for phasing of nucleic acids, in which the heavy atom is synthetically incorporated into the molecule. For example, Drew *et al.* (1980) determined the structure of $d(CGCG)_2$ by incorporating 5-bromocytosine into the synthesis of their nucleic acid and then using the bromine atoms for phasing.

## 6.6 Assessment of derivative formation

As might be expected, not every crystal soaked in a solution containing a heavy-atom reagent will be a derivative. Recently, Garman and Murray (2003) have summarized many of the techniques used in evaluating derivative formation; these include mass spectrometry, gel electrophoresis, and microPIXE (particle-induced X-ray emission). However, it is important to note that most of these techniques are just a guide that can help in evaluating derivative formation, but the ultimate method is to identify intensity changes between native and derivative crystals and to be able to confirm the significance of these changes by calculating the position of ordered heavy-atom sites. This can be achieved by comparing the different statistics calculated from the native and putative derivative data or between Friedel mates within the derivative data; significant differences should indicate successful derivative formation.

## 6.7 Determination of heavy-atom positions

By far, the most common procedure for the determination of heavy-atom positions is the difference Patterson method; it is often used in combination with the difference Fourier technique to locate sites in second and third derivatives.

### 6.7.1 Difference Patterson

The Patterson function (Patterson, 1934) is a phaseless Fourier summation similar to that in Eq. 4 but employing $F^2$ as coefficients, thus it can be calculated directly from the experimentally measured amplitudes ($F_P$) without the need to determine the phase angle. In the case of macromolecules, $(F_{PH} - F_P)^2$ are used as coefficients in Eq. 4 to produce a Patterson map (hence the name difference Patterson). Such a map contains peaks of vectors between atoms (interatomic vectors). Thus in the case of a difference Patterson of macromolecules, it is a heavy-atom vector map. For example if a structure has an atom at position (0.25, 0.11, 0.32) and another atom at position (0.10, 0.35, 0.15), there will be a peak in the Paterson map at position (0.25–0.10, 0.11–0.35, 0.32–0.15), meaning a peak at (0.15, −0.24, 0.17).

The interpretation of Patterson maps requires knowledge of crystallographic symmetry and space groups. Chapter 4 of Blundell and Johnson (1976) offers a concise review of these topics. The ease of interpretation of these maps depends on the quality of the data, the degree of isomorphism, the number of heavy-atom sites per macromolecule and the degree of substitution for each heavy atom. An ideal case is one in which: (a) the native and derivative data are of very good quality, (b) the derivative shows a high degree of isomorphism, (c) only one highly substituted heavy atom is present per macromolecule, and (d) the heavy atom is of sufficiently high atomic number to give significant differences. A simple example of how to interpret a Patterson map is described by Abdel-Meguid (1996).

### 6.7.2 Difference Fourier

As can be seen from Eq. 4, a Fourier synthesis requires phase angles as input, thus it cannot be used to locate heavy-atom positions in a derivative if no phase information exists. However, it can be used to determine such positions in a derivative, if phases are already available from one or more other derivatives. As in the case of a difference Patterson, the Fourier synthesis here also employs difference coefficients. They are of the form:

$$m(F_{PH} - F_P)e^{i\alpha_P} \quad (7)$$

where $F_{PH}$ and $F_P$ are the structure factor amplitude of the derivative and the native, respectively; $\alpha_P$ the protein phase angle calculated from other derivatives; and $m$ (figure of merit; whose value is between zero and one) is a weighting factor related to the reliability of the phase angle.

The success of this technique is highly dependent on the correctness of $\alpha_P$, since it has been clearly demonstrated that Fourier summations with correct phases but wrong amplitudes can result in the correct structure, while having incorrect phases even with correct amplitudes results in the wrong structure.

Difference Fourier techniques are most useful in locating sites in a multisite derivative, when a Patterson map is too complicated to be interpretable. The phases for such a Fourier must be calculated from the heavy-atom model of other derivatives in which a difference Patterson map was successfully interpreted, and should not be obtained from the derivative being tested, in order not to bias the phases. Also, difference Fourier techniques can be used to test the correctness of an already identified heavy-atom site by removing that site from the phasing model and seeing whether it will appear in a difference Fourier map. Again the success of this feed-back technique depends on the correctness of the phasing model.

## 6.8 Refinement of heavy-atom positions

Once the positions of the heavy atoms have been determined, their accuracy can be significantly improved through least squares refinement. This is achieved by allowing the heavy-atom positions, thermal parameters, and occupancy to vary, while minimizing the difference between the structure factor amplitudes calculated from the heavy-atom model and experimentally measured amplitudes. It is important to note that thermal parameters and occupancy are often correlated and should not be refined in the same cycle. A good strategy in heavy-atom refinement is to obtain the best possible heavy-atom model for each derivative alone, and then add one derivative at a time while refining, to insure that the best derivative does not dominate phasing to the exclusion of others. Also, it is better to omit minor sites from the initial refinement if there is doubt about their existence. These sites can be included in later steps of refinement when it becomes more certain that they are real.

## 6.9 Conclusion

Although the technique of isomorphous replacement was the dominant crystallographic technique for determination of *de novo* protein structures since its development in the early 1950s to the early 1990s, today it is but one of several techniques used routinely by those interested in elucidating the three-dimensional structures of macromolecules. Now those interested in crystallographic structure determination can choose from a number of methods to solve the phase problem, including single and multiple isomorphous replacement each with or without anomalous scattering, single- and multi-wavelength anomalous diffraction (SAD/MAD), and molecular replacement. The choice of which method to use can depend on a number of factors, such as the availability of synchrotron radiation, atomic coordinates of homologous protein, protein expressed in the presence of selenomethionine, etc.

## References


Abdel-Meguid, S. S. (1996). Structure determination using isomorphous replacement. *Method Mol. Biol.* **56**, 153–171.

Abdel-Meguid, S. S., Grindley, N. D. F., Templeton, N. S. and Steitz, T. A. (1984). Cleavage of the site-specific recombination protein γδ resolvase: the smaller of the two fragments binds DNA specifically. *Proc. Natl. Acad. Sci. USA* **81**, 2001–2005.

Ban, N., Nissen, P., Hansen, J., Moore, P. B. and Steitz, T. A. (2000). The complete atomic structure of the large ribosomal subunit at 2.4A resolution, *Science* **289,** 905–920.

Blundell, T. L. and Johnson, L. N. (1976). *Protein Crystallography*. Academic Press, London.

Crick, F. H.C. and Magdoff, B. S. (1956). The theory of the method of isomorphous replacement for protein crystals. *Acta Crystallogr.* **9**, 901–908.

Drenth, J. (1999). *Principles of Protein X-ray Crystallography*, 2nd edn. Springer-Verlag, Berlin.

Drew, H., Takano, T., Takana, S., Itakura, K. and Dickerson, R. E. (1980). High-salt d(CpGpCpG), a left-handed Z′ DNA double helix. *Nature (London)* **286**, 567–573.

Garman, E. and Murray, J. W. (2003). Heavy-atom derivatization. *Acta Crystallogr. D* **59**, 1903–1913.

Green, D. W., Ingram, V. M. and Perutz, M. F. (1954). The structure determination of heamoglobin: IV. Sign determination by the isomorphous replacement method. *Proc. R. Soc. London* **A225**, 287–307.

Hatfull, G. F., Sanderson, M. R., Freemont, P. S., Raccuia, P. R., Grindley, N. D. F. and Steitz, T. A. (1989). Preparation of heavy-atom derivatives using site-directed mutagenesis: introduction of cysteine residues into γδ resolvase. *J. Mol. Biol.* **208**, 661–667.

Holbrook, S. R. and Kim, S.-H. (1985). Crystallization and heavy-atom derivatives of polynucleotides. *Method Enzymol.* **114**, 167–176.

Kendrew, J. C., Bodo, G., Dintzis, H. M., Parrish, R. G., Wyckoff, H. W. and Phillips, D. C. (1958). *Nature* **181**, 662.

Kendrew, J. C., Dickerson, R. E., Strandberg, B. E., Hart, R. G., Davies, D. R., Phillips, D. C. and Shore, V. (1960). *Nature* **185**, 422.

Kim, S.-H., Shin, W. C. and Warrant, R. W. (1985). Heavy metal ion-nucleic acid interaction. *Method Enzymol.* **114**, 156–167.

Patterson, A. L. (1934). A Fourier series method for the determination of the components of interatomic distances in crystals. *Phys. Rev.* **46**, 372–376.

Petsko, G. A. (1985). Preparation of isomorphous heavy-atom derivatives. *Method Enzymol.* **114**, 147–156.

Petsko, G. A., Phillips, D. C., Williams, R. J. P. and Wilson, I. A. (1978). On the protein crystal chemistry of chloroplatinite ions: general principles and interactions with triose phosphate isomerase. *J. Mol. Biol.* **120**, 345–359.

Ringe, D., Petsko, G. A., Yanahura, F., Suzuki, K. and Ohmori, D. (1983). Structure of iron superoxide dismutase from *Pseudomonas ovalis* at 2.9 Å resolution. *Proc. Natl. Acad. Sci. USA* **80**, 3879–3883.

Schitz, M., Prange, T. and Fourme, R. (1994). *J. Appl. Cryst.* **27**, 950–960.

Sigler, P. B. (1970). Iodination of a single tyrosine in crystals of alpha-chymotrypsin. *Biochem.* **9**, 3609.

Taylor, G. L. (2003). The phase problem. *Acta Crystallogr. D* **59**, 1881–1890.

Wang, B. C. (1985). Resolution of phase ambiguity in macromolecular crystallography. *Method Enzymol.* **115**, 90–112.

CHAPTER 7

# Molecular replacement techniques for high-throughput structure determination

Marc Delarue

## 7.1 Introduction

In the context of structural genomics projects, there are two main routes to consider for solving efficiently and rapidly the three-dimensional (3D) structure of the target gene products by crystallography. The first is the Multiple-wavelength Anomalous Diffraction (MAD) technique (reviewed in Chapter 8 of this volume), which necessitates growing SeMet-substituted protein crystals. The second is molecular replacement (MR), which requires X-ray data for the native protein as well as the structure of a related homolog.

MR is an ensemble of techniques that aims to place and orientate an approximate molecular model in the unit cell of the crystal being studied. This will provide the starting phases needed to calculate the initial electron density map from which the protein model can be built, either manually by iterative use of reconstruction with molecular graphics packages (Jones *et al.*, 1991) followed by refinement (Murshudov *et al.*, 1997), or automatically if diffraction data up to 2.3 Ångstroms or better are available (ARP/wARP (Perrakis *et al.*, 2001), Solve/Resolve (Terwilliger, 2003)).

In this article, we will not focus on recent developments in refinement techniques, which benefited recently from better statistical treatments such as maximum likelihood targets for refinement (Adams *et al.*, 1999), but rather will describe in detail some of the newest developments in MR to get the best possible set of phases to initialize refinement and reconstruction in the best possible conditions.

As more and more structures are deposited in the Protein Data Bank (PDB) (Berman *et al.*, 2000), the chances of finding a (remote) homolog structure in the PDB have become higher and, therefore, molecular replacement techniques are increasingly useful. This is reflected, for instance, in the number of hits one gets by doing a search for the keywords 'molecular replacement' in one of the leading journals in the protein crystallography community, for example

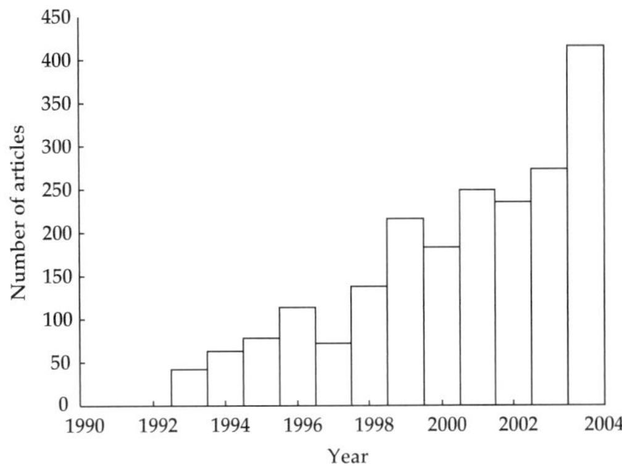


**Figure 7.1** Histogram of the number of articles in *Acta Crystallographica D* containing 'Molecular Replacement' in the title or abstract, year by year. The score for 2004 is a projection based on the first 6 months.

**Table 7.1** List of web sites for MR packages with a short description of their distinctive features

| Package name | Search space | NCS | Web site or e-mail address of the author | Flexib. sampling | Post-ref |
|---|---|---|---|---|---|
| AMoRe | 2x3D | Yes | jorge.navaza@gv.cnrs-gif.fr | No | RigBod |
| CNS | 2x3D | Yes | http://cns.csb.yale.edu/v1.1/ | No | CNS |
| MolRep | 2x3D | Yes | http://www.ysbl.york.ac.uk/~alexei/molrep.html | Some | RigBod |
| EPMR | 6D | Yes | http://www.msg.ucsf.edu/local/programs/epmr/epmr.html | No | Conj.Grad. |
| Qs | 6D | Yes | http://www.mbg.duth.gr/~glykos/Qs.html | No | Sim.Anneal |
| SoMore | 6D | No | http://www.caam.rice.edu/~djamrog/somore.html | No | BFGS |
| Phaser | 2x3D | Yes | http://www-structmed.cimr.cam.ac.uk/phaser | No | No |
| CaspR | 2x3D | Yes | http://igs-server.cnrs-mrs.fr/Caspr2/index.cgi | Yes | CNS |
| @TOME | 2x3D | Yes | http://bioserv.cbs.cnrs.fr/HTML_BIO/frame_meta.html | Yes | No |

*Acta Crystallographica D* (Fig. 7.1). Also, the growing number of sequences in the sequence databases have made the detection of remote homologs a more reliable task, as multialignments of several dozens of sequences allow for building position-dependent mutation matrices (profiles), which are much more sensitive than pairwise comparisons with a standard mutation matrix (Altschul *et al.*, 1997). Having detected a homolog of known 3D structure, a model can be quickly derived using standard and automatic comparative modelling techniques (Sali and Blundell, 1993). If several models are available, this may, in turn, result in an increase of the signal/noise ratio of MR searches.

One innovation, which owes much to bioinformatics, is the assembly of a web-interfaced suite of programs that makes use of all these new developments to perform a multialignment of a set of sequences, using of all the homologs found in the PDB to scan many different models through the MR procedure (Claude *et al.*, 2004). Refinement of the best solutions can also be performed to assess the correctness of the solution(s). In addition, remote structural homologs can be detected by threading methods through a suite of programs installed as a metaserver, which sends requests to several web servers and uses the results of each task to feed in the next request (Douguet and Labesse, 2001). Thus, there is a trend towards integrating different servers and/or packages in automatic protocols of MR, with the aim of obtaining the best possible model, which together greatly enhances the chances of finding the solution quickly.

At the same time, it is apparent that the MR method itself continues to be developed and improved by a number of new ideas. This is, in turn, reflected in the number of hits of the same keywords 'molecular replacement' in *Acta Crystallographica A*, a journal used mainly to describe methodological advances in crystallography: the cumulated number of hits between 1985 and 2004 is 534, an average of 27 articles per year. With the development of web-based interfaces, many of these new ideas are implemented in programs that are available on-line (together with the manuals) and it has become easier to test them rapidly, as they readily accept standard formats for the coordinates of the model and the X-ray diffraction data. Recently, another very interesting integrated approach called Mr Bump has been published (Keegan and Wynn, 2007).

We will review, here, most of these new methods, as well as the 'classical' ones, and will list their website address (Table 7.1), with a brief mention of their main characteristics. We have tested all of them for the purpose of this review, using the same input data files, and will describe the protocols that have been used throughout the text.

The rule of thumb for a successful application of molecular replacement is that the model should have a root-mean-square deviation (RMSD) on C-alpha coordinates ~2.0–2.5 Ångstroms with the target structure, corresponding to a sequence identity with the target of 25–35%. In practice, however, there are many more structures solved by MR in the PDB using models with sequence identity of 60% or higher than otherwise.

**Protocol 7.1 Checking your data**

- Check for completeness and redundancy at the desired resolution: usually a complete data set at 10–4 Ångstroms is fine.
- Check for possible twinning; read carefully the output of TRUNCATE in CCP4 (Collaborative Computational Project, 1994) or submit your data to the scrutiny of Todd Yeates web site (http://www.doe-mbi.ucla.edu/Services/Twinning).
- Check your space group (!):
  - for instance, if P2(1)2(1)2(1), check extinctions carefully;
  - for a polar space group, remember to try both enantiomorphic possibilities in the translation search, because they cannot be distinguished through Patterson methods used in the rotation function: P4(1) and P4(3), P6(1), and P6(5).
- Check for possible non-crystallographic symmetry (NCS):
  - check for a possible pure translational non-crystallographic symmetry (pseudo-symmetry) by calculating a native Patterson function and sorting its peaks;
  - check for two-fold axes, three-fold axes... using self-rotation function;
  - calculate the percentage of solvent in the crystal, for different hypothesis as to the number nmol of molecules in the asymmetric unit:

$$100 * (1. - \text{nmol} * (\text{MW} * \text{Vm})/\text{Va.u.})$$

where MW is the molecular weight of a monomer, Vm is the density of a protein ($\sim$0.73 cm$^3$/g), and Va.u. is the volume of the asymmetric unit. The percentage of solvent should be in the range 20–80%.

*Note*: It might be of interest to read in detail a recent 'tour de force' success story using MR in a very difficult case, with both high NCS (12 copies in the asymmetric unit) and twinning of the data (as discovered quite late in the process of structure solution) (Lee *et al.*, 2003).

There are, essentially, two issues in the application of MR automatic protocols to structural genomics:

**1.** Can one push the limits of the method so as to use models with less and less sequence identity with the target, i.e. models obtained by threading methods, with sequence identity levels between 15 and 25%? The answer seems to be yes (Jones, 2001), with newer methods capable of solving problems with sequence identity around 20% (Claude *et al.*, 2004; G. Labesse, personal communication; Abstract in the GTBio Meeting, June 2004, Lyon, France; Keegan and Wynn, 2007).

**2.** How far should one pursue efforts to solve the molecular replacement problem once the automatic protocols have failed; in other words, how much time should one spend in trying to solve difficult cases before deciding to go back to the bench and grow SeMet crystals, or use the SAD method with crystals soaked with one anomalous scatterer? This is actually a difficult question that depends on a lot of different issues, such as the expertise already present in the lab. in MR techniques, the solubility of the protein(s), which might or might not change upon selenomethionylation. This question cannot be answered in general but this is clearly an issue that one should keep in mind in defining the strategy for structure solution.

## 7.2 Test data used throughout this study with the different MR packages

The same data set at 2.5 Ångstrom resolution, collected at the European Synchrotron Research Facility (ID14-EH2), for a T. brucei 6-phosphoglucono-lactonase (6PGL) (Delarue *et al.*, 2007) was used throughout this study. The cell parameters are 70.3, 80.8, 90.3 in P2(1)2(1)2(1). There are two molecules in the asymmetric unit, related by a pseudotranslation, discovered by sorting the native Patterson peaks (Protocol 7.1). There are two possible models: 6PGL from *T. maritima* (1PBT) (about 40% sequence identity) and the glucosamine-6-phosphate deaminase (1DEA), which was detected by BLAST (25% sequence identity).

## 7.3 The standard molecular replacement method

### 7.3.1 Historical background: Patterson methods

The possibility and feasibility of molecular replacement was demonstrated by Rossmann and colleagues in the 1960s, as part of an effort to use non-crystallographic symmetry to solve the phase problem for macromolecules (Rossmann, 1990).

In this article, we will restrict ourselves to the case of finding the orientation and position of a known model in another unit cell, to help solve the molecular structure contained in this unit cell (Fig. 7.2). Traditionally, this six-dimensional (6D) search (three orientation angles and three translations are to be found) is divided into two separate and consecutive 3D search problems.

The first one consists of finding the orientation of the model through the so-called rotation function, which is defined in such a way that it should be maximum for the true orientation. In its real-space formulation, it consists of the convolution of the experimental Patterson function with the computed Patterson of the model in every possible orientation. If restricted to the molecular volume, the use of Patterson functions allows the superposition of intramolecular vectors, which are independent of relative translations between the model and the target. In its reciprocal-space formulation (Rossmann and Blow, 1962), it was demonstrated that the maximum of this function is indeed the expected solution. Later on, the formula was rearranged using the plane-wave expansion and spherical harmonics so as to use the powerful technique of Fast Fourier Techniques (FFT) (Crowther, 1972), whose numerical implementation was further refined and stabilized by Navaza (Navaza, 2001).

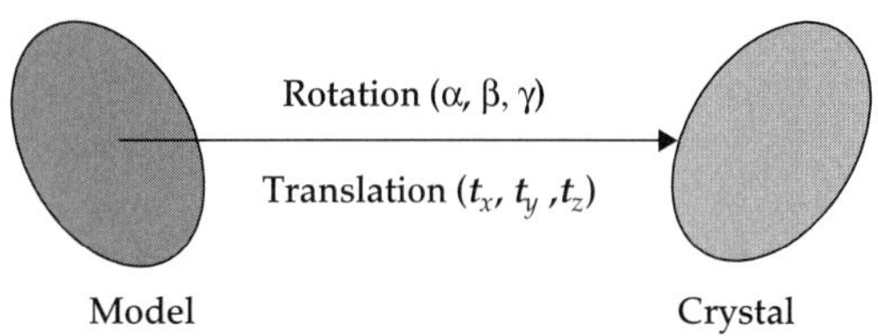


**Figure 7.2** Definition of the molecular replacement problem and the six degrees of freedom needed to describe it.

The second step consists of calculating a convolution of interatomic vectors between symmetry-related molecules of the correctly oriented model placed at different origins, with the experimental Patterson function. The reciprocal version of the

**Protocol 7.2 Test case on the same model, the same space group, with calculated data**

**1.** Rotate model by kappa around axis given by (phi, psi) using the following jiffy code:

```
ck = cosd(kappa)
sk = sind(kappa)
cp = cosd(psi)
sp = sind(psi)
cf = cosd(phi)
sf = sind(phi)
a(1, 2) = cp*sk + cf*sf*sp*sp*(1 – ck)
a(2, 1) = –cp*sk + cf*sf*sp*sp*(1 – ck)
a(1, 3) = –sf*sp*sk + cf*cp*sp*(1 – ck)
a(3, 1) = sf*sp*sk + cf*cp*sp*(1 – ck)
a(2, 3) = cf*sp*sk + sf*cp*sp*(1 – ck)
a(3, 2) = –cf*sp*sk + sf*cp*sp*(1 – ck)
a(1, 1) = ck + cf*cf*sp*sp*(1 – ck)
a(2, 2) = ck + sf*sf*sp*sp*(1 – ck)
a(3, 3) = ck + cp*cp*(1 – ck)
```

c... now rotate and translate, knowing the coordinates of the centre of gravity xg

```
do 100 k = 1, 3
xnew(k) = xg(k) + a(k, 1)*(x(1) – xg(1))
   + a(k, 2)*(x(2) – xg(2)) + a(k, 3)*(x(3) – xg(3))
100 continue
```

**2.** Calculate structure factors (by CCP4 sfall) for the transformed coordinates *xnew in your own space group.*

**3.** Run your favourite molecular replacement program using: (i) the unrotated model as a search model; and (ii) calculated data as experimental data.

**4.** Analyse results: make sure you understand the symmetry of the rotation function group, the translation solution (the y coordinate is undetermined in P2(1) etc.). Get a feeling of the (maximum) height of the signal you can expect.

**5.** Convince yourself that (kappa, phi, psi) applied in (1) is the same as the solution of MR.

c... Hint: here is the rotation matrix using eulerian angles alpha, beta, gamma (Urzhumtseva and Urzhumtsev, 1997).

c... Caveat: AMoRe first rotates the model so that the inertial axes coincide with *x*, *y*, and *z*.

calculation lends itself particularly well to Fourier (Crowther and Blow, 1967) and FFT techniques and is, in general, very rapid.

Usually, the top peaks of the translation search are then submitted to a low resolution quick rigid-body refinement, for which quick algorithms have been devised (Huber and Schneider, 1985; Navaza and Saludjian, 1997).The resolution is usually taken to be 12–4 Ångstroms or so; if one wants to use the low resolution terms, one should use a solvent effect correction technique (Fokine and Urzhumtsev, 2002).

The following issues are crucial for the success of the method:

- completeness and quality of the data
- accuracy of the model(s)
- completeness of the modelling of the unit cell
- type of score used as an indicator of the quality of the agreement.

This is reflected in the following input parameters:

- quality of the X-ray data (see Protocol 7.1)
- quality of the model(s), e.g. poly-Ala versus full model (see below, structural diversity)
- number of molecules per asymmetric unit
- target functions.

While the first point was addressed directly in Protocol 7.1, we will address the other points successively in the next paragraphs.

Before executing the real MR protocol, we strongly recommend running a test case. Understanding the conventions of the rotation function will be greatly facilitated by reading a recent

**Protocol 7.3 A typical molecular replacement session using AMoRe (Navaza, 2001)**

**1.** Read the Manual!
**2.** Move to an empty directory and type : *csh $AMORE/setup*
**3.** Then:
In ./d/:
– Create data.d (as data.example created by setup), and give cell parameters, symmetry related positions, resolution limits, and the number of molecule(s) (and their type) to search in the asymmetric unit
** 6PGL ** Title
70.315 80.852 90.31 90. 90. 90.
x,y,z * 1/2 + x, 1/2 − y,−z * 1/2 − x,−y, 1/2 + z * −x, 1/2 + y, 1/2 − z * end
0; orthogonalizing code
95. 0.0; % reflections, B-add
12.0 4.0; resolution range
1 2; NTYP, (nmol(n),n = 1,NTYP); a homodimer in the asymmetric unit
– Data files with diffraction data and model coordinates must be named hkl.d and xyzN.d, respectively; N = 1 for the first model, N = 2 for the second one...
Insert FORMAT card (upper case) in hkl.d and xyz{#}.d files (as hkl.example and xyz1.example created by setup).
– In ./i/:
Edit dato.i3 file which will look like this:
job + * + * + * + * + * + * + * + * + *
xyz model type (could ne a map)
1. 2 2 0.5 2. rot: %rad, 1min's, rotation function cutoff, step in degrees
c-c 50 0.5 50 1-body trans: option, nb. orient. to try, cutoff, nb. peaks
p-t 10 0.5 50 n-body trans: option, nb. orient. to try, cutoff, nb. peaks
10 20 fitting: nb. trans. to fit, nb. of iterations
0. packing: CoM cutoff for -crude- packing function
**4.** Finally, in ./:
*csh ./e/job dato*
*csh ./job* (protocol created by ./e/job for automated runs).
**5.** In our test case (6PGL), there was one solution well detached with a correlation coefficient of about 30%. However, we were unable to bring the R-factor below 52%, even after refinement, until we collected another data set with crystal soaked in a mercury derivative. This derivative turned out to be highly isomorphous and refinement with CNS produced an R-factor around 48%. Phases from the model were able to pull out the heavy atom sites through both isomorphous and anomalous Fourier difference maps. The resulting solvent-flattened 2.8 Ångstrom SIRAS map was interpretable.

article by Urzhumtseva and Urzhumtsev (1997) and Protocol 7.2.

In Protocol 7.3 we show a typical command file for the AMoRe package (Navaza, 2001), one of the most widely used packages for MR.

## 7.4 How does one know the solution has been found?

Sometimes, it is not so easy to convince oneself that the solution of the molecular replacement problem has, in fact, been found, even after rigid-body refinement; indeed, the first solution is not always well detached and different scores may produce different rankings. The most commonly used scores are correlation coefficients on either intensities or structure-factor amplitudes, and R-factors. Even though these criteria are formally related (Jamrog *et al.*, 2004), they can produce different rankings, especially if no solution is clearly detached. Some other criterion is then needed to discriminate between the potential solutions.

One possibility is to run simulated annealing refinement in torsion angle space as implemented in CNS (Brünger *et al.*, 1998). As this is one of the most powerful programs in terms of radius of convergence, it is especially useful to look for the decrease of the free-R-factor (Adams *et al.*, 1999), but this is a rather cpu-intensive task if several possible solutions are to be tested.

In difficult cases, there is actually a way to bring the R-work and R-free down by means other than just rigid-body or dihedral angle molecular dynamics, while still exploring just a few degrees of freedom, using the Normal Mode Analysis protocol (Delarue and Dumas, 2004). This is implemented in http://lorentz.immstr.pasteur.fr and will be described in the paragraph on 'exploring structural diversity' below (Lindahl *et al.*, 2006).

Another possibility is to look at the packing of the top 10 solutions, because this can prove a very discriminating criterion. Even if the solution is well detached, it is still mandatory (and reassuring) to examine carefully the corresponding packing arrangement. This necessary (but not sufficient) condition can actually be implemented in the translation function using analytical expression that can be evaluated using FFT (Harada *et al.*, 1981; Stubbs and Huber, 1991) and this is available in several MR packages, such as AMoRe (Navaza, 2001).

## 7.5 The need to find a better score for the rotation function: the Phaser euphoria

Many people have recognized that the rotation function suffers from some drawbacks and have tried to improve the score by using origin-removed Patterson functions, normalized structure factors E-values, etc. (Brünger, 1997).

Brünger and colleagues developed a 'direct rotation function', which is just a correlation coefficient between $E_{obs}$ and $E_{mod}$(Omega), the normalized structure factors of the crystal and of the rotated model, respectively. However, in this case the model is placed in a P1 cell of dimensions and angles identical to the ones of the crystal being studied. This works well but requires quite a lot of cpu as it is not amenable to FFT (DeLano and Brünger, 1995).

Another idea is to use different runs of the same program with slightly different models; it is most aptly described as an application of a 'consistency principle', namely it is required that the solution should appear consistently in all runs, even with a rather low score. Special algorithms have been developed to cluster similar solutions in eulerian angles space and convincing results have shown that it is indeed possible to increase the signal-to-noise ratio of the rotation function in this way (Urzhumtsev and Urzhumtseva, 2002).

Recently, Read extended the maximum likelihood formalism to the molecular replacement problem. Maximum likelihood puts on analytical grounds the notion that the best model is the one which maximizes the probability of having measured the actual experimental data at hand (structure factors). In his first implementation (BEAST), Read could show that he would get a much clearer signal in difficult cases, but the program was pretty slow (Read, 2001). Recently, he and his team developed a much more rapid version of his algorithm (PHASER), based on an approximation which can be evaluated by FFT, and this was incorporated in a single package that also contains the translation function (Storoni *et al.*, 2004). Roughly speaking, the rotation score is now based on a *weighted* origin-removed convolution of experimental and calculated Patterson functions. However, the authors stress that their program does not work very well in cases where pseudo-translational symmetry is present. This is because the formalism assumes that the translation vector between the molecules in the unit cell is randomly distributed so that the relative phases between pairs of symmetry-related molecules are sampled randomly. This assumption is obviously violated if there is pseudo-translational symmetry; however, the authors mention that they have found a way around this problem and that it will be fixed soon.

Once released, the programers received immediate praise from the crystallographic community, with messages of thanks posted to the CCP4 Bulletin Board, reporting how several structures that had resisted molecular replacement traditional methods for years were solved in less than a day by PHASER. So, this is certainly one the packages to try first when dealing with a tough MR problem.

## 7.6 Screening many solutions (*multi sunt vocati, pauci vero electi*)

The need for automated protocols is apparent from the strategy adopted by AMoRe to circumvent the problem that the score of the rotation function (RF) is far from being perfect and does not always rank the solutions correctly (Navaza, 2001). Indeed, it is often observed that the true solution is not the top solution, with many false positives. Hence, AMoRe runs a translation function (TF) for each of, typically, the top 50 or 100 solutions of the rotation function. This is actually quite rapid as TF is based on FFT; then, the first 10 solutions of each of these TF runs is in turn refined using a very effective implementation of rigid-body refinement (Navaza, 2001).

One complication occurs with polar space groups, where all the possibilities must be tried in the TF. If there is an ambiguity in the extinctions, for instance in the Laue group Pmmm, again all possibilities must be searched. This is usually done 'by hand', going through all different possibilities one by one.

## 7.7 Non-crystallographic symmetry protocols and six-dimensional search programs

If there is NCS in the crystal, all molecules of the asymmetric unit must be searched in turn; every time a potential solution has been found, it is possible to use this information to increase the signal-to-noise ratio of the searches for the other molecules. But then, the combinatorics of testing the 50 top solutions of the rotation function and then the 10 top solutions of each associated translation function for rigid-body refinement cannot be done 'by hand' as in the previous case, as soon as there is more than two molecules in the asymmetric unit. In NCS-MR, depending on the number of molecules present in the asymmetric unit, there are thousands of possibilities to be searched. Also, as one is searching with only a fraction of the asymmetric unit, the signal to be expected is intrinsically lower.

AmoRe (Navaza, 2001), and other programs such as MolRep (Vagin and Teplyakov, 1997, 2000) and, in fact, most MR programs handle this quite effectively in an expert fashion. MolRep is a very versatile program that has many different options implemented and is part of CCP4 (Collaborative Computational Project, 1994). Obviously, if one wants to try different possible models to increase the chance to find the correct solution, this again makes the search more computer intensive and the best way to deal with this is to follow a given, sensible protocol (see below).

As we mentioned already, the problem of the rotation function is its score, leading to a difficult energy landscape to be searched; we can now describe another way to tackle this problem. Since the Translation Function score is much more sensitive, one might try to run a translation for every possible rotation angle, therefore exploring the 6D space exhaustively. The space to be searched in eulerian angles depends on the space group of the crystal and can be found in Rao *et al.* (1980). It turns out that it is doable in most cases within reasonable cpu time with a 'normal' workstation.

There are at least two implementations of this protocol that appeared recently:

- SoMore, which performs a full 6D search with low-resolution data (usually 8 Ångstrom), followed by conjugate-gradient minimization of the best solutions (Jamrog *et al.*, 2003).

- Systematic rotation followed by translation searches, as generated by a script (Sheriff *et al.*, 1999) (see also Protocol 7.6).

Alternatively, one might want to search the N-6D space (alpha, beta, gamma, *tx*, *ty*, *tz* for each one of the *N* molecules of the asymmetric unit) using stochastic or *Monte Carlo* methods, as exhaustive searches are out of question. In this case, as all molecules are searched simultaneously, the problem of the low signal is less severe than with traditional MR methods. However, the process is quite cpu intensive. This has been implemented with success by Glykos and Kokkinidis (Glykos and Kokkinidis, 2000, 2001, 2003), who later included a simulated annealing protocol to increase the radius of convergence of the method (Queen of Spades or Qs method).

Other methods have used *genetic algorithms* to search the 6D space: one of them was originated by Lewis and colleagues (Chang and Lewis, 1997) and the other resulted in the popular program EPMR (Kissinger *et al.*, 1999, 2001). The EPMR program is now widely used in the crystallographic community and is especially simple to use (see Protocol 7.4). This makes it a very attractive candidate for use in a suite of programs integrated in a web site that goes all the way from model generation to the refinement of the final model (Rupp *et al.*, 2002).

All these methods exploit the fact that there is no need to recalculate the structure factors of the model each time it is rotated or translated; it is sufficient to be able to sample the structure factors at the rotated Miller indices, with or without a phase shift coming from the translation, and this can be done quite effectively by interpolation in reciprocal space (see Protocol 7.4).

## 7.8 How to choose the best model

Even the best possible MR package will fail if the model is not good. Hence, a good deal of

### Protocol 7.4 Stochastic search methods

**Use of EPMR (Kissinger *et al.*, 1999, 2001)**

*epmr -m 2 -h 4. -l 12. -n 50 example.cell example.pdb example.hkl >example.log &*

where example.cell contains the cell parameters and the space group number:

70.315 80.852 90.31 90. 90. 90. 19

and the model and X-Ray data are in example.pdb and example.hkl, respectively.

The options *-l* and *-h* define the low and high resolutions limits of the data, respectively, while the *-m* option defines the number of molecules to be searched.

The number of different starts is controlled by the *-n* option.

It is clear from this input lines that EPMR is very easy to use. Indeed, it is the MR package used by Rupp's automated protocol (Rupp *et al.*, (2002). It was successful in finding the solution of 6PGL, using default options.

**Use of Qs (Glykos and Kokkinidis, 2000, 2001, 2003)**

*Qs example.in >example.log &*

where example.in will look like this:

```
TARGET CORR-1
CYCLES 10
STEPS 1000000
STARTING_T 0.0150
FINAL_T 0.0050
INFO 1000
NOISE_ADDED 0.10
RESOLUTION 12.0 4.0
AMPLIT_CUTOFF 5.0
SIGMA_CUTOFF 0.0
RANDOM_SELECT 1.0
FREE 0.10
MODEL example.pdb
DATA example.hkl
GLOBAL_B 20.0
MOLECULES 2
SEED 357539
SCALECELL 4.0
MAXGRIDSPACING 1.0
SCMODE wilson
INTERPOLATION linear
CELL 70.315 80.852 90.31 90. 90. 90.
GROUP 19
```

efforts should be put into the generation of the best possible model.

### 7.8.1 The best starting model

Choosing the best model, even when only one possible homolog has been detected through alignment and/or threading methods, is not an easy task. This is usually dealt with by running MR, in turn, with different versions of the same model. Up to now, the general accepted rule of thumb was to remove those parts of the model that are suspected to be different in the target protein. In other words, it is believed to be better to have an incomplete model with no error than a complete one with errors. For instance, one could truncate all side chain atoms downstream of the CB (except for glycine) effectively changing the sequence into a poly-Ala; or one could choose to keep the side-chain atoms coordinates of only those residues that are strictly conserved between the template and the target and mutate all the others into alanine; or into a serine (changing the CG atom into an OG atom) if there is a conservative substitution.

There are all sorts of possibilities, including the one to keep all atoms inside the core of the molecule and truncating atoms with an accessibility to the solvent larger than a given criterion (Delarue *et al.*, 1990). Truncating loops with high B-factors or non-conserved loops is also possible, although it is only recently that automatic protocols have been devised to do the latter, using information contained in the multialignement of the sequences (Claude *et al.*, 2004). The richest diversity of approaches is the one available in MrBump (Keegan and Wynn, 2007). When there are several possible models (templates), the situation becomes more difficult to handle; if there are only two or three, they can be tried individually, say as poly-Ala models. If there are more, as in structures solved by Nuclear Magnetic Resonance (NMR) and deposited in the PDB, which usually contain 20 *a priori* 'equivalent' models, specialized protocols have been devised and tested (Chen, 2001), some of which are available on Gerard Kleywegt's web site (http://xray.bmc.uu.se/usf/factory_6.html). One idea is to use an 'average' structure; another idea is to weight each atom by a pseudo B-factor, which is calculated by an empirical formula that is a function of the RMSD of the position of this atom in the 20 different models of the PDB (see Protocol 7.5). This usually works well (Wilmanns and Nilges, 1999).

### 7.8.2 Using homology-modelling derived models

Up to now, models of the protein derived by homology-modelling techniques were not heavily used, because people were reluctant to use models which contain some errors. In these models, all side-chains have been reconstructed, as well as insertions and deletions. So, in a sense, the model is more complete, but it is not certain that this will facilitate the search for the solution of the molecular replacement solution. One fearful feature of homology modelling is that the refinement of the model using standard force-field tends to worsen the model, rather than improve it, at least in test cases. Also, as there are several homology modelling programs available on the web, all of which use different methods (distance geometry: Modeller (Sali and Blundell, 1993), mean-field optimization techniques (Koehl and Delarue, 1994, 1995)) and sometimes depend on the generation of random numbers, the question arises as to which should be used.

It is well known that the result of homology modelling is highly dependent on the quality of the alignment between the template and the target. Obviously, the success of MR will be highly dependent on the accuracy of the alignment between the template and the target sequences

**Protocol 7.5 How to handle NMR models (Chen, 2001; Wilmanns and Nilges, 1999)**

Go to Gerard Kleywegt's website (Uppsala Software Company):
http://xray.bmc.uu.se/usf/factory_6.html

Get the script in:
ftp://xray.bmc.uu.se/pub/gerard/omac/multi_probe
and follow the advice given by Y. W. Chen (2001).

(Schwarzenbacher *et al.*, 2004). Hence the need to examine critically the multialignement and, in some cases, to modify it manually. Luckily, it is now possible to couple the multialignement refining process to the building of models, that is process simultaneously 1D and 3D information, and this can be done on the fly. This allows visualization directly on the 3D level the effect of a modification of the alignment and makes it possible to avoid meaningless alignments (ViTo, (Catherinot and Labesse, 2004)). However, it demands human intervention and cannot be made automatic. It should be stressed that homology modelling techniques can handle effectively the case where several structural models (templates) are available.

### 7.8.3 Detecting low-homology models (using fold-recognition algorithms)

Recently, as became apparent during the last Critical Assessment of Techniques for Protein Structure Prediction (CASP5) competition (CASP5, 2003), methods to detect structural homology with virtually no sequence homology have become more reliable and convincing. They are often based on so-called metaservers, which address requests to web-based servers of various sorts (secondary prediction methods, threading methods, etc.) and then issue some sort of a consensus score more reliably than any of the separate methods used (Douguet and Labesse, 2001).

One might wonder, then, if it would be possible to use such methods to pick up remote homologs and use them in MR problems (Jones, 2001). Obviously, the obtained models will contain a lot of errors, so why use them? In particular, homology modelling programs rely heavily on the original backbone coordinates of the template; as there is a well-known exponential law relating the lack of sequence identity between two proteins and the RMSD of their coordinates (Chothia and Lesk, 1986), this is indeed worrisome. So at first sight, it might appear that this kind of model would be useless. However, recent re-examination of the same data show that, if one filters out the outliers in the paired atoms of the structural alignment, the relation between RMSD and lack of sequence identity is no longer exponential but simply linear (Martin and Labesse, personal communication). Therefore, by truncating the model judiciously, keeping only the conserved core and removing all the variable loops, one might indeed have a useful model.

So altogether, it seems that the trend of conservatively using only models of high sequence identity is now changing, as homology modelling techniques and low-sequence identity structural homolog detection methods are being refined, stimulated by the increase in both the number of sequences and the number of structures, and also because automatic protocols allow for the testing of many different models. Indeed, given the number of possibilities to generate plausible models, clearly there is room for an automatic method trying different things in turn, and then ranking the different solutions. This is precisely what has been done recently in a suite of programs such as the one called CaspR (Claude *et al.*, 2004), showing very promising results. This is described in more details in the following section.

## 7.9 The integrated molecular replacement method: comparison of automatic protocols

### 7.9.1 CaspR (Claude *et al.*, 2004)

CaspR is a combination of well-established, standalone software tools; for a general flowchart of the program see Fig. 7.3. First it reads an MTZ file (CCP4), and extracts from it the unit cell parameters and space group number. Then it runs T-Coffee (Notredame *et al.*, 2000) and 3D-Coffee (O'Sullivan *et al.*, 2004) to get the best possible alignment of the template and the target sequences; in doing so, it identifies variables regions that are likely to be non-conserved in the target structure. Then it runs Modeller (Sali and Blundell, 1993) to construct 10 different models. All these different models are then subjected to AMoRe (Navaza, 2001) molecular replacement protocols, with all sorts of different modifications: either truncated from the unreliable regions or not, either as poly-Ala or not. Then CNS (Brünger *et al.*, 1998) is used to subject the models with the best MR scores to a round of simulated annealing protocol in internal coordinates space.

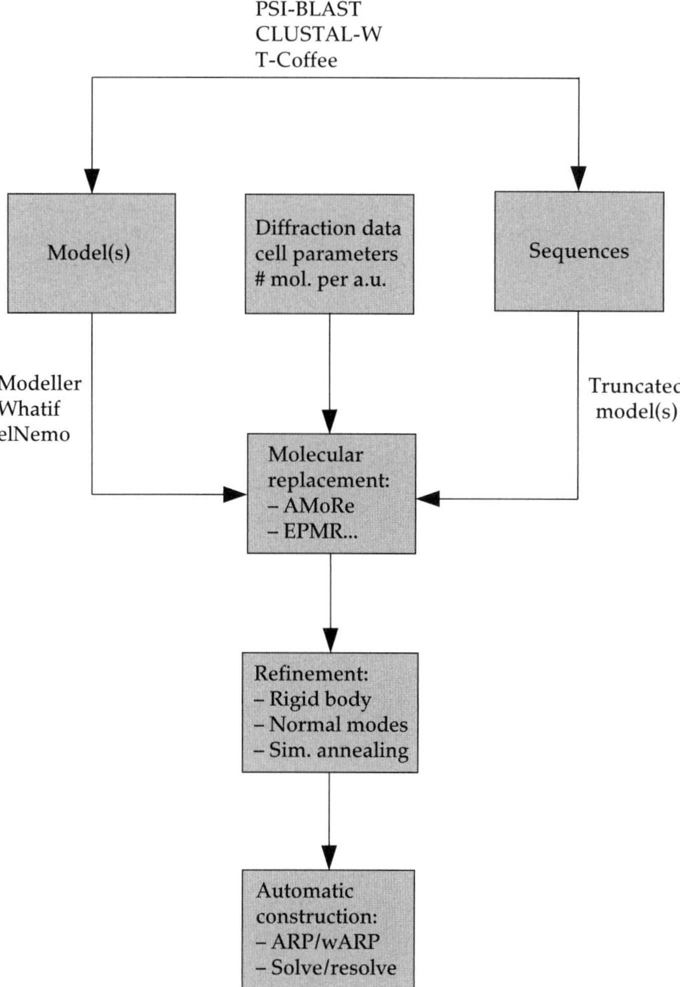


**Figure 7.3** General flow chart for an integrated molecular replacement structure solution protocol.

The status of the procedure can be checked on line at any time. Once submitted, a job gets a reference number that is sent to the user by e-mail and that has to be quoted for consulting the job status. The data are erased after one week.

It is our experience that even if the solution can be obtained with more traditional MR packages, the final R-factor and R-free obtained with CaspR are usually much lower, thereby effectively reducing the time spent in manual reconstruction in front of the graphics display terminal. For instance, in the 6PGL case, the best score of the MR itself was obtained with the truncated Model #2 of Modeller (Corr. Coeff. = 38%), but the best refined model was Model #23 with Rwork = 43.8% and Rfree = 52%; this kind of score could never be obtained with only Rigid Body refinement.

### 7.9.2 TB Structural Genomics Consortium (Rupp *et al.*, 2002)

In the case of the TB Structural Genomics Consortium (Kissinger *et al.*, 2001), PSI-BLAST is used to find homolog(s) in the PDB, CLUSTAL-W is used for obtaining a multialignment, AL2TS to build the model(s), EPMR (Kissinger *et al.*, 1999, 2001) to do molecular replacement and CNS (Brünger *et al.*, 1998) to refine the model(s). In addition, the authors use their own local version of ARP/wARP, called Shake-and-wARP, to attempt automatic building of the model in the electron density map. Clearly, many combinations of such protocols are possible, by using different software at each step of the whole process.

### 7.9.3 G. Labesse and L. Martin (GTBio, Lyon, June 2004)

In this case, the @TOME meta server (Douguet and Labesse, 2001) is used to detect low structural homology three-dimensional models, as a consensus between six different 'threading servers'. Then, after model building by Modeller (Sali and Blundell, 1993), MolRep (Vagin and Teplyakov, 1997; 2000) is run, followed by refinement using Refmac (Murshudov *et al.*, 1997). In some instances, this suite of programs was able to find the MR solution using a model with sequence identity as low as 20% (G. Labesse, personal communication).

### 7.9.4 MrBump (Keegan and Wynn, 2007)

This is the most complete program, downloadable from CCP4 web site www.ccp4.ac.uk/MrBUMP. Sequence multialignment is performed by MAFFT or ClustalW. The model is edited using four different protocols, including CHAINSAW and the one available in MOLREP. Then either Phaser or MolRep is used for MR and finally Refmac is used for refinement. The approach is intended more to be exhaustive than fast, but this is not a problem with current computers. The philosophy is summarized by the author's following statement: 'It is concluded that exploring a range of search models automatically can be valuable in many cases'.

## 7.10 Making the most of your model using normal modes: pushing towards the systematic exploration of structural diversity

A new way to take into account the structural variability of protein structures is being actively explored, with very promising results. It is well known that many proteins can exist in several structural states, such as open or closed, depending on the presence of one of their substrates. The RMSD between the two forms is such that if one tries to solve the MR with the wrong form, MR will fail, while it might succeed if one uses the other form. Of course, one might try to use both, but it may well be that all accessible structural states of a given protein are not deposited in the PDB.

Hinge-bending as well as shear movements have been documented over the years, with RMSD between the two forms of the same proteins spanning a very large range. While hinge-bending movements can, in principle, be tackled by dividing the protein into individual domains and then solving the MR problem for each of the domains, other types of movements are very difficult to deal with, especially as they are very collective and involve the coordinated movement of many atoms. Recently, however, it has become apparent that most structural transitions observed in the PDB can be modelled quite accurately by the low-frequency Normal Modes derived from a simplified representation of the protein, namely the Elastic Network Model (Krebs *et al.*, 2002). The Elastic Network Model (Fig. 7.4) is a simplified representation of a protein where each residue is represented by a single point that is linked to its spatial neighbours (within a given radius, usually taken as 8–12 Ångstroms) by a spring of constant strength (Tirion, 1996); it works surprisingly well despite its simplicity and is by construction most apt to model collective movements (Delarue and Sanejouand, 2000). In fact, it has been shown that most movements can be modelled with, on average, only two normal modes (Krebs *et al.*, 2002). As a result of advances in the computation of such modes, there is virtually no limit to the size of the macromolecular assembly for which low-frequency normal modes can be calculated (Tama and Sanejouand, 2001).

Recently, two independent applications of these ideas to MR have been described (Fig. 7.5):

- One is intended to produce structural diversity in the model by grid-sampling the amplitudes of (at most) two of the lowest-frequency normal modes;

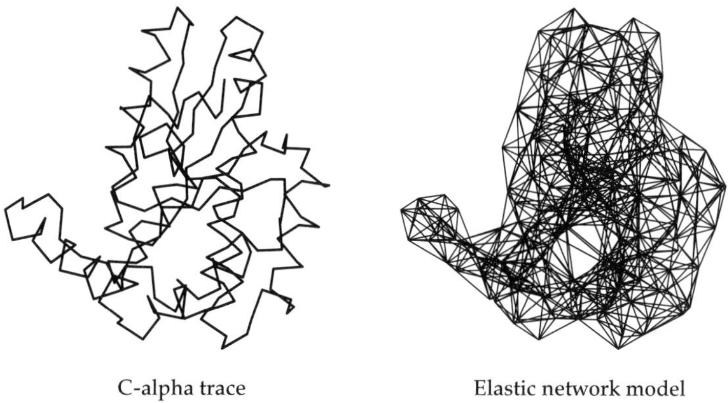


**Figure 7.4** Illustration of the Elastic Network Model for thymidylate kinase (cut-off = 8 Ångstroms).

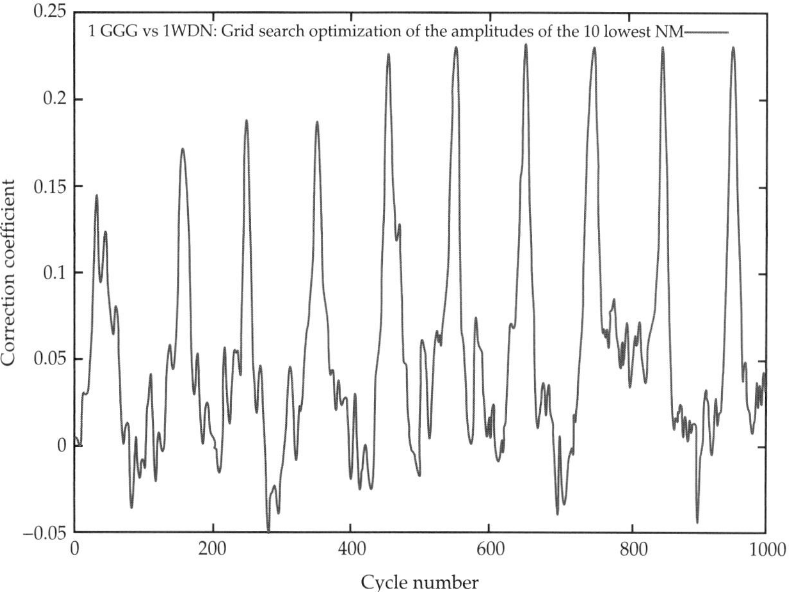


**Figure 7.5** Grid sampling and refinement of the amplitudes along, successively, the first 10 Normal Modes for the glutamine-binding protein. The open structure (1WDN) is used to calculate Normal Modes and then deformed along these Normal Modes to fit X-ray data of the closed form (1GGG PDB code). The amplitudes are scanned from −500 to 500, using 100 grid points for each normal mode. Once the best amplitude is found for a given Normal Mode, it remains fixed for the following searches along the next available Normal Modes. It is apparent from this plot that no further increase of the correlation coefficient is obtained after the 5th mode has been added. The RMSD between 1WDN and 1GGG is 5.5 Ångstroms. After finding the amplitudes of the 10 lowest frequency Normal Mode using this grid-based approach, Conjugate Gradient refinement is used to further improve the correlation coefficient, reaching ultimately 26%. The initial score was 6%. The same example is also treated in Suhre and Sanejouand (2004), with similar results.

all the produced models are subjected to a standard MR protocol, leading to solutions in cases where all the other methods failed. This is now available on line as a web server called elNemo (Suhre and Sanejouand, 2004a, b).

- The other is a refinement program using a correlation coefficient with the experimental data as a target and conjugate-gradient algorithm as the engine, while the only degrees of freedom are the amplitudes of the lowest frequency Normal Modes

(Delarue and Dumas, 2004). It is intended to be used as a generalization of Rigid Body refinement, as the six degrees of freedom of global rotation and translation are also included as Normal Modes of null frequency. Convincing test cases, as well as real cases, have shown that it works well and, although it was originally described to work with C-alpha-only models, it has now been modified to include all atoms in the refinement and has also been put on line as a web-server at http://lorentz.immstr.pasteur.fr (Lindahl *et al.*, 2006).

## 7.11 How to get the best (least-biased) starting phases

Once the best possible molecular replacement solution has been found and refined by, for example, the CNS simulated annealing protocol in internal coordinates space, there is a need to use the least possible biased phases, as recently emphasized in Rupp *et al.* (2002). This is usually done by using SIGMAA-weighted maps, originally due to Read (Read, 1986, 1990) and implemented in most crystallographic packages (Refmac5 (Murshudov *et al.*, 1997), CNS (Brünger *et al.*, 1998), etc.). Then, one can go on and try to use directly automatic construction methods such as ARP/wARP (Perrakis *et al.*, 2001) or Resolve (Terwilliger, 2003). However, we wish to point out here that there exists a very general formalism based on mean-field theory and statistical mechanics (Delarue and Orland, 2000) which, in principle, should allow most of the errors contained in the model phases to be eliminated and which points to the use of other kinds of weighted coefficients, calculated in a self-consistent manner, as inputs to the Fourier transforms to calculate the map. Even though the method still remains to be implemented in a space-group general program, preliminary results in the space group P2(1)2(1)2(1) show great promise (Delarue, unpublished work).

## 7.12 Special cases: phased TF and locked RF

If, for some reason, MR failed, leaving no option but to look for heavy atom derivatives and if, for lack of isomorphism, the phasing power of these derivatives does not allow for a straightforward interpretation of the map, it is still possible to use the experimental phases in conjunction with the model at hand. Indeed, MR in even a poor experimental map will in general give a much stronger signal-to-noise ratio than regular MR based on structure factor moduli only. This can be done in reciprocal space using an analog of the translation function, called the phased translation function (Bentley, 1997). This is simply a correlation coefficient between maps of the model placed at different origins and the experimental one. This can be carried out in reciprocal space, where all the products of type ($|\mathbf{F}_{obs}|.|\mathbf{F}_{model}|$) have been replaced by $Re(\mathbf{F}_{obs}.\mathbf{F}^*_{model})$, where $\mathbf{F}^*_{model}$ is the complex conjugate of $\mathbf{F}_{model}$. Once the model has been placed correctly in the unit cell of the crystal, experimental phases can be combined with phases from the model, followed by solvent flattening using a molecular envelope derived directly from the model. This usually results in a much better map. However, the rotation function is still, in general, no easier to solve than without the experimental phases, so one might think that all of this is of little use. There is, however, a way out of it, which is to scan the entire rotation function space with the phased translation search as a score (see Protocol 7.6). The present author has used it in a number of cases with success: it is, in general, computationally doable and leads to a very clear signal (Li de la Sierra *et al.*, 2001; Delarue *et al.*, 2002).

Another special case where the signal of the rotation function could be enhanced concerns crystals where the self-rotation function can be interpreted without ambiguity; in this case, the so-called locked cross-rotation function (Tong and Rossmann, 1997) allows to search for cross rotations which are compatible with the self rotation function. This usually results in a much better signal-to-noise ratio.

## 7.13 Concluding remarks

It seems pretty clear that easy-to-use and web-interfaced, automated protocols will be more and more useful in the near future. The possibility of combining different tools at different steps of the integrated process makes it inevitable that more of them will be developed and open to the public. What is good about this kind of approach is that

**Protocol 7.6 Full 6D search with the phased translation function**

**1.** Determine the asymmetric unit of the space group of the rotation function of your space group (Urzhumtsev and Urzhumtseva, 2002).
**2.** Write and execute jiffy code to create the (formatted) input file to explore exhaustively rotation space:

```
c...
  delta0 = 5.
  delta1 = delta0/2.
  do i = -imax,imax
     do j = 0,jmax
          do k = -kmax, kmax
                  alpha = alpha0 + delta0*i
                  beta = beta0 + delta1*j
                  gamma = gamma0 + delta0*k
                  write(2,'(3f10.5,4x,2f11.3,i10)')alpha,beta,gamma,r,t,iu
          enddo
     enddo
  enddo
c...
```

The resulting formatted file will be fed to AMoRe as m1.rts, the input file to Phased Translation Function (PTF).
**3.** Run AMoRe PTF in all the possible space groups (see Protocol 7.1), after having run a test case (as in Protocol 7.2).
**4.** Use perl to extract the information on the solutions with the highest score from the (enormous) log file. Sort the solutions.
**5.** Plot results using gnuplot.

the authors can keep track of both the successes and failures of jobs submitted by the crystallographic community and use these as bench marks and opportunities to improve their protocols. So their performances should keep growing, provided that crystallographers don't shy away from them because of confidentiality problems.

Finally, by way of setting perspectives, we wish to point out that the only alternative to the use of many different models successively, with the same protocol, until one of them gives a better signal, would be to use all of them simultaneously, weighted by some linear prefactor that remains to be determined. The sum of all these prefactors for all the different copies of the model should of course be 1. This is actually reminiscent of some recent approaches to the refinement of macromolecules using a multicopy strategy, so as to take into account some inherent flexibility of the model (Burling and Brünger, 1994; Pellegrini *et al.*, 1997). We suggest that refining the weights of the different models at each orientation of the model in the rotation function should improve the signal-to-noise ratio of the whole procedure. Translation function searches should then proceed normally using the best combination of weighted model(s) identified in this way.

## Acknowledgments

We gratefully acknowledge financial support from ACI-IMPB045 from the Ministere de la Recherche et de la Technologie and from GDR 2417 du CNRS. We thank G. Labesse (CBS, CNRS, Montpellier) for discussions and communication of results prior to publication and P. Dumas (IBMC, CNRS, Strasbourg) for comments on the manuscript.

## References

Adams, P., Pannu, N. S., Read, R. J., and Brünger, A. T. (1999) Extending the limits of molecular replacement through combined simulated annealing and maximum-likelihood refinement. *Acta Crystallogr. D* **55**, 181–190.

Altschul, S. F, Madden, T. L., Schaffer, A. A., Zhang, J., Zhang, Z., Miller, W., and Lipman, D. J. (1997) Gapped BLAST and PSI-BLAST: a new generation of protein database search programs. *Nucleic Acids Res.* **25**, 3389–3402.

Bentley, G. (1997) Phased translation function. *Method Enzymol.* **276**, 611–619.

Berman, H. M., Westbrook, J., Feng, Z., Gilliland, G., Bhat, T. N., Weissig, H., Shindyalov, I. N., and Bourne, P. E. (2000) The Protein Data Bank. *Nucleic Acids Res.* **28**, 235–242.

Brünger, A. T. (1997) Free R values: Cross-validation in crystallography. *Method Enzymol.* **277**, 366–396.

Brünger, A. T., Adams, P. D., Clore, G. M., DeLano, W. L., Gros, P., Grosse-Kunstleve, R. W., *et al.* (1998) Crystallography and NMR system: A new software suite for macromolecular structure determination. *Acta Crystallogr. D* **54**, 905–921.

Burling, F. T. and Brünger, A. T. (1994) Thermal motion and conformational disorder in protein crystal structures. *Isr. J. Chem.* **34**, 165–175.

CASP5 (2003). *Proteins*, **53**, Suppl. 6.

Catherinot, V. and Labesse, G. (2004) ViTO: tool for refinement of protein sequence-structure alignments. *Bioinformatics*, 20: 3694–3696.

Chang, G. and Lewis, M. (1997) Molecular replacement using genetic algorithms. *Acta Crystallogr. D* **53**, 279–289.

Chen, Y. W. (2001) Solution solution: using NMR models for molecular replacement. *Acta Crystallogr. D* **57**, 1457–1461.

Chothia, C. and Lesk, A. (1986) The relation between the divergence of sequence and structure in proteins. *EMBO J.* **5**, 823–826.

Claude, J. B., Suhre, K., Notredame, C., Claverie, J. M., and Abergel, C. (2004) CaspR: a web server for automated molecular replacement using homology modelling. *Nucleic Acids Res.* **32**, W606–609.

Collaborative Computational Project (1994) The *CCP4* suite: programs for protein crystallography. *Acta Crystallogr. D* **50**, 760–763.

Crowther, R. A. (1972). The fast rotation function. In: *The Molecular Replacement Method*, Rossmann, M. G., ed. Gordon and Breach, New York, pp. 173–185.

Crowther, R. A. and Blow, D. M. (1967) A method of positioning a known molecule in an unknown crystal structure. *Acta Crystallogr.* **23**, 544–548.

DeLano, W. L. and Brünger, A. T. (1995) The direct rotation function: Patterson correlation search applied to molecular replacement. *Acta Crystallogr. D* **51**, 740–748.

Delarue, M., Samama, J. P., Mourey, L., and Moras, D. (1990) Crystal structure of bovine antithrombin III. *Acta Crystallogr. B* **46**, 550–556.

Delarue, M. and Orland, H. (2000). General formalism for phase combination and phase refinement: a statistical thermodynamics approach in reciprocal space. *Acta Crystallogr. A* **56**, 562–574.

Delarue, M. and Sanejouand, Y. H. (2002) Simplified normal mode analysis of conformational transitions in DNA-dependent polymerases: the elastic network model. *J. Mol. Biol.* **320**, 1011–1024.

Delarue, M., Boule, J. B., Lescar, J., Expert-Bezancon, N., Jourdan, N., Sukumar, N., Rougeon, F., and Papanicolaou, C. (2002) Crystal structures of a template-independent DNA polymerase: murine terminal deoxynucleotidyltransferase. *EMBO J.* **21**, 427–439.

Delarue, M. and Dumas, P. (2004) On the use of low-frequency normal modes to enforce collective movements in refining macromolecular structural models. *Proc. Natl. Acad. Sci. USA* **101**, 6957–6962.

Delarue, M., Duclert-Savatier, N., Miclet, E., Haouz, A., Giganti, D., Ouazzani, J., Lopez, P. Nilges, M. and Stoven, V. (2007) Three dimensional structure and implications for the catalytic mechanism of 6-phosphogluconolactonase from *Trypanosoma brucei. J. Mol. Biol.* **366**, 868–881.

Douguet, M. and Labesse, G. (2001) Easier threading through web-based comparisons and cross-validations. *Bioinformatics* **17**, 752–753.

Fokine, A. and Urzhumtsev, A. (2002) On the use of low-resolution data for translation search in molecular replacement. *Acta Crystallogr. D* **58**, 72–74.

Glykos, N. M. and Kokkinidis, M. (2000) A stochastic approach to molecular replacement. *Acta Crystallogr. D* **56**, 169–174.

Glykos, N. M. and Kokkinidis, M. (2001) Multi-dimensional molecular replacement. *Acta Crystallogr. D* **57**, 1462–1473.

Glykos, N. M. and Kokkinidis, M. (2003) Structure determination of a small protein through a 23-dimensional molecular-replacement search. *Acta Crystallogr. D* **59**, 709–718.

Harada, Y., Lifchitz, A., Berthou, J., and Jolles, P. (1981) A translation function combining packing and diffraction information: an application to lysozyme (high-temperature form). *Acta Crystallogr. A* **37**, 398–406.

Huber, R. and Schneider, M. (1985) A group refinement procedure in protein crystallography using Fourier transforms. *J. Appl. Crystallogr.* **18**, 165–169.

Jamrog, D. C., Zhang, Y., and Phillips, G. N., Jr. (2003) *SOMoRe*: a multi-dimensional search and optimization approach to molecular replacement. *Acta Crystallogr. D* **59**, 304–314.

Jamrog, D. C., Zhang, Y., and Phillips, G. N., Jr. (2004) On the equivalence between a commonly used correlation coefficient and a least-squares function. *Acta Crystallogr. A* **60**, 214–219.

Jones, D. T. (2001) Evaluating the potential of using fold-recognition models for molecular replacement. *Acta Crystallogr. D* **57**, 1428–1434.

Jones, T. A., Zhou, J. Y., Cowan, S. W., and Kjelgaard, M. (1991) Improved methods for building protein models in electron density maps and the location of errors in these models. *Acta Crystallogr. A* **47**, 110–119.

Keegan, R. M. and Wynn, M. D. (2007) Automated search-model discovery and preparation for structure solution by molecular replacement. *Acta Crystallogr. D* **63**, 447–457.

Kissinger, C. R., Gelhaar, D. K., and Fogel, D. B. (1999) Rapid automated molecular replacement by evolutionary search. *Acta Crystallogr. D* **55**, 484–491.

Kissinger, C. R., Gelhaar, D. K., Smith, B. A., and Bouzida, D. (2001) Molecular replacement by evolutionary search. *Acta Crystallogr. D* **57**, 1474–1479.

Koehl, P. and Delarue, M. (1994) Application of a self-consistent mean field theory to predict protein side-chains conformation and estimate their conformational entropy. *J. Mol. Biol.* **239**, 249–275.

Koehl, P. and Delarue, M. (1995) A self consistent mean field approach to simultaneous gap closure and side-chain positioning in homology modelling. *Nature Struct. Biol.* **2**, 163–170.

Krebs, W. G., Alexandrov, V., Wilson, C. A., Echols, N, Yu, H., and Gerstein, M. (2002) Normal mode analysis of macromolecular motions in a database framework: developing mode concentration as a useful classifying statistic. *Proteins* **48**, 682–695.

Lee, S., Sawaya, M. R., and Eisenberg, D. (2003) Structure of superoxide dismutase from *Pyrobaculum aerophilum* presents a challenging case in molecular replacement with multiple molecules, pseudo-symmetry and twinning. *Acta Crystallogr. D* **59**, 2191–2199.

Li de la Sierra, I., Munier-Lehmann, H., Gilles, A. M., Barzu, O., and Delarue, M. (2001) X-ray structure of TMP kinase from *Mycobacterium tuberculosis* complexed with TMP at 1.95 A resolution. *J. Mol. Biol.* **311**, 87–100.

Lindahl, E., Azuara, C., Koehl, P. and Delarue, M. (2006) NOMAD-Ref: visualization, deformation and refinement of macromolecular structures based on all-atom normal mode analysis. *Nucleic Acid Res.* **34**, W52–56.

Murshudov, G. N., Vagin, A. A., and Dodson, E. J. (1997) Refinement of macromolecular structures by the maximum-likelihood method. *Acta Crystallogr. D* **53**, 240–255.

Navaza, J. (2001) Implementation of molecular replacement in *AmoRe*. *Acta Crystallogr. D* **57**, 1367–1372.

Navaza, J. (2001). Rotation functions. In: *International Tables of Crystallography*, vol. F, p. 269–274. KluwerAcademic Publishers. Dordrecht, Boston, London.

Navaza, J. and Saludjian, P. (1997) AmoRe: an automated molecular replacement package. *Method Enzymol.* **276**, 581–594.

Notredame, C., Higgins, D., and Heringa, J. (2000) T-Coffee: A novel method for fast and accurate multiple sequence alignment. *J. Mol. Biol.* **302**, 205–217.

O'Sullivan, O., Suhre, K., Abergel, C., Higgins, D. G., and Notredame, C. (2004) 3DCoffee: combining protein sequences and structures within multiple sequence alignment. *J. Mol. Biol.* **340**, 385–395.

Pellegrini, M., Gronbech-Jensen, N., Kelly, J. A., Pfluegl, G., and Yeates, T. O. (1997) Highly constrained multiple-copy refinement of protein crystal structures. *Proteins* **29**, 426–432.

Perrakis, A., Harkiolaki, M., Wilson, K. S., and Lamzin, V. S. (2001) ARP/wARP and molecular replacement. *Acta Crystallogr. D* **57**, 1445–1450.

Rao, S. N., Hih, J. H., and Hartsuck, J. A. (1980) Rotation-function space groups. *Acta Crystallogr. A* **36**, 878–884.

Read, R. J. (1986) Improved Fourier coefficients for maps using phases from partial structures with errors. *Acta Crystallogr. A* **42**, 140–149.

Read, R. J. (1990) Structure-factor probabilities for related structures. *Acta Crystallogr. A* **46**, 900–912.

Read, R. J. (2001) Pushing the boundaries of molecular replacement with maximum likelihood. *Acta Crystallogr. D* **57**, 1373–1382.

Rossmann, M. G. (1990) The molecular replacement method. *Acta Crystallogr. A* **46**, 73–82.

Rossmann, M. G. and Blow, D. M. (1962) The detection of sub-units within the crystallographic asymmetric unit. *Acta Crystallogr.* **15**, 24–31.

Rupp, B., Segelke, B.W., Krupka, H.I., Lekin, T., Shafer, J., Zemla, A., Topani, D., Snell, G., and Earnest, G. (2002).The TB structural genomics consortium crystallization facility: towards automation from protein to electron density. *Acta Crystallogr. D* **58**, 1514–1518.

Sali, A. and Blundell, T. L. (1993) Comparative protein modelling by satisfaction of spatial restraints. *J. Mol. Biol.* **234**, 779–815.

Schwarzenbacher, R., Godzik, A., Grzechnik, S. K., and Jaroszewski, L. (2004) The importance of alignment accuracy for molecular replacement. *Acta Crystallogr. D* **60**, 1229–1236.

Sheriff, S., Klei, H. E., and Davis, M. E. (1999) Implementation of a six-dimensional search using the *AMoRe* translation function for difficult molecular-replacement problems. *J. Appl. Crystallogr.* **32**, 98–101.

Storoni, L. C., McCoy, A. J., and Read, R. J. (2004) Likelihood-enhanced fast rotation functions. *Acta Crystallogr. D* **60**, 432–438.

Stubbs, M. T. and Huber, R. (1991) An analytical packing function employing Fourier transforms. *Acta Crystallogr. A* **47**, 521–526.

Suhre, K. and Sanejouand, Y. H. (2004a) On the potential of normal-mode analysis for solving difficult molecular-replacement problems. *Acta Crystallogr. D* **60**, 796–799.

Suhre, K. and Sanejouand, Y. H. (2004b) ElNemo: a normal mode web server for protein movement analysis and the generation of templates for molecular replacement. *Nucleic Acids Res.* **32**, W610–614.

Tama, F. and Sanejouand, Y. H. (2001) Conformational change of proteins arising from normal mode calculations. *Protein Engng.* **14**, 1–6.

Terwilliger, T. (2003a) Automated main-chain model building by template matching and iterative fragment extension. *Acta Crystallogr. D* **59**, 38–44.

Terwilliger, T. (2003b) Automated side-chain model building and sequence assignment by template matching. *Acta Crystallogr. D* **59**, 45–49.

Tirion, M. (1996) Large Amplitude Elastic Motions in Proteins from a Single-Parameter, Atomic Analysis. *Phys. Rev. Lett.* **77**, 1905–1908.

Tong, L. and Rossmann, M. G. (1997) Rotation function calculations with GLRF program. *Method Enzymol.* **276**, 594–611.

Urzhumtsev, A. and Urzhumtseva, L. (2002) Multiple rotation function. *Acta Crystallogr. D* **58**, 2066–2075.

Urzhumtseva, L. M. and Urzhumtsev, A. G. (1997) Tcl/Tk-based programs. II. *CONVROT*: a program to recalculate different rotation descriptions. *J. Appl. Crystallogr.* **30**, 402–410.

Vagin, A. and Teplyakov, A. (1997) *MOLREP*: an automated program for molecular replacement. *J. Appl. Crystallogr.* **30**, 1022–1025.

Vagin, A. and Teplyakov, A. (2000) An approach to multi-copy search in molecular replacement. *Acta Crystallogr. D* **56**, 1622–1624.

Wilmanns, M. and Nilges, M. (1999) Molecular replacement with NMR models using distance-derived pseudo B factors. *Acta Crystallogr. D* **52**, 973–982.

CHAPTER 8

# MAD phasing

H. M. Krishna Murthy

## 8.1 Introduction

From a relatively small beginning, nearly two decades ago (Guss *et al.*, 1988; Murthy *et al.*, 1988; Hendrickson *et al.*, 1988, 1989), phase calculation using Multiple-wavelength Anomalous Diffraction (MAD) data has become more widespread and generally used. There are many reviews that cover diverse aspects of the process (Hendrickson, 1991; Smith, 1991; Ogata, 1998; Ealick, 2000). This review will concentrate on the most recent developments and experimental details, and is essentially an update to the author's earlier review (Krishna Murthy, 1996).

## 8.2 Theoretical background

Phasing in macromolecular structure determination entails the independent determination of the sine and cosine functions of the phase angle. Traditionally, this has been accomplished through measurement of diffraction data on one or more heavy atom derivatives of the macromolecule, and using the Multiple Isomorphous Replacement (MIR) methodology. In favourable cases, a single derivative that has measurable anomalous diffraction can also be used, in a variation of MIR termed single isomorphous replacement supplemented with anomalous scattering (SIRAS). MAD phasing exploits the signals generated from atoms in crystals that have absorption frequencies close to that of the wavelength of incident X-rays. The theoretical underpinnings of anomalous diffraction are covered in detail in several standard works (Blundell and Johnson, 1976; James, 1982). A more elaborate outline of the derivation of the formulae is given in references Hendrickson (1985, 1991) and Krishna Murthy (1996). Briefly, assuming that the wavelength dependent structure factor expression may be written as:

$$^{\lambda}F_T(\mathbf{h}) = {}^{0}F_T(\mathbf{h}) + \sum_k \left[\left(\frac{f'_k}{f^0_k}\right) + \left(\frac{if''_k}{f^0_k}\right)\right] {}^{0}F_{Ak}(\mathbf{h})$$

Where

$^{\lambda}F_T(\mathbf{h})$ is the structure factor at wavelength $\lambda$

$^{0}F_T(\mathbf{h})$ is the structure factor contribution from all normal scatterers

$^{0}F_{Ak}(\mathbf{h})$ is the structure factor contribution from the $k^{\text{th}}$ kind of anomalous scatterer

$f'_k$ and $f''_k$ = dispersive and Bijvoet components for the $k^{\text{th}}$ kind respectively, of the anomalous diffractor at wavelength $\lambda$, and

$f^0_k$ = normal scattering factor,

then, the experimentally measurable quantity, namely, wavelength dependent intensity, assuming for simplicity a single kind of anomalous diffractor ($k = 1$), may be written as:

$$\begin{aligned}|^{\lambda}F_T(\mathbf{h})|^2 &= |^{0}F_T(\mathbf{h})|^2 + a(\lambda)|^{0}F_A(\mathbf{h})|^2 \\ &\quad + b(\lambda)|^{0}F_T(\mathbf{h})||^{0}F_A(\mathbf{h})| \cos[^{0}\phi_T(\mathbf{h}) - {}^{0}\phi_A(\mathbf{h})] \\ &\quad + c(\lambda)|^{0}F_T(\mathbf{h})||^{0}F_A(\mathbf{h})| \sin[^{0}\phi_T(\mathbf{h}) - {}^{0}\phi_A(\mathbf{h})]\end{aligned}$$

where

$$a(\lambda) = \left(\frac{f^{\Delta}}{f^0}\right)^2, \quad b(\lambda) = 2\left(\frac{f'}{f^0}\right),$$

$$c(\lambda) = 2\left(\frac{f''}{f^0}\right) \quad \text{and} \quad f^{\Delta} = (f'^2 + f''^2)^{1/2}.$$

Also $^{0}\phi_T(\mathbf{h})$ is the phase of $^{0}F_T(\mathbf{h})$ and $^{0}\phi_A(\mathbf{h})$ is the phase of $^{0}F_A(\mathbf{h})$.

There are three unknowns in these equations:

$$|{}^0F_T(\mathbf{h})|, |{}^0F_A(\mathbf{h})| \text{ and } {}^0\phi_T(\mathbf{h}) - {}^0\phi_A(\mathbf{h}) = \Delta\phi$$

The quantities, $a(\lambda)$, $b(\lambda)$, and $c(\lambda)$ are, by definition, scalar functions of $f'$ and $f''$ and can be estimated through direct measurements of these quantities, without knowledge of the structure. Historically, in early work on MAD phasing, these three unknowns were determined through an algebraic method due to Karle (1980). Measurements at several wavelengths provided the over determination needed for a least squares determination of these unknown parameters (Hendrickson, 1985). Although the above relations treat the case of a single anomalous scatterer, the methodology can, in principle, be generalized to several anomalous scatterers.

Most of the later applications of MAD phasing have, however, followed the treatment of MAD phasing as a special case of the MIR phasing approach. Following a suggestion by Hendrickson (Hendrickson, 1991), the first practical application of this approach was in the determination of the structure of a fragment of histone H5 (Ramakrishnan *et al.*, 1993). This approach has the advantage that the significant body of theoretical knowledge accumulated in the application of the MIR procedure, as well as the numerous programs that implement it, can directly be used in analysis of MAD data. The basic tenet of the MAD as a special case of MIR approach is to treat the data measured at one of the wavelengths as 'native'. Data measured at the other wavelengths are then treated as a set of derivatives; dispersive differences play the role of isomorphous differences while the Bijvoet differences provide the orthogonal, anomalous information; their traditional role in MIR. The theoretical basis of this approach is detailed in a review by Ramakrishnan and Biou (1997). However, a limitation of this approach is that it cannot be easily generalized to the case of multiple types of anomalous scatterers within a unit cell or to multiple MAD data sets. In addition, one set of data, the 'native', is arbitrarily treated as not containing an anomalous component. This latter assumption is, in most MAD data measurement regimes, quite unjustified. A modification, which approximates a pseudo SIRAS situation, is based on the suggestion that the magnitude of anomalous diffraction is small, compared to the total diffraction from a unit cell. This approach permits ready generalization to the simultaneous presence of many kinds of anomalous scatterers in the asymmetric unit (Terwilliger, 1994a). Briefly, the pseudo SIRAS formalism may be summarized as below.

Denoting by $F^+(\lambda_j)$ and $F^-(\lambda_j)$, the two components of a Bijvoet pair, at wavelength $\lambda_j$, the average structure factor amplitude at that wavelength is given by

$$\overline{F}(\lambda_j) = \frac{1}{2}[F^+(\lambda_j) + F^-(\lambda_j)]$$

and the Bijvoet difference by

$$\Delta_{\text{Ano}}(\lambda_j) = F^+(\lambda_j) - F^-(\lambda_j)$$

To achieve analogy with MIR, the normal and anomalous diffraction parts for a reflection are expressed as separate quantities,

$$\overline{F}(\lambda_j) = |\mathbf{F}_0 + \mathbf{F}_\text{H}(\lambda_j)|$$

where $\mathbf{F}_0$ and $\mathbf{F}_\text{H}(\lambda_j)$ are the structure factor for all the non-anomalously scattering atoms and that for just the anomalous scatterers, respectively. The method then goes on to derive estimates for $F_0$, $\Delta_{\text{Ano}}(\lambda_0)$ and $\overline{F}(\lambda_0)$, where the subscript '0' indicates that one of the wavelengths that the data have been measured at can be arbitrarily chosen for this evaluation. An estimate for $\mathbf{F}_\mathbf{H}(\lambda_j)$, the structure factor at any of the wavelengths other than the chosen one, can be derived from the observation that anomalous scattering changes in magnitude with wavelength although the phase is wavelength independent,

$$\mathbf{F}_\mathbf{H}(\lambda_j) = \mathbf{F}_\mathbf{H}(\lambda_0)\frac{f_0 + f'(\lambda_j)}{f_0 + f'(\lambda_0)}$$

where the terms containing the $f$'s represent the real part of the scattering factor for the anomalous scatterers.

An approximation to $\Delta_{\text{Ano}}(\lambda_0)$ is also estimated from similar arguments (Terwilliger and Eisenberg, 1987),

$$\Delta_{\text{Ano}}(\lambda_j) \approx \Delta_{\text{Ano}}(\lambda_0)\frac{f''(\lambda_j)}{f''(\lambda_0)}$$

If one assumes that the anomalous component is small relative to the normal scattering from the unit cell contents, then

$$\overline{F}(\lambda_j) \approx F_0 + F_H(\lambda_j)\cos\alpha$$

Where $\alpha$ is the difference in the phase angle between the anomalous and non-anomalous structure factors, the equivalent of $\Delta\phi$ in the Hendrickson formulation above. From these isomorphous equivalents from the MAD data can be formulated

$$\Delta_{\text{Iso}}(\lambda_0) \approx F_H(\lambda_0)\cos\alpha$$
$$\approx [f_0 + f'(\lambda_j)]\frac{\overline{F}(\lambda_i) - \overline{F}(\lambda_j)}{f'(\lambda_i) - f'(\lambda_j)}$$

From this, one might estimate approximate values for $F_0$ and $\overline{F}(\lambda_0)$,

$$F_0 \approx \overline{F}(\lambda_j) - \Delta_{\text{Iso}}(\lambda_0)\frac{f_0 + f'(\lambda_j)}{f_0 + f'(\lambda_0)} \quad \text{and}$$

$$\overline{F}(\lambda_0) \approx F_0 + \Delta_{\text{Iso}}(\lambda_0)$$

Setting up a formal equivalence between MAD and MIR data representations is very useful for combining information from two different sources, for cases in which each by itself is insufficient for structure solution. These sources could, for example, be MAD data sets determined from crystals with different types of anomalous scatterers; or a MAD and a MIR data set. The assumption, that anomalous signals are small compared to normal scattering, leads to some errors, primarily in the estimation dispersive differences. These errors, however, have been estimated to be no larger than about 4% (Terwilliger, 1994a). An alternative formulation, that does not depend on either assuming a special 'error free' data set or that the anomalous signal be small, has also been described (Bella and Rossmann, 1998). The method is based on an earlier, theoretical approach developed for MIR which treats all isomorphous data sets as equivalent, in terms of error analysis (Cullis *et al.*, 1961). It is also easily extensible to multiple kinds of anomalous scatterers, as well as MAD experiments made on different crystals.

More recently, efforts have also been made to develop direct approaches to the calculation of phases from MAD data. A method that augments other MAD phasing approaches, through improving poorly estimated phases, has been suggested (Fan *et al.*, 1990). The method decomposes the MAD data into a series of one-wavelength anomalous scattering (OAS) subsets. A subset of MAD phases with the highest figures of merit are used as a starting point, and improved phases for the more poorly determined phases computed through the OAS phase-ambiguity resolution procedure (Fan *et al.*, 1990). The direct methods phase probabilities for each phase are determined from each of the OAS subsets, through a modified tangent formula, and an improved phase set is generated through combination with the starting MAD phases, weighted with their associated figures of merit (Gu *et al.*, 2001). Estimation of phases from MAD data through computation of the joint probability distribution functions for each phase, based on earlier direct methods approaches, has also been reported. A probability distribution function approach had earlier been developed for estimation of $|{}^0F_A(\mathbf{h})|$ values (Giacovazzo and Siliqi, 2001; Burla *et al.*, 2003). It has been combined with a procedure for estimation of the probabilities for $|{}^0F_T(\mathbf{h})|$ and ${}^0\phi_T(\mathbf{h})$, using normalized structure factors, derived from measured MAD data, to provide a complete solution to the MAD phasing problem. The theoretical principles and application to a number of test data sets have been described (Giacovazzo and Siliqi, 2004).

## 8.3 Experimental considerations

The following sections address some of the choices that need to be made in the design and implementation of MAD experiments. Examples are taken from the work in the author's laboratory since he is most familiar with those.

### 8.3.1 Incorporation of anomalous scatterers

The best atomic types to include as anomalous scatterers are those that produce the largest absorption and dispersive signals. Electrons in the L and M shells of atoms are less strongly held by their nuclei and generate larger $f'$ and $f''$ signals. Most reported MAD experiments have been performed at either a K or L edge, although at least one has used an M edge (Liu *et al.*, 2001). Absorption curves for the selenium K-edge and holmium L-edge are shown in Fig. 8.1. Note that the $f'$ and $f''$ magnitudes for selenium are in the 4–5 electrons, while the corresponding values for holmium) are in the 18–20 electron range (Merritt, 1998). Techniques for the introduction of anomalous scatterers can be

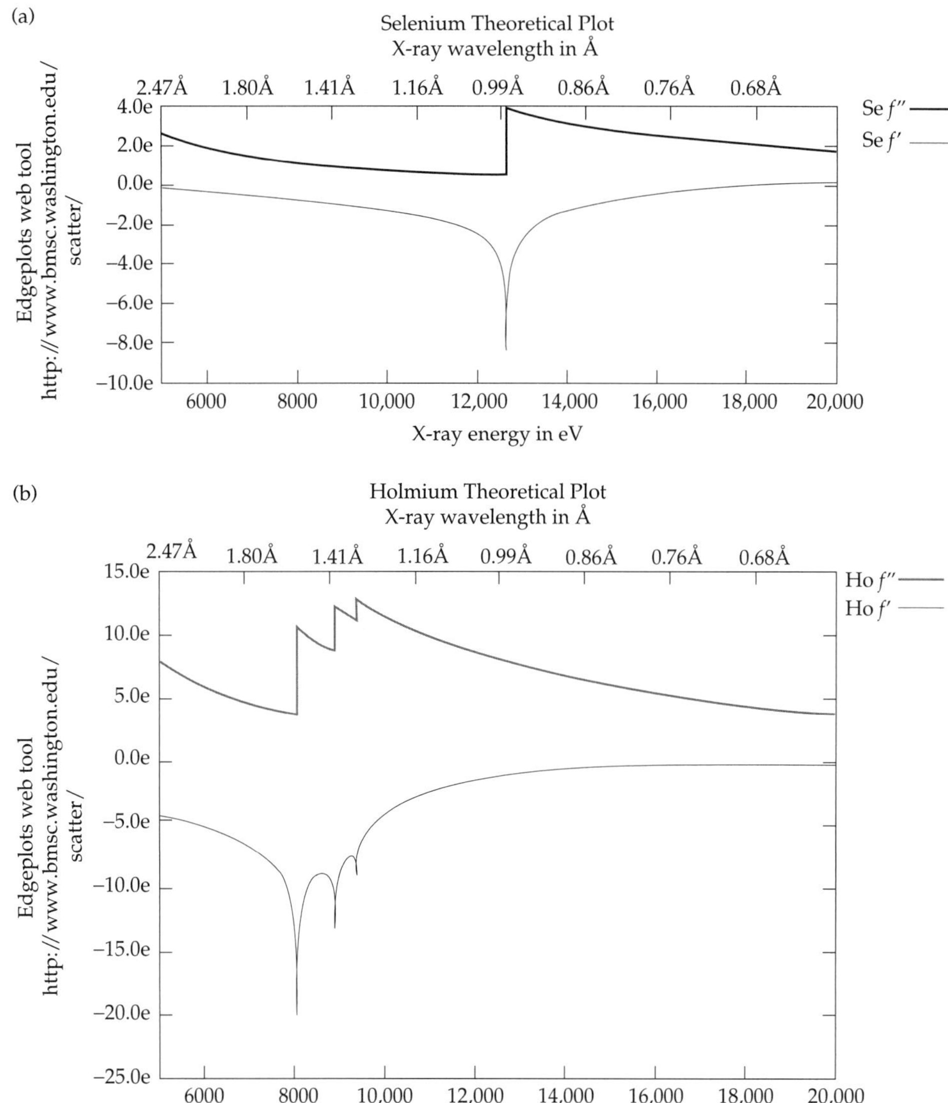


**Figure 8.1** Theoretically calculated $f'$ and $f''$ values as a function of wavelength in the vicinity of the K absorption edge of selenium (a) and the L edge of holmium (b).

broadly divided into two kinds; those that are specific to a particular macromolecule and the more general techniques. Many proteins, for instance, have metal binding sites and are, in their physiological states, associated with metal ions. In many cases these metal ions have absorption edges that are either at very long wavelengths or do not produce very strong signals. They can however be replaced by metal ions that have larger signals, if necessary by first chelating out the naturally associated ions. For example, in the solution of the structure of Taq DNA polymerase the native $Mg^{++}$ were replaced by $Ho^{+++}$ ions, by first soaking the crystals in *o*-phenanthroline/EDTA to extract the $Mg^{++}$, followed by soaking in $HoC1_3$. Although MAD phasing was not used in this particular structure determination, it illustrates the potential for incorporation of atomic species which can provide larger anomalous signals (Urs, 1999). Other types of specialized techniques include labelled substrates or inhibitors that carry atoms or groups that scatter anomalously.

The most broadly used, general technique is the replacement of methionines in proteins by selenomethionines. The technique, initially introduced by Hendrickson and coworkers (Hendrickson *et al.*, 1990), has been widely used. Essentially, it involves expression of the cloned protein of interest in a methionine auxotrophic strain with exogenously provided selenomethionine as the sole source of this amino acid. Several reviews that describe the technology and the problems involved (Hendrickson, 1998), including methods for replacing both methionine and cysteine residues by their seleno analogs (Strub *et al.*, 2003), and replacement of other residues by methionines to generate sites for selenium atoms (Gassner and Matthews, 1999), are available. Although achieving incorporation of selenomethionines into proteins expressed in non-bacterial cell cultures is still in many cases a challenge, it has been accomplished in some cases (Lustbader *et al.*, 1995; Wu *et al.*, 1994). Techniques for introduction of selenium into nucleic acids and oligonucleotides have also been described (Teplova *et al.*, 2002).

A second set of general methods is the classical approach used for derivatization of crystals in MIR phasing (Garman and Murray, 2003). One of the advantages of MAD methods over MIR is that isomorphism is not a requirement. The structure of a derivatized protein that is non-isomorphous with the native can be determined directly from MAD data, and the native structure subsequently determined through molecular replacement, if necessary. Among the more recent developments in techniques of anomalous scatterer incorporation are those that involve binding of anomalous scatterers to macromolecular surfaces using short soaks. Compounds that can be used include both negative ions such as bromide (Dauter, 2000, 2002) as well as positive ions such as rubidium (Korolev *et al.*, 2001). In addition, binding cavities in proteins for inert gas atoms such as argon and krypton have been utilized for this purpose (Cohen *et al.*, 2001; Quillin and Matthews, 2003). For larger proteins, clusters of heavy atoms have been successfully used (Rudenko *et al.*, 2003). Bijvoet ($S_{Bij-Calc}$) and dispersive ($S_{Dis-Calc}$) signal levels expected from the incorporated anomalously scattering species can be estimated through use of the following formulae:

$$S_{Bij-Calc} = \left(\frac{1}{2}\right)^{1/2} \left(\frac{N_A}{N_P}\right)^{1/2} \left(\frac{f''_{Peak}}{Z_{eff}}\right) \text{ and}$$

$$S_{Dis-Calc} = \left(\frac{1}{2}\right)^{1/2} \left(\frac{N_A}{N_P}\right)^{1/2} \left(\frac{|f'_{Edge} - f'_{Remote}|}{Z_{eff}}\right)$$

where

$N_A$ = number of anomalous scatterers
$N_P$ = number of non-hydrogen atoms in the macromolecule
$Z_{eff}$ = effective scattering factor of an atom (6.7 for protein atoms).

### 8.3.2 Choice of wavelengths

Generally, data at three wavelengths is preferred for determination of both Bijvoet and dispersive differences, in case of a single anomalous species in the unit cell. Although in principle, the remote wavelength can double as the peak wavelength, as long as it is on the short wavelength side of the absorption edge (Peterson *et al.*, 1996), use of an additional wavelength improves accuracy and precision of the determined phases (Hendrickson, 1991). In particular for methods that use least squares calculations in estimation or refinement of parameters, it provides valuable additional observations. Bijvoet differences are determined by measurements of either Bijvoet or Friedel mates at the wavelength corresponding to the absorption peak, or on its short wavelength side. Dispersive differences require the measurement of intensities at two different wavelengths, one at a wavelength close to the inflexion point of the absorption spectrum (Fig. 8.1), and a second one far enough away to enable as a large a signal as practicable. It is important, in many cases, to measure the absorption spectrum from the crystals under study since the environment of the anomalous scatterer might significantly change the position of the infection point of the $f'$ curve from its theoretically calculated position (Fanchon and Hendrickson, 1990). It is also clear, from Fig. 8.1, that the $f'$ curve varies very rapidly, as a function of wavelength, near the inflection point. Since the magnitude of the measured dispersive signal is dependent on the difference in $f'$ values between the inflection point and the 'remote' wavelength chosen, prior accurate determination and representation of the $f'$ curve is important. The $f'$ and $f''$ values are generally determined from fluorescence measurements that are then transformed into $f'$ and $f''$ values through use of programs such as CHOOCH (Evans and Pettifer, 2001) that use standard transformations (Cromer and Lieberman, 1970) for the purpose. Wavelengths at which to measure data can then be directly read off from $f'$ and $f''$ curves plotted as a function of wavelength. For measurement of Bijvoet signals the choice of wavelength is less critical, at least if one ascertains that the monochromator is positioned on the lower wavelength side of the absorption curve, as the absorption continues to be high at longer distances from the peak value (Fig. 8.1). Acquisition of fluorescence scans, derivation of the absorption and dispersion curves, and suggestion of wavelengths for data measurement are completely automated at most synchrotron beam lines. There are also websites, such as that at the University of Washington (Merritt, 1998), which permit theoretical estimates of $f'$ and $f''$ values for potential anomalous scatterers to be obtained, as well providing other useful information for experimental design. In addition, other prescriptions for making optimal wavelength choices are available (Narayan and Ramaseshan, 1981; Burla *et al.*, 2004).

As a practical example, the theoretical spectrum and the wavelength choices made for measurement of MAD data for vaccinia complement-control protein (VCP) are shown in Fig. 8.2. There were two crystal forms of the protein, in space groups $P2_12_12_1$ and C222. One heavy atom derivative, $Eu^{+++}$ and $Au^{+++}$, respectively, for each of them was obtained and MAD data were collected on each of them on the X4A beam line at NSLS. The structure was eventually determined, through MAD phasing for the $P2_12_12_1$crystal form and the structure in C222 determined using molecular replacement (Murthy *et al.*, 2001).

### 8.3.3 Data measurement and processing

Data precision and accuracy are generally important, but acquire additional importance due to the small size of MAD signals. Although some multiwavelength data have been measured on laboratory sources (Hendrickson *et al.*, 1986), synchrotron beam lines are the most practical sources for such measurements. Signals at K absorption edges tend to be numerically smaller than those at the L and M edges (Fig. 8.1) and somewhat larger errors can be tolerated in the latter cases. When data measurements were usually made with crystals mounted within capillaries, it was customary to enhance precision in measured Bijvoet mates by orienting crystals such that reflections related through Friedel symmetry were measures on the same frames, at nearly the same time (Murthy *et al.*, 1988, 1999). However, with flash-frozen crystals, orientation along real or reciprocal crystal axes is not practicable. Thus in generally for data measurement on flash-frozen crystals the inverse-beam geometry is preferred (Hendrickson, 1991). The time required for movements of the goniometer motor to measure data in the inverse-beam geometry would worsen potential radiation damage to the crystal. Although measurement of data on flash-cooled crystals has reduced the severity of radiation damage it is still a significant problem, even at –100°C (Garman, 2003). In most cases this conundrum is addressed by measuring data in blocks of 30°– 45° before swinging the goniometer to the corresponding inverse orientation, and alternating until all the data needed are measured. If needed, even smaller blocks can be used for crystals that show little decay. A review by Gonzalez (2003) has greater detail on data measurement principles and practices, including measurement of MAD data. Processing of data at each wavelength is done as it is for single wavelength experiments, with the proviso that the Bijvoet differences are not merged, at least at the peak wavelength.

The most popular processing programs in current use are probably DENZO/SCALEPACK (Otwinowski and Minor, 1997) and MOSFLM (Project, 1994). Careful scaling is particularly important to preserve the relatively weak anomalous signals, and minimize the consequences of

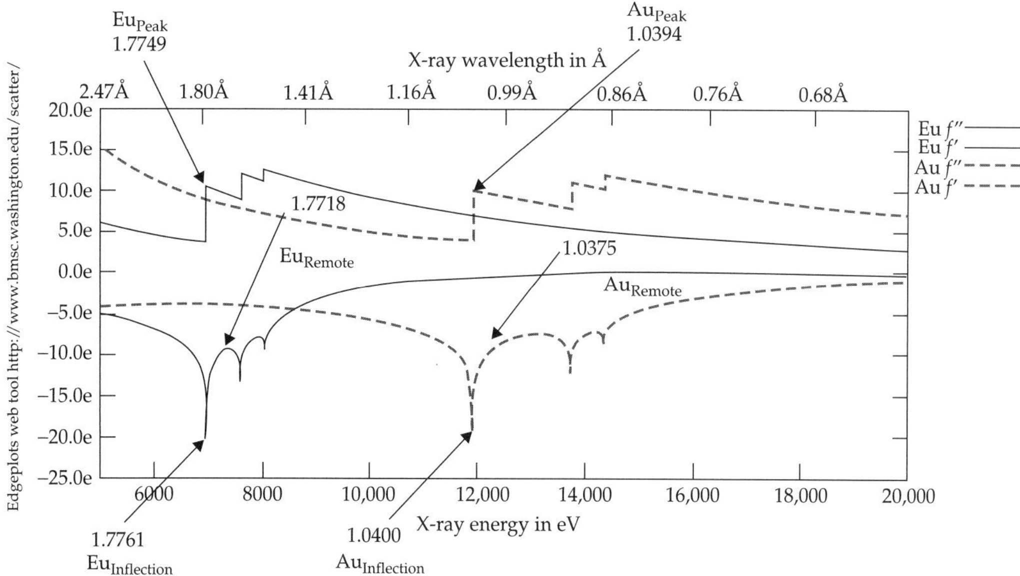


**Figure 8.2** Theoretically calculated $f'$ and $f''$ curves in the neighbourhood of the L edges of europium (solid line) and gold (dashed line) are shown. Wavelengths that were chosen for measurement of peak, inflection point and remote data in each case are marked by arrows and labelled.

measurement errors (Borek *et al.*, 2003). Both scaling of multiple measurements within a wavelength as well scaling equivalent observations across different wavelengths are equally important. Most successful MAD experiments have used local scaling procedures that divide the data into small segments based on resolution and the time at which they were measured (usually parameterized as detector frame number). Although we have found that SCALEPACK is somewhat less flexible in this regard, SCALA of the CCP4 system (Project, 1994) gives excellent results. Multiparameter scaling has also been incorporated into SCALEPACK (Otwinowski *et al.*, 2003). The program system MADSYS (Hendrickson, 1991) also has a very comprehensive set of local scaling capabilities. In addition, the SOLVE/RESOLVE (Terwilliger, 2003) system also provides a variety of multiparameter scaling options. A set of statistics for MAD data, measured for the $Eu^{+++}$ derivative of VCP, scaled using a resolution shell based local scaling option in SCALA is shown in Table 8.1. A set of theoretically estimated diffraction ratios assuming the presence of one to four $Eu^{+++}$ ions in the asymmetric unit is also given at the bottom of the table. The numbers in the $B_{sig}$ column are measured diffraction ratios calculated using the formula at the foot of the table, with those computed for the centrosymmetric pairs in that resolution bins in parentheses. The theoretically estimated ratios are computed at zero scattering angle, and their comparison with the Bijvoet signal in the lowest resolution bin would lead one to expect three to four $Eu^{+++}$ in the asymmetric unit. Comparison of the dispersive signals listed in the $D_{Sig}$ column is consistent with this expectation. The measured signals increase in both absolute magnitude and relative to the expected signals. This is expected since the B factors that engineer a fall off in intensity affect the inner shell electrons, which are responsible for the anomalous effects, less severely. However, some of this advantage is offset by the larger relative error levels in intensity measurements at higher resolutions.

### 8.3.4 Phase calculation and refinement

The first task is the determination of the positions of the anomalous scatterers.

$$|{}^0F_T(\mathbf{h})|, |{}^0F_A(\mathbf{h})| \text{ and } {}^0\phi_T(\mathbf{h}) - {}^0\phi_A(\mathbf{h}) = \Delta\phi$$

**Table 8.1** Observed and expected MAD signals for VCP

| Resolution (Å) | $B_{sig}$ (centric) | $D_{sig}$ ($R_{sym}$) |
|---|---|---|
| 4.20 | 0.092 (0.016) | 0.097 (0.011) |
| 3.33 | 0.097 (0.019) | 0.086 (0.027) |
| 2.91 | 0.114 (0.030) | 0.131 (0.024) |
| 2.85 | 0.173 (0.047) | 0.147 (0.051) |
| 2.46 | 0.191 (0.049) | 0.169 (0.066) |
| 2.31 | 0.215 (0.060) | 0.182 (0.083) |

| Expected signals for $N_A$ = | 1 | 2 | 3 | 4 | ($f'' = 10.7, f' = -18.3$ $N_P = 1400, Z_{eff} = 6.7$) |
|---|---|---|---|---|---|
| Bijvoet | 0.037 | 0.052 | 0.064 | 0.074 | |
| Dispersive | 0.032 | 0.045 | 0.055 | 0.063 | |

$$B_{sig} = \frac{\sum_{Nref} |F^{+Peak} - F^{-Peak}|}{1/2 \sum_{Nref} (F^{+Peak} + F^{-Peak})} \quad D_{sig} = \frac{\sum_{Nref} |F^{Edge} - F^{Remote}|}{1/2 \sum_{Nref} (F^{Edge} + F^{Remote})}$$

$$R_{sym} = \frac{\sum_{Nref} |I - \langle I \rangle|}{\sum_{Nref} I} \quad \text{where } \langle I \rangle = \sum_{Nref} I_i/\text{Nobs}$$

In the original, algebraic implementation, this was done by determination of the three unknown quantities through least squares minimization of the MAD equation:

$$\begin{aligned} |^{\lambda}F_T(\mathbf{h})|^2 &= |^0F_T(\mathbf{h})|^2 + a(\lambda)|^0F_A(\mathbf{h})|^2 \\ &+ b(\lambda)|^0F_T(\mathbf{h})||^0F_A(\mathbf{h})| \cos\left[^0\phi_T(\mathbf{h}) - {}^0\phi_A(\mathbf{h})\right] \\ &+ c(\lambda)|^0F_T(\mathbf{h})||^0F_A|(\mathbf{h}) \sin\left[^0\phi_T(\mathbf{h}) - {}^0\phi_A(\mathbf{h})\right] \end{aligned}$$

Making the following substitutions,

$$p_1 = |^0F_T(\mathbf{h})|^2, p_2 = |^0F_A(\mathbf{h})|^2,$$

$$p_3 = |^0F_T(\mathbf{h})||^0F_A(\mathbf{h})| \cos\left[^0\phi_T(\mathbf{h}) -^0 \phi_A(\mathbf{h})\right]$$

and

$$p_4 = |^0F_T(\mathbf{h})||^0F_A(\mathbf{h})| \sin\left[^0\phi_T(\mathbf{h}) -^0 \phi_A(\mathbf{h})\right]$$

the MAD equation can written as

$$|^{\lambda}F_T(\mathbf{h})|^2 = p_1 + a(\lambda)p_2 + b(\lambda)p_3 + c(\lambda)p_4 = Y_c$$

From which the $p$'s can be determined by fitting the measured Bijvoet and dispersive differences, along with the knowledge of the $f'$ and $f''$ values, as described in Krishna Murthy (1996) and Hendrickson (1985). It follows that:

$$|^0F_T(\mathbf{h})| = \sqrt{p_1}, \ |^0F_A(\mathbf{h})| = \sqrt{p_2} \quad \text{and}$$

$$\Delta\phi =^0 \phi_T(\mathbf{h}) -^0 \phi_A(\mathbf{h}) = \tan^{-1}(p_4/p_3)$$

The $|^0F_A(\mathbf{h})|$ values thus derived can be used to determine the positions of anomalous scatterers through computation of a Patterson synthesis, or by other methods. This step leads to the knowledge of $^0\phi_A(\mathbf{h})$, and since $\Delta\phi =^0 \phi_T(\mathbf{h}) -^0\phi_A(\mathbf{h})$, the $^0\phi_T(\mathbf{h})$ for each reflection can be computed. All of these steps are implemented in the MADSYS (Hendrickson, 1991) system of programs.

In implementation of MAD as a special case of MIR, the well developed theoretical foundation for MIR (Blundell and Johnson, 1976), and the program system that has long been in use can directly be used (Ramakrishnan and Biou, 1997). The anomalous scatterer positions are determined either from Patterson functions or from the use of direct methods, and the phases calculated and refined using a robust maximum likelihood target (La Fortelle and Bricogne, 1997). For example, the CCP4 program MLPHARE can be used for this purpose (Project, 1994). Some details from the structure determination of the dengue virus serotype 2 (Den2) protease complexed with the mung bean Bowman–Birk type inhibitor (MbBBI) will be used as an example (Murthy, 2000). The asymmetric unit of the crystals, space group $P2_12_12_1$, consists of two molecules of the enzyme, in complex with one molecule of the inhibitor. The asymmetric unit, overall has a two-fold non-crystallographic symmetry which is very approximate for the inhibitor and somewhat better for the two protease molecules. The MbBBI sequence shows evidence of having evolved through gene duplication, leading to inexact, but discernible, sequence symmetry (Chen *et al.*, 1992). One heavy atom derivative, by soaking in $HoC1_3$, was obtained, and turned out to be non-isomorphous with the native. MAD data were measured at three wavelengths at APS on the IMCA beam line at three wavelengths bracketing the LIII edge of $Ho^{+++}$ to a resolution of 2.1 Å (Terwilliger, 2002). Four $Ho^{+++}$ sites in the asymmetric unit were determined through Patterson functions, using SOLVE (Terwilliger, 2003), although the automated structure solution capabilities of the program were not used. The estimated $f'$ values at the three wavelengths were −18.4 (inflection point), −12.0 (Peak), and −9.0 (Remote). The positions, occupancies, and B factors of the sites determined were refined in MLPHARE (Project, 1994). The data at the inflection point were treated as the native data, because they had the most negative estimated $f'$, which would give positive real occupancies for the heavy atom sites in the MIR formulation

$$Q \propto f\,(F_{\text{PHi}}) - f(F_{\text{P}}) = f'_{\text{inflection}} - f'_{\text{Peak/Remote}}$$

of the problem. The anomalous occupancies are determined by the $f''$ values that are estimated at the different lengths. The heavy atom refinement produced an improvement from the initial figure of merit of 0.47 to a final one of 0.53. It was further improved using density modification and averaging, by exploiting the two-fold symmetry within the asymmetric unit, to 0.66. An electron density map using these phases could be readily interpreted in terms of the complete structure (Murthy *et al.*, 2000).

SOLVE/RESOLVE is a program system that permits automation of all the steps between processed data and interpretation of phased maps. These include scaling of data measured at multiple wavelengths, location of anomalous scatterers, phase calculation and refinement, phase improvement through density modification algorithms, and interpretation of resulting maps in terms of a molecular model (Terwilliger, 1997a, 2003, 2004). The approach implemented in this program system to the determination of anomalous scatterer positions is based on Bayesian statistics that takes into account correlated errors between measurements at different wavelengths. The theoretical basis of the approach is described in Terwilliger (1994a, 1997b, 1997c). The $|{}^0F_A(\mathbf{h})|$ and a quantity $\alpha$, which is closely related to $\Delta\phi$, are computed through a Bayesian probability estimate from the input MAD data (Terwilliger, 1994b). The anomalous scatterer positions implied by the $|{}^0F_A(\mathbf{h})|$ and $\alpha$ are verified by correlation of the calculated Patterson function and the observed functions, that are computed from the Bijvoet and Dispersive differences. Additional verification of the correctness of the solution is done through cross-validation difference Fourier maps. The quality of the obtained solutions is expressed as Z-scores, which compare the figures of merit for a particular solution with those for random values of estimated parameters. The score thus provides an estimate of the probability of having obtained the solution by chance. In the final calculation of phases, an error formulation, broadly based on the Blow and Crick error model (Blundell and Johnson, 1976), explicitly considers errors that are propagated to structure factor calculation at each wavelength due to inaccuracies in the anomalous scatterer model, and leads to maximum likelihood estimates for the $|{}^0F_T(\mathbf{h})|$ and ${}^0\phi_T(\mathbf{h})$. Native maps computed using these quantities are improved through density modification procedures. The SOLVE/RESOLVE system was used for the solution of the VCP structure, although solvent flattening, symmetry averaging, and building of the atomic model was done outside of it. The MAD data for the $Eu^{+++}$ derivative, statistics for which are listed in Table 8.1, were used. SOLVE suggested the presence of four $Eu^{+++}$ sites in the asymmetric unit, which is in substantial agreement with the estimate derived from comparison of expected signals with measured diffraction ratios detailed above (Table 8.1). The solution obtained had an overall Z-score of 29.3, with a figure of merit of 0.65 to a resolution of 2.2 Å. This improved to 0.78 on solvent flattening and non-crystallographic symmetry averaging of the density due to the two copies in the asymmetric unit. A model of the complete molecules could easily be built into a map computed with these phases (Murthy, 2001).

Although heavy atom or anomalous scatterer positions have traditionally determined from Patterson maps, alternative techniques are available. Especially in cases where the number of anomalous scatterers is large, direct methods have been used. In fact, in the determination of the low resolution structure of *Clostridium acdi-urici* ferredoxin through MAD phasing (Murthy, 1988), the positions of the iron-sulphur clusters was located using MULTAN (Germain *et al.*, 1970). Because the data measured from the crystals were limited to about 5 Å, no attempt could be made to obtain individual Fe and S atomic positions. A selected set of $|{}^{0}F_A(\mathbf{h})|$ values were input to the program for computation of E values and to obtain a very straightforward solution for the centroids of the two clusters in the asymmetric unit by using the phase set with the highest combined figure of merit. Although Patterson methods have been used to determine substructures containing as many as 50 atoms (Terwilliger, 1997a), Direct Methods are becoming increasingly popular for this purpose. For example, the 15 Se positions in selenomethionyl T7 DNA polymerase were determined (Doublie *et al.*, 1998) using SHELXS86 (Uson and Sheldrick, 1999). Newer programs such as SnB (Howell, 2000) have also had remarkable success in determining large substructures of anomalous scatterers in macromolecules. Thirty Se positions in selenomethionyl substituted *S*-adenosyl-homocysteine hydrolase (Turner *et al.*, 1998), 48 Se positions in the selenomethionyl EphB2 receptor (Thanos *et al.*, 1999), and 65 Se positions in selenomethionyl ADP-L-glycero-D-mannoheptose 6-epimerase (Deacon and Ealick, 1999) have all been successfully determined using SnB. The CCP4 program system also offers direct methods based techniques for determination of anomalous scatterer positions through the RANTAN (Yao, 1981) and ACORN (Yao, 2002) programs.

In most cases, structures are determined from MAD data measured on a single crystal. This makes the processed data inherently more precise and accurate, because scaling across crystals can be avoided. A second circumstance that contributes to the quality of MAD phases is the fact that the data come crystals that approximate true isomorphism more closely than MIR data. Nevertheless, there might be cases in which the anomalous signal might be too weak to provide a complete solution, although approximate phases can be obtained. Classically, it has been possible to combine structural information from more than one source through combination of suitably weighted phases. This is generally accomplished by multiplication of the probability distribution function for the phase from each source (Hendrickson and Lattman, 1970). A probability density function representation for MAD phases has been derived (Pahler *et al.*, 1990) and such functions have long been available for MIR and model phases (Blundell and Johnson, 1976). Currently the most widely used improvement technique for initial phases from any source is that of statistical density modification (Cowtan and Main, 1998; Cowtan and Zhang, 1999). SOLVE/RESOLVE has algorithms that implement such techniques (Terwilliger, 2001, 2002), as do programs in the CCP4 system (Project, 1994). Non-crystallographic symmetry can also be used to improve the quality of initial phases obtained from any source including MAD phases.

MAD phasing has, over the past decade, become one of the most widely used techniques to determine macromolecular structures. Its increasing appeal can be attributed to a succession of advances in the theoretical foundations, data measurement, processing, and analysis as well as parallel developments in refinement of initially obtained phases. Currently ongoing work, in extending data measurement at synchrotrons to longer wavelengths, is likely to permit the use of K absorption edges of native sulphur atoms in proteins and phosphorous in nucleic acids as potential MAD signal generators.

## Acknowledgements

Thanks are due to Abdul Ajees for help in preparation of the manuscript and to Larry DeLucas for support and encouragement. The research in my laboratory is supported by NIH grants AI45623 and AI51615 and by a Focused Giving grant from the Johnson & Johnson department of Corporate Science and Technology.

## References


Bella, J. and Rossmann, M. G. (1998). A general phasing algorithm for multiple MAD and MIR data. *Acta Crystallogr. D* **54** , 159–174.

Blundell, T.J. and Johnson, L. N. (1976). *Protein Crystallography*. Academic Press, New York.

Borek, D., Minor, W. and Otwinowski, Z. (2003). Measurement errors and their consequences in protein crystallography. *Acta Crystallogr. D* **59**, 2031–2038.

Burla, M. C., *et al.* (2003). SAD or MAD phasing: location of the anomalous scatterers. *Acta Crystallogr. D* **59**, 662–669.

Burla, M. C., *et al.* (2004). MAD phasing: choosing the most informative wavelength combination. *Acta Crystallogr. D* **60**, 1683–1686.

Chen, P., *et al.* (1992). Reactive sites of an anticarcinogenic Bowman-Birk proteinase inhibitor are similar to other trypsin inhibitors. *J. Biol. Chem.* **267**, 1990–1994.

Cohen, A., *et al.* (2001). MAD phasing with krypton. *Acta Crystallogr. D* **57**, 233–238.

Cowtan, K. and Main, P. (1998). Miscellaneous algorithms for density modification. *Acta Crystallogr. D* **54**, 487–493.

Cowtan, K. D. and Zhang, K. Y. (1999). Density modification for macromolecular phase improvement. *Prog. Biophys. Mol. Biol.* **72**, 245–270.

Cromer, D. T. and Lieberman, D. (1970). Relativistic calculation of anomalous scattering factors for X-rays. *J. Chemical Physics* **53**, 1891–1898.

Cullis, R., *et al.* (1961). The structure of haemoglobin, VIII. A three-dimensional Fourier synthesis at 5.5 Å resolution: determination of the phase angles. *Proceed. Royal Soc. London Series A* **265**,15–38.

Dauter, Z. (2002). New approaches to high-throughput phasing. *Curr. Opin. Struct. Biol.* **12**, 674–678.

Dauter, Z., Dauter, M. and Rajashankar, K. R. (2000). Novel approach to phasing proteins: derivatization by short cryo-soaking with halides. *Acta Crystallogr. D* **56**, 232–237.

Deacon, A. M. and Ealick, S. E. (1999). Selenium-based MAD phasing: setting the sites on larger structures. *Structure Fold Des.* **7**, R161–166.

Doublie, S., *et al.* (1998). Crystal structure of a bacteriophage T7 DNA replication complex at 2.2 A resolution. *Nature* **391**, 251–258.

Ealick, S. E. (2000). Advances in multiple wavelength anomalous diffraction crystallography. *Curr. Opin. Chem. Biol.* **4**, 495–499.

Evans, G. and Pettifer, R. F. (2001). CHOOCH: A program for deriving anomalous scattering factors from X-ray flourescence spectra. *J. Appl. Crystallogr.* **34**, 82–86.

Fan, H. F., *et al.* (1990). Combining direct methods with isomorphous replacement or anomalous scattering data. VII. Ab initio phasing of one-wavelength anomalous scattering data from a small protein. *Acta Crystallogr. A* **46**, 935–939.

Fanchon, E. and Hendrickson, W. A. (1990). Effect of the anisotropy of anomalous scattering on the MAD phasing method. *Acta Crystallogr. A* **46**, 809–820.

Garman, E. (2003). 'Cool' crystals: macromolecular cryocrystallography and radiation damage. *Curr. Opin. Struct. Biol.* **13**, 545–551.

Garman, E. and Murray, J. W. (2003). Heavy-atom derivatization. *Acta Crystallogr. D* **59**, 1903–1913.

Gassner, N. C. and Matthews, B. W. (1999). Use of differentially substituted selenomethionine proteins in X-ray structure determination. *Acta Crystallogr. D* **55**, 1967–1970.

Germain, G., Main, P. and Woolfson, M. M. (1970). On the application of phase relationships to complex structures. II. Getting a good start. *Acta Crystallogr. B* **26**, 274–285.

Giacovazzo, C. and Siliqi, D. (2001). The method of joint probability distribution functions applied to MAD techniques. The two-wavelength case for acentric crystals. *Acta Crystallogr. A* **57**, 700–707.

Giacovazzo, C. and Siliqi, D. (2004). Phasing via SAD/MAD data: the method of the joint probability distribution functions. *Acta Crystallogr. D* **60**, 73–82.

Gonzalez, A. (2003). Optimizing data collection for structure determination. *Acta Crystallogr. D* **59**, 1935–1942.

Gu, Y. X., *et al.* (2001). Direct-method-aided phasing of MAD data. *Acta Crystallogr. D* **57**, 250–253.

Guss, J. M., *et al.* (1988). Phase determination by multiple-wavelength x-ray diffraction: crystal structure of a basic 'blue' copper protein from cucumbers. *Science* **241**, 806–811.

Hendrickson, W. (1998). Teaching an old bug new tricks. *Nature Biotechnol.* **16**, 910–911.

Hendrickson, W. A. (1985). Analysis of protein structure from diffraction measurements at multiple wavelengths. *Trans. Am. Crystallogr. Assoc.* **21**, 11–21.

Hendrickson, W. A. (1991). Determination of macromolecular structures from anomalous diffraction of synchrotron radiation. *Science* **254**, 51–58.

Hendrickson, W. A. and Lattman, E. E. (1970). Representation of phase probability distributions for simplified combination of independent phase information. *Acta Crystallogr. B* **26**, 136–143.

Hendrickson, W. A., Horton, J. R. and LeMaster, D. M. (1990). Selenomethionyl proteins produced for analysis by multiwavelength anomalous diffraction (MAD): a vehicle for direct determination of three-dimensional structure. *EMBO J.* **9**, 1665–1672.

Hendrickson, W. A., *et al.* (1986). Structure of D-selenolanthionine determined directly from multiwavelength anomalous diffraction of bremstrahlung. *Am. Crystallogr Assoc Series* **2** (Abstract) **14**, 48.

Hendrickson, W. A., *et al.* (1988). Crystallographic structure analysis of lamprey hemoglobin from anomalous dispersion of synchrotron radiation. *Proteins* **4**, 77–88.

Hendrickson, W. A., *et al.* (1989). Crystal structure of core streptavidin determined from multiwavelength anomalous diffraction of synchrotron radiation. *Proc. Natl. Acad. Sci. USA* **86**, 2190–2194.

Howell, P. L., *et al.* (2000). Optimizing DREAR and SnB parameters for determining Se-atom substructures. *Acta Crystallogr. D* **56**, 604–617.

James, R. W. (1982). *The Optical Principles of the Diffraction of X-rays*. Oxbow Press, Woodbridge, CT.

Karle, J. (1980). Some developments in anomalous dispersion for the structural investigation of macromolecular systems in biology. *Int. J. Quantum Chem.* (Quantum Biol. Symp.) **7**, 357–367.

Korolev, S., *et al.* (2001). Using surface-bound rubidium ions for protein phasing. *Acta Crystallogr. D* **57**, 1008–1012.

Krishna Murthy, H. M. (1996). The use of multiple wavelength anomalous diffraction in ab initio phase determination. *Method Mol. Biol.* **56**, 127–152.

Krishna Murthy, H. M., *et al.* (1999). Crystallization, characterization and measurement of MAD data on crystals of dengue virus NS3 serine protease complexed with mung-bean Bowman-Birk inhibitor. *Acta Crystallogr D* **55**, 1370–1372.

La Fortelle, E. D. and Bricogne, G. (1997). Maximum-likelihood heavy-atom parameter refinement for multiple isomorphous replacement and multiwavelength anomalous diffraction methods. *Method Enzymol.* **276**, 472–494.

Liu, Y., Ogata, C. M. and Hendrickson, W. A. (2001). Multiwavelength anomalous diffraction analysis at the M absorption edges of uranium. *Proc. Natl. Acad. Sci. USA* **98**, 10648–10653.

Lustbader, J. W., *et al.* (1995). The expression, characterization, and crystallization of wild-type and selenomethionyl human chorionic gonadotropin. *Endocrinology* **136**, 640–650.

Merritt, E. A. (1998). X-ray anomalous scattering. http://skuld.bmsc.washington.edu/scatter/

Murthy, H. M., *et al.* (1988). Crystal structure of Clostridium acidi-urici ferredoxin at 5-A resolution based on measurements of anomalous X-ray scattering at multiple wavelengths. *J. Biol. Chem.* **263**, 18430–18436.

Murthy, H. M., *et al.* (2000). Crystal structure of Dengue virus NS3 protease in complex with a Bowman-Birk inhibitor: implications for flaviviral polyprotein processing and drug design. *J. Mol. Biol.* **301**, 759–767.

Murthy, H. M., Clum, S. and Padmanabhan, R. (1999). Dengue virus NS3 serine protease. Crystal structure and insights into interaction of the active site with substrates by molecular modeling and structural analysis of mutational effects. *J. Biol. Chem.* **274**, 5573–5580.

Murthy, K. H., *et al.* (2001). Crystal structure of a complement control protein that regulates both pathways of complement activation and binds heparan sulfate proteoglycans. *Cell* **104**, 301–311.

Narayan, R. and Ramaseshan, S. (1981). Optimum choice of wavelengths in the anomalous scattering technique with synchrotron radiation. *Acta Crystallogr. A* **37**, 636–641.

Ogata, C. M. (1998). MAD phasing grows up. *Nat. Struct. Biol.* **5** (Suppl.), 638–640.

Otwinowski, Z. and Minor, W. (1997). Processing of X-ray Diffraction Data Collected in the Oscillation Mode. *Method Enzymol.* **276A**, 307–326.

Otwinowski, Z., *et al.* (2003). Multiparametric scaling of diffraction intensities. *Acta Crystallogr. A* **59**, 228–234.

Pahler, A., Smith, J. L. and Hendrickson, W. A. (1990). A probability representation for phase information from multiwavelength anomalous dispersion. *Acta Crystallogr. A* **46**, 537–540.

Peterson, M. R., *et al.* (1996). MAD Phasing strategies explored with a brominated oligonucleotide crystal at 1.65Å resolution. *J. Synchrotron Radiation* **3,** 24–34.

Project, C. C. (1994). The CCP4 suite: programs for protein crystallography. *Acta Crystallogr. D* **50**, 760–763.

Quillin, M. L. and Matthews, B. W. (2003). Selling candles in a post-Edison world: phasing with noble gases bound within engineered sites. *Acta Crystallogr. D* **59**, 1930–1934.

Ramakrishnan, V. and Biou, V. (1997). Treatment of multiwavelength anomalous diffraction data as a special case of multiple isomorphous replacement. *Method Enzymol.* **276**, 538–557.

Ramakrishnan, V., *et al.* (1993). Crystal structure of globular domain of histone H5 and its implications for nucleosome binding. *Nature* **362**, 219–223.

Rudenko, G., *et al.* (2003). 'MAD'ly phasing the extracellular domain of the LDL receptor: a medium-sized protein, large tungsten clusters and multiple non-isomorphous crystals. *Acta Crystallogr. D* **59**, 1978–1986.

Smith, J. L. (1991). Determination of three-dimensional structure by multiwavelength anomalous diffraction. *Current Opin. Struct. Biol.* **1**, 1002–1011.

Strub, M. P., *et al.* (2003). Selenomethionine and selenocysteine double labeling strategy for crystallographic phasing. *Structure* (Camb) **11**, 1359–1367.

Teplova, M., *et al.* (2002). Covalent incorporation of selenium into oligonucleotides for X-ray crystal structure determination via MAD: proof of principle. Multiwavelength anomalous dispersion. *Biochimie* **84**, 849–858.

Terwilliger, T. C. (1994a). MAD phasing: Bayesian estimates of F(A). *Acta Crystallogr. D* **50**, 11–16.

Terwilliger, T. C. (1994b). MAD phasing: treatment of dispersive differences as isomorphous replacement information. *Acta Crystallogr. D* **50**, 17–23.

Terwilliger, T. (1997a). *SOLVE/RESOLVE*. http://www.solve.lanl.gov/

Terwilliger, T. C. (1997b). Bayesian correlated MAD phasing. *Acta Crystallogr. D* **53**, 571–579.

Terwilliger, T. C. (1997c). Multiwavelength anomalous diffraction phasing of macromolecular structures: analysis of MAD data as single isomorphous replacement with anomalous scattering data using the MADMRG Program. *Method Enzymol.* **276**, 530–537.

Terwilliger, T. C. (2001). Maximum-likelihood density modification using pattern recognition of structural motifs. *Acta Crystallogr. D* **57**, 1755–1762.

Terwilliger, T. C. (2002). Automated structure solution, density modification and model building. *Acta Crystallogr. D* **58**, 1937–1940.

Terwilliger, T. C. (2003). SOLVE and RESOLVE: automated structure solution and density modification. *Method Enzymol.* **374**, 22–37.

Terwilliger, T. (2004). SOLVE and RESOLVE: automated structure solution, density modification and model building. *J. Synchrotron Radiat.* **11**, 49–52.

Terwilliger, T. and Eisenberg, D. (1987). Isomorphous replacement: effects of errors on the phase probability distribution. *Acta Crystallogr A* **43**, 6–13.

Thanos, C. D., Goodwill, K. E. and Bowie, J. U. (1999). Oligomeric structure of the human EphB2 receptor SAM domain. *Science* **283**, 833–836.

Turner, M. A., *et al.* (1998). Structure determination of selenomethionyl S-adenosylhomocysteine hydrolase using data at a single wavelength. *Nat. Struct. Biol.* **5**, 369–376.

Urs, U. K., Murali, R. and Krishna Murthy, H. M. (1999). Structure of taq DNA polymerase shows a new orientation for the structure-specific nuclease domain. *Acta Crystallogr. D* **55**, 1971–1977.

Uson, I. and Sheldrick, G. M. (1999). Advances in direct methods for protein crystallography. *Curr. Opin. Struct. Biol.* **9**, 643–648.

Wu, H., *et al.* (1994). Structure of human chorionic gonadotropin at 2.6 A resolution from MAD analysis of the selenomethionyl protein. *Structure* **2**, 545–558.

Yao, J. X. (2002). ACORN in CCP4 and its applications. *Acta Crystallogr. D* **58**, 1941–1947.

Yao, J. X. (1981). On the application of phase relationships to complex structures. XVIII. RANTAN-random MULTAN. *Acta Crystallogr. A* **37**, 642–644.

CHAPTER 9

# Application of direct methods to macromolecular structure solution

**Charles M. Weeks and William Furey**

## 9.1 Introduction

The phasing technique known as direct methods is a powerful tool for determining protein structures. It has been shown that these methods, as implemented in computer programs such as *SnB* (Miller *et al.*, 1994; Weeks and Miller, 1999), *SHELXD* (Schneider and Sheldrick, 2002), and *SIR2000* (Burla *et al.*, 2000), are capable of solving small protein structures provided that a single set of unique diffraction data has been measured accurately to atomic resolution (1.2 Å or better). Such applications have been made to structures containing as many as 2000 independent non-hydrogen atoms (Frazão *et al.*, 1999). Successful applications in the 1.2–1.7 Å range have been reported occasionally (Chowdhury *et al.*, 2005; Burla *et al.*, 2003) and are more likely when a number of sulphur (or heavier) atoms are present. In addition, by using helical fragments as a starting point, the direct-methods program *ACORN* (Foadi *et al.*, 2000) has been able to phase small proteins at 1.45 Å (Rajakannan *et al.*, 2004b) and 1.9 Å (Rajakannan *et al.*, 2004a) resolution.

In most cases, however, the protein molecules are larger or the resolution of the data is lower, and phasing becomes a two-stage process. If two or more intensity measurements are available for each reflection with differences arising only from some property of a small substructure, the positions of the substructure atoms can be found first, and then the substructure can serve as a bootstrap to initiate the phasing of the complete structure. Suitable substructures may consist of heavy atoms soaked into a crystal in an isomorphous replacement experiment, or they may consist of the set of atoms that exhibit resonant or anomalous scattering when the wavelength of the incident radiation is near the absorption edge of a particular element. In the case of single isomorphous replacement (SIR) or single anomalous dispersion (SAD) data, the application of direct methods to difference structure-factor amplitudes $|\Delta F|$ iso $(= ||F_{PH}| - |F_P||)$ and $|\Delta F|$ ano $(= ||F+| - |F-||)$, respectively, is an excellent way to locate the heavy-atom sites. (The notation used for the structure factors is $F_P$ (native protein), $F_{PH}$ (derivative), $\Delta F$ or $F_A$ (substructure), $F+$ and $F-$ (for respectively, in the presence of anomalous dispersion)). Multiple isomorphous replacement (MIR) data can be accommodated simply by treating the data separately for each derivative, and multiple-wavelength anomalous dispersion (MAD) data can be handled by examining the anomalous differences for each wavelength individually or by combining them together in the form of $F_A$ structure factors (Karle, 1989; Hendrickson, 1991). The dispersive differences between two wavelengths of MAD data also can be treated as pseudo-SIR differences. Since heavy atoms are rarely closer than 3–4 Å, the resolution of the data typically collected for isomorphous replacement or MAD experiments is sufficient for direct-methods determination of substructures. At least a dozen proteins with selenomethionine (SeMet) substructures containing more than 50 Se sites have been solved by direct methods, and the largest of these, ketopantoate hydroxymethyltransferase, has 160 Se sites (von Delft *et al.*, 2003).

Since the great majority of direct-methods applications to macromolecules involve the solution of a substructure as an intermediate step, this is the type

**Table 9.1** Test data: methylmalonyl-coenzyme A epimerase from *Propionibacterium shermanii*

| | |
|---|---|
| PDB accession code | 1JC4 |
| Derivative | Selenomethionine |
| Space group | $P2_1$ |
| Cell constants | $a = 43.60$, $b = 78.62$, $c = 89.43$, $\beta = 91.95$ |
| Asymmetric unit contents: | |
| number of chains | 4 (identical) |
| residues per chain | 148 |
| S containing residues per chain | 7 Met and 2 Cys |
| substructure | Se24 |
| Matthews coefficient ($V_m$) | 2.01 Å/Dalton |
| Solvent fraction | 0.36 |
| MAD data: | |
| maximum resolution | 2.1 Å |
| wavelengths* | IP (0.9793), PK (0.9792), HR (0.9184) |
| $f'$ | −9.16, −7.49, −0.92 |
| $f''$ | 8.20, 8.22, 4.23 |

*Wavelength abbreviations are IP (inflection point), PK (peak), and HR (high-energy remote).

of application that will be described in this chapter. Although any of the common direct-methods programs could be used to phase substructures, certain programs are more convenient because they are part of a program pipeline that makes it possible not only to determine the substructure, but also to phase the protein itself and possibly to perform other downstream operations. Pipelines increase the potential for automation and, therefore, higher throughput. Examples of program packages that have pipelines involving direct methods are the following: (a) the authors' own program *BnP* (Weeks *et al.*, 2002) consisting of the subprograms *DREAR* (Blessing and Smith, 1999), *SnB* (Weeks and Miller, 1999), and components of the PHASES suite (Furey and Swaminathan, 1997); (b) the set of programs *XPREP* (Bruker AXS, 2005), *SHELXD* (Schneider and Sheldrick, 2002), and *SHELXE* (Sheldrick, 2002) written by George Sheldrick; (c) the *PHENIX* package (Adams *et al.*, 2004) which includes the direct-methods program *HySS* (Grosse-Kunstleve and Adams, 2003); and (d) *autoSHARP* (Bricogne *et al.*, 2003).

In the following sections, the steps required to carry out the two-stage phasing process for proteins are described in detail and illustrated through the application of *BnP* to the MAD data set for the selenomethionine derivative of methylmalonyl-coenzyme A epimerase from *P. shermanii* (PDB accession code 1JC4) (McCarthy *et al.*, 2001). Background information about this enzyme and its data set is summarized in Table 9.1. The *BnP* program has two operational modes, automatic and manual. In automatic mode, which is geared to routine high-throughput applications, the user needs only to specify a few parameters, and the entire two-stage phasing process from substructure determination through phase refinement and solvent flattening is chained together and started by clicking a single button. On the other hand, manual mode is available for large structures or difficult problems with marginal data, and it allows the user to control many parameters and to execute the major steps in the phasing process sequentially.

## 9.2 Data preparation

As will be described in Section 9.3, direct methods are techniques that use probabilistic relationships among the phases to derive values of the individual phases from the experimentally measured amplitudes. In order to take advantage of these relationships, a necessary first step is the replacement of the usual structure factors, $F$, by the normalized structure factors (Hauptman and Karle, 1953),

$$|E_{\mathbf{H}}| = |F_{\mathbf{H}}| \Big/ \left[ \varepsilon_{\mathbf{H}} \sum_{j=1}^{N} f_j^2 \right] \tag{1}$$

where the $f_j$ are the scattering factors for the $N$ atoms in the unit cell and the integers $\varepsilon_{\mathbf{H}}$ (Shmueli and Wilson, 1996) correct for the space-group-dependent higher average intensities of some groups of reflections. The quantity $\langle |E|^2 \rangle$ is always unity for the whole data set, hence the term 'normalized'. The $F$s express the scattering from real atoms with a finite size whereas $E$s represent scattering from point atoms at rest, and the effect of dividing by a function of $f_j$ is to eliminate any fall-off of intensity as a function of $\sin(\theta)/\lambda$. The distribution of $|E|$ values is, in principle, independent of the unit cell size and contents, but it does depend on whether a centre of symmetry is present. Statistics describing the distribution of $|E|$ values for the peak-wavelength data

**Table 9.2** Normalized structure-factor magnitude statistics for the peak-wavelength data for methylmalonyl-coA epimerase (1JC4)

| | Experimental | | Theoretical | |
|---|---|---|---|---|
| | Acentric | Centric | Acentric | Centric |
| $\langle\|E\|\rangle$ | 0.885 | 0.729 | 0.866 | 0.798 |
| $\langle\|E\|^2\rangle$ | 1.002 | 0.834 | 1.000 | 1.000 |
| $\langle\|\|E\|^2-1\|\rangle$ | 0.757 | 0.903 | 0.736 | 0.968 |
| Fraction $\|E\| \geq 1$ | 0.351 | 0.258 | 0.368 | 0.320 |
| Fraction $\|E\| \geq 2$ | 0.023 | 0.034 | 0.018 | 0.050 |
| Fraction $\|E\| \geq 3$ | 0.0003 | 0.002 | 0.0001 | 0.003 |

for methylmalonyl-coenzyme A epimerase (1JC4) are shown in Table 9.2. Comparison of the observed experimental values to the theoretical values for centric and acentric data shows how closely the observed distribution matches the expected.

Normalization can be accomplished simply by dividing the data into concentric resolution shells, taking the epsilon factors into account, and applying the condition $\langle|E|^2\rangle = 1$ to each shell. Alternatively, a least-squares-fitted scaling function can be used to impose the normalization condition. The procedures are similar regardless of whether the starting information consists of $|F|$, $|\Delta F|$ (iso or ano), or $|F_A|$ values and leads to $|E|$, $|E_\Delta|$, or $|E_A|$ values. Mathematically precise definitions of the SIR and SAD difference magnitudes, $|E_\Delta|$, that take into account the atomic scattering factors $|f_j| = |f_j^0 + f_j' + if_j''|$ have been derived and are implemented in the program *DIFFE* (Blessing and Smith, 1999) that is distributed as part of the *DREAR* component of the *SnB* and *BnP* packages. Alternatively, $|E_A|$ values can be derived from $|F_A|$ values using the *XPREP* program (Sheldrick) or the *MADBST* component of *SOLVE* (Terwilliger and Berendzen, 1999).

Direct methods are notoriously sensitive to the presence of even a small number of erroneous measurements. This is especially problematical for difference data where the quantities used involve small differences between two much larger measurements such that errors in the measurements can easily disguise the true signal. When using MAD or SAD data to locate anomalous scatterers, it is important not to include high-resolution data that lack a significant anomalous or dispersive signal. Since the use of normalized structure factors emphasizes high-resolution data, direct methods are especially sensitive to noise in this data. Fortunately, very high-resolution data are generally not required to find substructures, and a high-resolution cut-off of 3 Å is typical. Since there is some anomalous signal at all the wavelengths in a MAD experiment, a good test is to calculate the correlation coefficient between the signed anomalous differences $\Delta F$ at different wavelengths as a function of the resolution. A good general rule is to truncate the data where this correlation coefficient falls below 25–30%.

One of the best ways to ensure accuracy is to measure highly redundant data. Care should be taken to eliminate outliers and observations with small signal-to-noise ratios before initiating the phasing process. Fortunately, it is usually possible to be stringent in the application of cut-offs because the number of difference reflections that are available from a protein-sized unit cell is typically much larger than the number of heavy-atom positional parameters that must be determined for a substructure. In fact, only 2–3% of the total possible reflections at 3 Å need be phased in order to solve substructures using direct methods, but these reflections must be chosen from those with the largest $|E_\Delta|$ values as will be discussed further in Section 9.3.

The *DIFFE* program (Blessing and Smith, 1999) rejects data pairs ($|E_1|$, $|E_2|$) [i.e. SIR pairs ($|E_P|$, $|E_{PH}|$), SAD pairs ($|E+|$, $|E-|$), and pseudo-SIR dispersive pairs ($|E_{\lambda1}|$, $|E_{\lambda2}|$)] or difference $E$ magnitudes ($|E_\Delta|$) that are not significantly different from zero or deviate markedly from the expected distribution. The following tests are applied where the default values for the cut-off parameters ($T_{MAX}$, $X_{MIN}$, $Y_{MIN}$, $Z_{MIN}$, and $Z_{MAX}$), are shown in parentheses and are based on empirical tests with known data sets (Smith *et al.*, 1998; Howell *et al.*, 2000).

1. Pairs of data are excluded if $|(|E_1| - |E_2|) - \text{median}(|E_1| - |E_2|)|/\{1.25 * \text{median}[|(|E_1| - |E_2|) - \text{median}(|E_1| - |E_2|)|]\} > T_{MAX}$ (6.0).
2. Pairs of data are excluded for which either $|E_1|/\sigma(|E_1|)$ or $|E_2|/\sigma(|E_2|) < X_{MIN}$ (3.0).
3. Pairs of data are excluded if $||E_1|-|E_2||/[\sigma^2(|E_1|)+\sigma^2(|E_2|)]^{1/2} < Y_{MIN}$ (1.0).

**Table 9.3** Number of peak-wavelength anomalous difference data pairs for 1JC4 remaining after successive application of the various DIFFE significance tests

| | Number |
|---|---|
| Unique reflection data ($\|E\|$) | 66, 122 |
| Total anomalous reflection pairs ($\|E_D\|$) | 28, 399 |
| Data pairs passing $T_{MAX}$ test | 28, 364 |
| Data pairs passing $X_{MIN}$ test | 26, 898 |
| Data pairs passing $Y_{MIN}$ test | 13, 790 |
| Data pairs passing $Z_{MIN}$ and $Z_{MAX}$ tests | 2474 |

**4.** Normalized $|E_\Delta|$ are excluded if $|E_\Delta|/\sigma(|E_\Delta|) < Z_{MIN}$ (3.0).

**5.** Normalized $|E_\Delta|$ are excluded if $[|E_\Delta| - |E_\Delta|_{MAX}]/\sigma(|E_\Delta|) > Z_{MAX}$ (0.0).

The parameter $T_{MAX}$ is used to reject data with unreliably large values of $||E_1| - |E_2||$ in the tails of the $(|E_1| - |E_2|)$ distribution. This test assumes that the distribution of $(|E_1| - |E_2|)/\sigma(|E_1| - |E_2|)$ should approximate a zero-mean unit-variance normal distribution for which values less than $-T_{MAX}$ or greater than $+T_{MAX}$ are extremely improbable. The quantity $|E_\Delta|_{MAX}$ is a physical least upper bound such that $|E_\Delta|_{MAX} = \Sigma|f|/[\varepsilon\Sigma|f|^2]^{1/2}$ for SIR data and $|E_\Delta|_{MAX} = \Sigma f''/[\varepsilon\Sigma(f'')^2]^{1/2}$ for SAD data. Table 9.3 shows the number of useable reflections remaining after applying the *DIFFE* significance tests to the peak-wavelength anomalous differences for 1JC4.

## 9.3 Substructure phasing

The phase problem of X-ray crystallography may be defined as the problem of determining the phases $\phi$ of the normalized structure factors $E$ when only the magnitudes $|E|$ are given. Since there are many more reflections in a diffraction pattern than there are independent atoms in the corresponding crystal, the phase problem is overdetermined, and the existence of relationships among the measured magnitudes is implied. Direct methods (Hauptman and Karle, 1953) are *ab initio* probabilistic methods that seek to exploit these relationships, and the techniques of probability theory have identified the linear combinations of three phases whose Miller indices sum to zero (i.e. $\Phi_{\mathbf{HK}} = \phi_{\mathbf{H}} + \phi_{\mathbf{K}} + \phi_{-\mathbf{H}-\mathbf{K}}$) as relationships useful for determining unknown structures. (The quantities $\Phi_{\mathbf{HK}}$ are known as structure invariants because their values are independent of the choice of origin of the unit cell.) The conditional probability distribution,

$$P(\Phi_{\mathbf{HK}}) = \left[2\pi I_0(A_{\mathbf{HK}})\right]^{-1} \exp(A_{\mathbf{HK}} \cos \Phi_{\mathbf{HK}}), \quad (2)$$

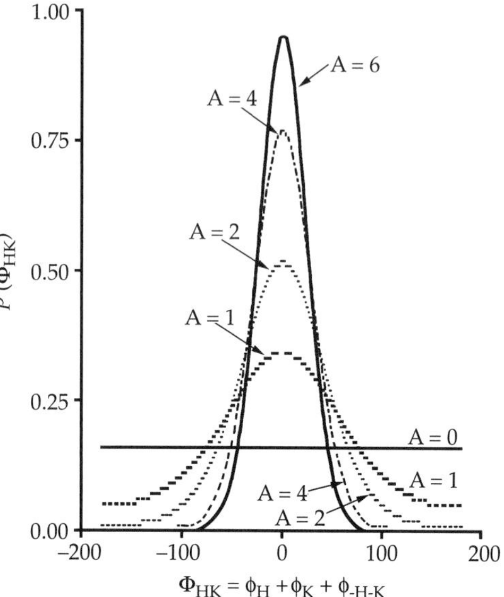


**Figure 9.1** The conditional probability distribution, $P(\Phi_{\mathbf{HK}})$, of the three-phase structure invariants, $\Phi_{\mathbf{HK}}$, having associated parameters $A_{\mathbf{HK}}$ with values of 0, 1, 2, 4, and 6. When $A \approx 0$, all values of $\Phi_{\mathbf{HK}}$ are equally likely, and no information useful for phase determination is available. However, the sum of the three phases for most invariants with $A \approx 6$ is close to 0°, and an estimate of one phase can be made if the other two are known.

of the three-phase or triplet invariants is illustrated in Fig. 9.1, and it depends on the parameter $A_{\mathbf{HK}}$ where

$$A_{\mathbf{HK}} = \left(2/N^{1/2}\right) |E_{\mathbf{H}} E_{\mathbf{K}} E_{\mathbf{H}+\mathbf{K}}| \quad (3)$$

and $N$ is the number of non-hydrogen atoms, here presumed to be identical, in the asymmetric unit (Cochran, 1955). When normalized difference magnitudes ($|E_\Delta|$) are being used, $N$ is the number of heavy or anomalously scattering atoms comprising the substructure. From Fig. 9.1, it is clear that the probabilistic estimate of an invariant value is most reliable when $A_{\mathbf{HK}}$ is large, and this occurs when the associated normalized magnitudes ($|E_{\mathbf{H}}|$, $|E_{\mathbf{K}}|$, and $|E_{-\mathbf{H}-\mathbf{K}}|$) are large or the number of atoms in the unit cell is small. Thus, it is the largest $|E_{\Delta}|$ or $|E_{\mathbf{A}}|$, remaining after the application of all appropriate cut-offs, that need to be phased in direct-methods substructure determinations. The triplet invariants involving these reflections are generated, and a sufficient number of those invariants with the highest $A_{\mathbf{HK}}$ values are retained to achieve an adequate degree of overdetermination of the phases by keeping the invariant-to-reflection ratio sufficiently large (e.g. 10:1). The inability to obtain a sufficient number of accurate invariant estimates is the reason why full-structure phasing by direct methods is possible only for the smallest proteins.

### 9.3.1 Multiple trial procedures

*Ab initio* phase determination by direct methods requires not only some information about the likely values of a set of invariants, but also a set of starting phases. Once the values for some pairs of phases ($\phi_{\mathbf{K}}$ and $\phi_{-\mathbf{H}-\mathbf{K}}$) are available, the triplet structure invariants can be used to generate further phases ($\phi_{\mathbf{H}}$ which, in turn, can be used iteratively to evaluate still more phases. The number of cycles of phase expansion or refinement that must be performed to get an adequate number of sufficiently accurate phases depends on the size of the structure to be determined. To obtain starting phases, a so-called multisolution or multiple trial approach is taken in which the reflections are each assigned many different starting values in the hope that one or more of the resultant phase combinations will lead to a solution (Germain and Woolfson, 1968). Typically, a random-number generator is used to assign initial values to all phases from the outset (Baggio *et al.*, 1978). A variant of this procedure employed in *SnB* is to use the random-number generator to assign initial coordinates to the atoms in the trial structures and then to obtain initial phases from a structure-factor calculation. In *SHELXD* (Schneider and Sheldrick, 2002), the percentage of successful trial structures is increased by using better-than-random sets of starting coordinates that are, in some way, consistent with the Patterson function.

### 9.3.2 Phase refinement

Once a set of initial phases has been chosen, it must be refined against the set of structure invariants whose values are presumed to be known. So far, two optimization methods (tangent refinement and parameter-shift reduction of the minimal function) have proven useful for extracting phase information from the structure invariants. The tangent formula (Karle and Hauptman, 1956),

$$\tan(\phi_{\mathbf{H}}) = \frac{\sum_{\mathbf{K}} |E_{\mathbf{K}}E_{\mathbf{H}-\mathbf{K}}| \sin(\phi_{\mathbf{K}} + \phi_{\mathbf{H}-\mathbf{K}})}{\sum_{\mathbf{K}} |E_{\mathbf{K}}E_{\mathbf{H}-\mathbf{K}}| \cos(\phi_{\mathbf{K}} + \phi_{\mathbf{H}-\mathbf{K}})} \quad (4)$$

can be used to compute the value of a phase, $\phi_{\mathbf{H}}$, given a sufficient number of pairs ($\phi_{\mathbf{K}}$, $\phi_{-\mathbf{H}-\mathbf{K}}$) of known phases. By treating each reflection in turn as $\phi_{\mathbf{H}}$, an entire set of phases can be refined. Although the tangent formula is a powerful tool that has been widely used in direct-methods programs for 40 years, it suffers from the disadvantage that, in space groups without translational symmetry, it is perfectly fulfilled by a false solution with all phases equal to zero, thereby giving rise to the so-called 'uranium-atom' solution with one dominant peak in the corresponding Fourier synthesis. This problem can be largely overcome for small-molecule structures by including carefully selected higher-order invariants (quartets) in a modified tangent formula (Schenk, 1974; Hauptman, 1974).

Constrained minimization of an objective function like the minimal function (Debaerdemaeker and Woolfson, 1983; DeTitta *et al.*, 1994)

$$R(\Phi) = \frac{\sum_{\mathbf{H},\mathbf{K}} A_{\mathbf{HK}} \left[\cos \Phi_{\mathbf{HK}} - \frac{I_1(A_{\mathbf{HK}})}{I_0(A_{\mathbf{HK}})}\right]}{\sum_{\mathbf{H},\mathbf{K}} A_{\mathbf{HK}}} \quad (5)$$

provides an alternative approach to phase refinement or phase expansion. $R(\Phi)$ (also known as $R_{\text{MIN}}$) is a measure of the mean-square difference between the values of the cosines of the triplets ($\mathbf{\Phi}_{\mathbf{HK}}$) calculated using the current set of phases and the expected probabilistic values of the same quantities as given by the ratio of modified Bessel functions, $I_1(A_{\mathbf{HK}})/I_0(A_{\mathbf{HK}})$. The minimal function is expected to have a constrained global minimum when the phases are equal to their correct values for some choice of origin and enantiomorph. An algorithm known as parameter shift (Bhuiya and

Stanley, 1963), has proved to be quite powerful and efficient as an optimization method when used to reduce the value of the minimal function. For example, a typical phase-refinement stage consists of three iterations or scans through the reflection list, with each phase being shifted a maximum of two times by 90° in either the positive or negative direction during each iteration. The refined value for each phase is selected, in turn, through a process that involves evaluating the minimal function using the original phase and each of its shifted values (Weeks *et al.*, 1994). The phase value that results in the lowest minimal-function value is chosen at each step. Refined phases are used immediately in the subsequent refinement of other phases.

### 9.3.3 Dual-space optimization

Conventional reciprocal-space direct methods, implementing tangent-formula refinement in computer programs such as *MULTAN* (Main *et al.*, 1980) and *SHELXS* (Sheldrick, 1990), can provide solutions for structures containing less than 100–150 unique non-hydrogen atoms. With few exceptions, however, phase refinement alone is not sufficient to solve larger structures. In such cases, successful applications require a dual-space optimization procedure that has come to be known as *Shake-and-Bake* (Weeks *et al* 1994; Miller *et al.*, 1993). *Shake-and-Bake* is also a powerful method for smaller structures and substructures, effectively avoiding most cases of false minima. The distinctive feature of this procedure is the repeated and unconditional alternation of reciprocal-space phase refinement (*Shaking*) with a complementary real-space process that seeks to improve phases by applying constraints (*Baking*). The *Shake-and-Bake* algorithm was implemented first in *SnB* and then independently in *SHELXD*. Although both reciprocal-space optimization methods are available in both programs,

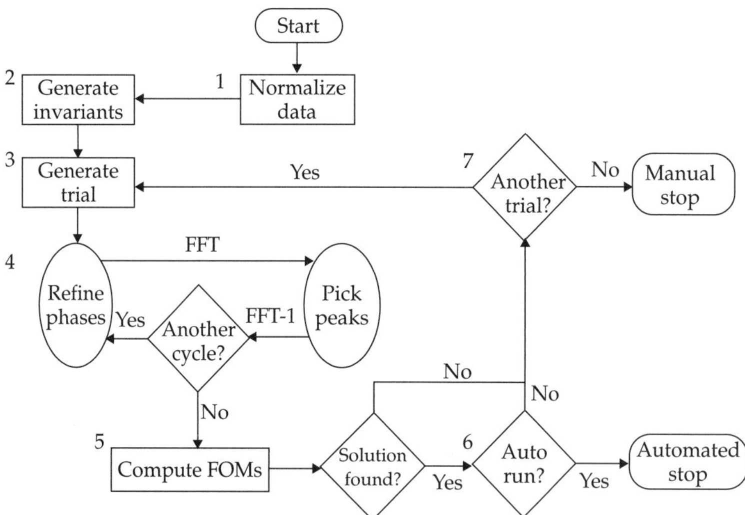


**Figure 9.2** A flow chart for structure solution using the direct-methods algorithm, *Shake-and-Bake*. Steps 1 and 2 involve computation of the normalized structure-factor magnitudes and generation of the triplet structure invariants, respectively. Step 3 is the start of the outer or trial structure loop in which trial structures consisting of randomly positioned atoms are generated and tested. Step 4 is the inner or *Shake-and-Bake* cycle which consists of a phase-refinement step involving either tangent-formula refinement or parameter-shift optimization of the minimal function, a Fourier transform resulting in an electron-density map, selection of the largest peaks on the map, and an inverse Fourier transform to generate new phases by means of a structure-factor calculation assuming these peaks to be atoms. The number of such cycles and the number of peaks used in the structure-factor calculation depend on the number of atoms to be located (Table 9.4). Step 5 is the evaluation of the refined trial structures based on the values of the figures of merit (see Section 9.4) to determine which trials, if any, are solutions. If solutions are being recognized automatically (Step 6), the substructure determination is terminated as soon as the first solution is found. Otherwise, processing continues (Step 7) until a predetermined number of trial structures are completed.

*SnB* uses the parameter-shift optimization of the minimal function by default whereas *SHELXD* uses the tangent formula. The complete direct-methods phasing process using the *Shake-and-Bake* algorithm is illustrated in flow-chart form in Fig. 9.2, and the main parameters are summarized in Table 9.4.

### 9.3.4 Peak picking

Peak picking is a simple but powerful way of imposing a real-space atomicity constraint. In each cycle of the *Shake-and-Bake* procedure, the largest peaks are used as an up-dated trial structure without regard to chemical constraints other than a minimum allowed distance between atoms (e.g. 1 Å for full structures and 3 Å for substructures). By analogy to the common practice in macromolecular crystallography of omitting part of a structure from a Fourier calculation in the hope of finding an improved position for the deleted fragment, a variant of the peak picking procedure in which approximately one-third of the largest peaks are randomly omitted in each cycle can also be used (Schneider and Sheldrick, 2002).

If markedly unequal atoms are present, appropriate numbers of peaks (atoms) can be weighted by the proper atomic numbers during transformation back to reciprocal space in a subsequent structure-factor calculation. Thus, *a priori* knowledge concerning the chemical composition of the crystal is used, but no knowledge of constitution is required or used during peak selection. It is useful to think of peak picking in this context as simply an extreme form of density modification appropriate when the resolution of the data is small compared to the distance separating the atoms. The imposition of physical constraints counteracts the tendency of phase refinement to propagate errors or produce overly consistent phase sets. For example, the ability to eliminate chemically impossible peaks at special positions using a symmetry-equivalent cut-off distance prevents the occurrence of most false minima (Weeks and Miller, 1999).

**Table 9.4** Recommended values of the major *Shake-and-Bake* parameters for substructure applications expressed in terms of $N$, the expected number of heavy atoms or anomalous scatterers

| Parameter | Recommended value | 1JC4 Value |
|---|---|---|
| Phases | $30N$ | 840 |
| Triplet invariants | $300N$ | 8400 |
| Peaks selected | $N$ | 28 |
| Cycles 2 | $N$ | 56 |

The values used for the 1JC4 example are based on the sequence information that there are seven methionines per chain and that the Matthews coefficient and solvent fraction are consistent with four chains in the asymmetric unit; therefore, $N = 28$.

## 9.4 Recognizing solutions

Since direct methods are implemented in the form of multiple-trial procedures, it is essential to have some means of distinguishing the refined trial structures that are solutions from those that are not. The *SnB* program computes three figures of merit that are useful for this purpose. These quantities are the minimal function or $R_{\mathrm{MIN}}$ (Eq. 5) calculated directly from the constrained phases corresponding to the final peak positions for a trial structure, a crystallographic $R$ value based on normalized magnitudes [i.e., $R_{\mathrm{CRYST}} = (\Sigma||E_{\mathrm{OBS}}| - |E_{\mathrm{CALC}}||)/\Sigma|E_{\mathrm{OBS}}|$], and a correlation coefficient,

$$CC = \left[\sum wE^2_{\mathrm{OBS}}E^2_{\mathrm{CALC}}\cdot\sum w - \sum wE^2_{\mathrm{OBS}}\cdot\sum w \cdot E^2_{\mathrm{CALC}}\right] \Big/ \left\{\left[\sum wE^4_{\mathrm{OBS}}\cdot\sum w - \left(\sum wE^2_{\mathrm{OBS}}\right)^2\right] \cdot \left[\sum wE^4_{\mathrm{CALC}}\cdot\sum w - \left(\sum wE^2_{\mathrm{CALC}}\right)^2\right]\right\}^{1/2}, \quad (6)$$

between $|E_{\mathrm{OBS}}|$ and $|E_{\mathrm{CALC}}|$ where the weight $w$ is usually unity (Fujinaga and Read, 1987). The minimal function and the crystallographic $R$ have relatively small values for solutions, but the correlation coefficient is relatively large.

The use of figures of merit to determine whether solutions have been found and, if so, which trials were successful can be illustrated with the methylmalonyl-coA epimerase (1JC4) test data. A manual *BnP* job for 100 trial structures was run using the normalized anomalous difference magnitudes ($|E_\Delta|$) for the peak wavelength data. A maximum resolution cut-off of 3 Å was applied, and only reflections with a signal-to-noise ratio (i.e. $|E_\Delta|/\sigma(|E_\Delta|)$) greater than 3.0 were used. A histogram of the final $R_{\mathrm{MIN}}$ values for the 100 trials is shown in Fig. 9.3. A clear bimodal distribution of $R_{\mathrm{MIN}}$ values is a strong indication that a solution(s) exists. Confirmation that this is true for Trial 5 in this example can be obtained by inspecting a trace of its $R_{MIN}$ value as a function of refinement cycle (Fig. 9.4). Solutions usually show an abrupt decrease in value over a few cycles followed by stability at the lower value. Further confirmation can be obtained by looking for corroboration from the other figures of merit as illustrated in Table 9.5 where low $R_{MIN}$ values are correlated with low $R_{CRYST}$ values and high $CC$ values.

For high-throughput operation, solutions need to be recognized quickly and automatically, the direct-methods-based substructure determination process terminated once a solution has been found, and protein phasing initiated. Since the ranges of figure-of-merit values for solutions are not the same for all structures and data sets, a careful analysis is required to avoid false solutions and to minimize the number of missed solutions. In automatic mode, *BnP* looks for a trial having $R_{MIN}$ and $R_{CRYST}$ values significantly less than their respective means. In this context, 'significant' means differing by more than three standard deviations. The standard deviations are computed based on the figure-of-merit values for a minimum of five trials which have been preliminarily classified as non-solutions because their $R_{CRYST}$ values are greater than 0.32. As a result, *BnP* always processes at least six trials. As shown by the statistics in Table 9.6, Trial 5 in the 1JC4 example is a solution that can be recognized automatically.

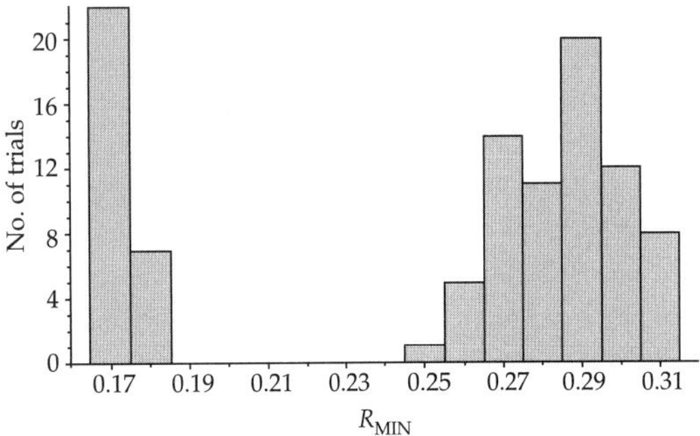


**Figure 9.3** A histogram of the final $R_{MIN}$ values for 100 trial structures for 1JC4 was made by dividing the range of the values (0.168–0.316) into 15 intervals (buckets) and counting the number of trials whose $R_{MIN}$ values fell into each bucket. The 29 trials in the two left buckets with the lowest $R_{MIN}$ values are solutions, and the trials in the buckets at the right are non-solutions.

## 9.5 Choosing correct sites

For each job involving an $N$-site substructure, the *SnB* component of *BnP* provides an output file of $1.5N$ peak positions for the best trial (based on its final $R_{MIN}$ value) sorted in descending order according to the corresponding electron density. It is then necessary to decide which, and how many, of these peaks correspond to actual atoms. The first $N$ peaks have the highest probability of being correct, and in

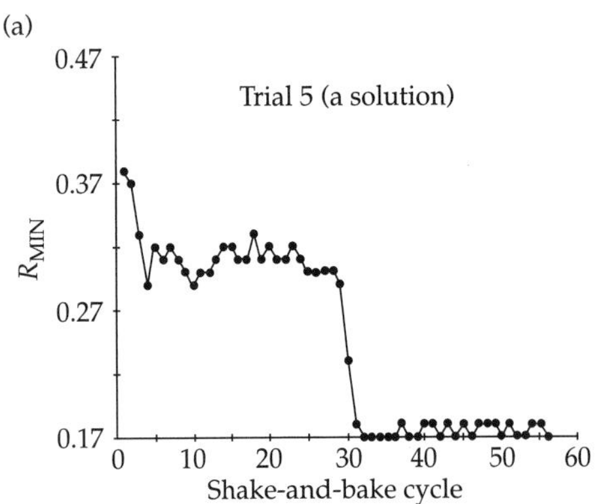


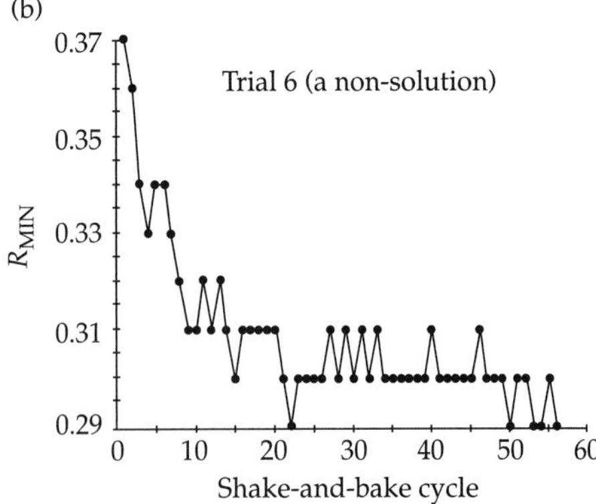


**Figure 9.4** Plots of the minimal-function value over 56 *Shake-and-Bake* cycles for (a) a solution and (b) a non-solution. Trial 5 illustrates the characteristic features of a solution – a small initial decrease in $R_{MIN}$ value followed by a plateau, a sharp decrease over a few cycles, and a second plateau at a lower value. The first plateau may be shortened significantly if convergence to a solution occurs quickly.

**Table 9.5** Figures of merit for Trials 1–10 in the 1JC4 example sorted in increasing order according to the value of $R_{MIN}$

| Trial | $R_{MIN}$ | $R_{CRYST}$ | CC |
|---|---|---|---|
| 7 | 0.174 | 0.27 | 0.42 |
| 8 | 0.174 | 0.25 | 0.44 |
| **5** | **0.177** | **0.27** | **0.45** |
| **4** | **0.265** | **0.39** | **0.38** |
| 3 | 0.275 | 0.38 | 0.39 |
| 10 | 0.286 | 0.42 | 0.39 |
| 6 | 0.295 | 0.40 | 0.35 |
| 9 | 0.297 | 0.40 | 0.30 |
| 2 | 0.306 | 0.40 | 0.34 |
| 1 | 0.309 | 0.41 | 0.30 |

Trials 5, 7, and 8 are solutions. Note the pronounced differences in figure-of-merit values between Trials 5 and 4.

**Table 9.6** Statistics of solution recognition for Trial 5

| | $R_{MIN}$ | $R_{CRYST}$ |
|---|---|---|
| $\langle R\rangle$ | 0.290 | 0.396 |
| $\sigma$ | 0.019 | 0.011 |
| $(\langle R\rangle - R_5)/\sigma$ | 5.85 | 11.05 |

The means and standard deviations were computed using Trials 1, 2, 3, 4, and 6.

some cases this simple guideline is adequate. Alternatively, a conservative approach is to accept the $0.8N$ to $0.9N$ top peaks.

Whether working in manual or automatic mode, occupancy refinement against one of the isomorphous, anomalous, or dispersive difference data sets is a good way to detect spurious sites. The occupancy of such sites will typically fall to very low values, and it has been found that, when refined against peak-wavelength (largest $f''$) anomalous differences, an occupancy cut-off of 0.2 distinguishes true sites from false sites and, in fact, most false sites refine to occupancies less than 0.05. This refinement is very rapid and has the advantage of being hand insensitive. Thus, it need be carried out only once prior to enantiomorph determination (see Section 9.6).

The results of occupancy refinement for Trial 5 in the 1JC4 example are given in Table 9.7. Of the 25 highest peaks, 24 pass the occupancy cut-off (peak 20 is spurious), and these 24 peaks do indeed correspond to correct Se sites. Although the sequence of this protein suggests that there may be seven SeMet residues in each of the four monomers in the asymmetric unit (Table 9.1), the four amino terminal SeMet residues are missing from the structure deposited in the Protein Data Bank. Thus, even though more sites were sought during substructure

**Table 9.7** Refined occupancy (Occ) values for the $1.5N$ peaks from Trial 5 sorted in decreasing order according to their height on the direct-methods electron density map

| Peak | Select? | Occ | Peak | Select? | Occ | Peak | Select? | Occ |
|---|---|---|---|---|---|---|---|---|
| 1 | Yes | 1.00 | 15 | Yes | 0.63 | 29 | No | 0.08 |
| 2 | Yes | 0.88 | 16 | Yes | 0.80 | 30 | No | 0.08 |
| 3 | Yes | 0.91 | 17 | Yes | 0.71 | 31 | No | 0.03 |
| 4 | Yes | 0.94 | 18 | Yes | 0.70 | 32 | No | 0.08 |
| 5 | Yes | 0.65 | 19 | Yes | 0.57 | 33 | No | 0.10 |
| 6 | Yes | 0.99 | 20 | No | 0.01 | 34 | No | 0.05 |
| 7 | Yes | 0.77 | 21 | Yes | 0.55 | 35 | No | 0.01 |
| 8 | Yes | 0.56 | 22 | Yes | 0.53 | 36 | No | 0.06 |
| 9 | Yes | 0.72 | 23 | Yes | 0.52 | 37 | No | 0.11 |
| 10 | Yes | 0.54 | 24 | Yes | 0.52 | 38 | No | 0.03 |
| 11 | Yes | 0.99 | 25 | Yes | 0.50 | 39 | No | 0.02 |
| 12 | Yes | 0.71 | 26 | No | 0.05 | 40 | No | 0.13 |
| 13 | Yes | 0.78 | 27 | No | 0.06 | 41 | No | 0.01 |
| 14 | Yes | 0.62 | 28 | No | 0.03 | 42 | No | 0.03 |

'Yes' indicates that the peak passes the cutoff of 0.2 and should be selected as an atom (Se site) for subsequent steps in the protein phasing procedure.

phasing, the direct-methods procedures are robust, and the occupancy refinement can detect successfully a large percentage of false sites.

Peaks consistently occurring in several independent trial solutions are the most likely to correspond to real atomic sites. Thus, while working under conditions (such as the *BnP* manual mode) where more than one solution may be present, additional information regarding which sites are likely to be correct can be obtained by comparing the peaks from different trials. The program *NANTMRF* (Smith, 2002), which takes into account the fact that different solutions may have different origins or enantiomorphs and then finds the best overall match between two sets of peak positions, is available within the *BnP* package and makes it easy to do the comparison. Table 9.8 shows the results of trial comparison for the 1JC4 example. Recall that the figures of merit (Table 9.5) indicate that Trials 7 and 8 are solutions but Trial 6 is not. Thus, the results of the peak comparison agree with this conclusion and confirm the site choices based on the occupancy refinement.

## 9.6 Determining the proper enantiomorph

Even if anomalous dispersion data are involved in the substructure determination process, it is the magnitude of the anomalous differences that are used, and the substructures of the biologically occurring macromolecule and its enantiomer are both consistent with the data. The probability that the substructure obtained by direct methods can be developed into a protein model with L-amino acids and right-handed $\alpha$ helices is 50%. Therefore, before proceeding further, other information must be used to determine the correct hand.

**Table 9.8** Comparison of the peak positions for 1JC4 Trial 5 to peaks for other trials

| Trial 5 | Trial 6 | | Trial 7 | | Trial 8 | |
|---|---|---|---|---|---|---|
| Peak | Peak | Distance | Peak | Distance | Peak | Distance |
| 1 | | | 1 | 0.05 | 1 | 0.04 |
| 2 | | | 2 | 0.16 | 4 | 0.09 |
| 3 | | | 3 | 0.06 | 2 | 0.09 |
| ... | | | | | | |
| 18 | 27 | 0.68 | 12 | 0.16 | 18 | 0.19 |
| 19 | 30 | 0.83 | 19 | 0.12 | 22 | 0.18 |
| 20 | | | | | | |
| 21 | | | 23 | 0.16 | 17 | 0.18 |
| 22 | | | 22 | 0.21 | 20 | 0.30 |
| 23 | | | 21 | 0.40 | 23 | 0.25 |
| 24 | | | 24 | 0.28 | 24 | 0.20 |
| 25 | | | 4 | 0.17 | 19 | 0.17 |
| 26 | | | | | | |
| 27 | | | | | | |
| ... | | | | | | |
| 40 | | | 34 | 0.52 | | |
| 41 | | | | | 36 | 0.63 |
| 42 | | | | | | |
| # Matches | 2 | | 25 | | 25 | |
| (Distance) | 0.75 | | 0.19 | | 0.17 | |

Distances less than 1 Å to peaks on the other maps are indicated. Peaks 1–19 as well as peaks 21–25 of Trial 5 (a presumed solution) have matching peaks on both Trials 7 and 8. There are no matches for peaks 28–39. The two 'matches' with Trial 6 as well as the matches with peaks 40 and 41 involve either weaker peaks or larger distances and, therefore, are likely to be spurious. The large number of matches between Trials 5 and 7 and between Trials 5 and 8 indicate that the corresponding maps are essentially identical. Peaks that match are the most likely to be Se sites.

In order to choose the enantiomorph, sets of protein phases must be computed from the set of heavy-atom sites found by direct methods and validated by occupancy refinement as well as the set of sites related to the first set by inversion. Electron-density maps must be generated for both sets of phases and examined for the presence of biologically correct stereochemistry. If no anomalous dispersion data are available, this is the only option. However, if anomalous scattering measurements are included when the maps are computed, it is possible to use other criteria to select the enantiomorph automatically. For example, in *BnP* protein-solvent envelope masks are created for each map using a protein–solvent boundary determination algorithm (Wang, 1985), and then the standard deviations of the electron densities in the protein and solvent regions are computed as well as the ratio σ(protein)/σ(solvent). This ratio should be higher for the correct enantiomorph since atomic sites and gaps between chains within the protein region are expected to show large variations whereas solvent regions should be relatively flat with little variation. In practice, this ratio is a robust discriminator even when challenged deliberately by solutions having both missing and false sites, and prior substructure or phase refinement is not a requirement (Weeks *et al.*, 2002). The results of applying this enantiomorph determination criterion to the 1JC4 example are shown in Table 9.9. It should be noted that the standard deviation ratio defined here is similar to, but not identical to, criteria used in the programs *SOLVE* (Terwilliger and Berendzen, 1999) and *SHELXE* (Sheldrick, 2002).

**Table 9.9** Enantiomorph discrimination for protein maps based on the Se sites for 1JC4 trial 5

| Enantiomorph | σ(solvent) | σ(protein) | σ(pro)/σ(sol) |
|---|---|---|---|
| Original | 18.65 | 30.18 | 1.62 |
| Alternate | 23.08 | 27.14 | 1.18 |

The σ values are a measure of the electron-density variation in the protein and solvent regions, and the ratio of these numbers is a measure of the 'contrast' between the two regions. Since anomalous dispersion data were used to phase the maps, the map for the correct hand will show greater contrast. In this case, the original direct-methods sites give rise to greater contrast thereby indicating that these sites do correspond to the correct enantiomorph.

## 9.7 From substructure to protein

Once the proper enantiomorph has been found, the substructure determination is complete, and the first of the two stages of protein phasing – the one in which direct methods plays a role – is finished. The second stage involves substructure/protein phase refinement, including the determination of optimum values for scaling, positional, and thermal parameters as well as expected lack-of-closure estimates and protein phases. Next, solvent flattening can be used to bring about further phase improvement. In automated programs such as *BnP*, these steps can be combined with the preceding substructure determination, and the entire process run as a single job in favourable cases. The resulting protein phases are then available for export to graphics programs such as *O* (Jones *et al.*, 1991) for manual examination of protein models or to programs such as *ARPwARP* (Morris *et al.*, 2003) and *RESOLVE* (Terwilliger, 2003) for automated chain tracing.

## Acknowledgment

The preparation of this chapter was supported by NIH grant EB002057.

## References

Adams, P. D., Gopal, K., Grosse-Kunstleve, R. W., Hung, L.-W., Ioerger, T. R., McCoy, A. J., *et al.* (2004). Recent developments in the *PHENIX* software for automated crystallographic structure determination. *J. Synchrotron Rad.* **11,** 53–55.

Baggio, R., Woolfson, M. M., Declercq, J. P., and Germain, G. (1978). On the application of phase relationships to complex structures. XVI. A random approach to structure determination. *Acta Crystallogr. A* **34,** 883–892.

Bhuiya, A. K. and Stanley, E. (1963). The refinement of atomic parameters by direct calculation of the minimum residual. *Acta Crystallogr.* **16,** 981–984.

Blessing, R. H. and Smith, G. D. (1999). Difference structure-factor normalization for heavy-atom or anomalous-scattering substructure determinations. *J. Appl. Cryst.* **32,** 664–670.

Bricogne, G., Vonrhein, C., Flensburg, C., Schiltz, M. and Paciorek, W. (2003). Generation, representation and flow of phase information in structure determination: recent developments in and around *SHARP* 2.0. *Acta Crystallogr. D* **59,** 2023–2030.

Bruker AXS (2005). XPREP, Version 2005/1. Bruker AXS Inc., Madison, Wisconsin, USA.

Burla, M. C., Camalli, M., Carrozzini, B., Cascarano, G. L., Giacovazzo, C., Polidori, G. and Spagna, R. (2000). *SIR2000*, a program for the automatic *ab initio* crystal structure solution of proteins. *Acta Crystallogr. A* **56**, 451–457.

Burla, M. C., Carrozzini, B., Cascarano, G. L., De Caro, L., Giacovazzo, C. and Polidori, G. (2003). *Ab initio* protein phasing at 1.4Å resolution. *Acta Crystallogr. A* **59**, 245–249.

Chowdhury, K., Bhattacharya, S. and Mukherjee, M. (2005). *Ab initio* structure solution of nucleic acids and proteins by direct methods: reciprocal-space and real-space approach. *J. Appl. Cryst.* **38**, 217–222.

Cochran, W. (1955). Relations between the phases of structure factors. *Acta Crystallogr.* **8,** 473–478.

Debaerdemaeker, T. and Woolfson, M. M. (1983). On the application of phase relationships to complex structures. XXII. Techniques for random phase refinement. *Acta Crystallogr. A* **39**, 193–196.

DeTitta, G. T., Weeks, C. M., Thuman, P., Miller, R. and Hauptman, H. A. (1994). Structure solution by minimal-function phase refinement and Fourier filtering. I. Theoretical basis. *Acta Crystallogr. A* **50**, 203–210.

Foadi, J., Woolfson, M. M., Dodson, E. J., Wilson, K. S., Yao, J.-X. and Zheng, C.-D. (2000). A flexible and efficient procedure for the solution and phase refinement of protein structures. *Acta Crystallogr. D* **56**, 1137–1147.

Frazão, C., Sieker, L., Sheldrick, G. M., Lamzin, V., LeGall, J. and Carrondo, M. A. (1999). *Ab initio* structure solution of a dimeric cytochrome c3 from *Desulfovibrio gigas* containing disulfide bridges. *J. Biol. Inorg. Chem.* **4,** 162–165.

Fujinaga, M. and Read, R. J. (1987). Experiences with a new translation-function program. *J. Appl. Cryst.* **20**, 517–521.

Furey, W. and Swaminathan, S. (1997). PHASES-95: a program package for processing and analyzing diffraction data from macromolecules. *Method Enzymol.* **277,** 590–620.

Germain, G. and Woolfson, M. M. (1968). On the application of phase relationships to complex structures. *Acta Crystallogr. B* **24**, 91–96.

Grosse-Kunstleve, R. W. and Adams, P. D. (2003). Substructure search procedures for macromolecular structures. *Acta Crystallogr. D* **59**, 1966–1973.

Hauptman, H. A. (1974). On the theory and estimation of the cosine invariants $\cos(\varphi_l + \varphi_m + \varphi_n + \varphi_p)$. *Acta Crystallogr. A* **30**, 822–829.

Hauptman, H. A. and Karle, J. (1953). *Solution of the Phase Problem. I. The Centrosymmetric Crystal.* ACA Monograph Number 3. Edwards Brothers, Ann Arbor, MI.

Hendrickson, W. A. (1991). Determination of macromolecular structures from anomalous diffraction of synchrotron radiation. *Science* **254,** 51–58.

Howell, P. L., Blessing, R. H., Smith, G. D. and Weeks, C. M. (2000). Optimizing *DREAR* and *SnB* parameters for determining Se-atom substructures. *Acta Crystallogr. D* **56**, 604–617.

Jones, T. A., Zou, J. Y., Cowtan, S. W. and Kjeldgaard, M. (1991). Improved methods for building protein models in electron density maps and the location of errors in these models. *Acta Crystallogr. A* **47**, 110–119.

Karle, J. (1989). Linear algebraic analyses of structures with one predominant type of anomalous scatterer. *Acta Crystallogr. A* **45**, 303–307.

Karle, J. and Hauptman, H. (1956). A theory of phase determination for the four types of non-centrosymmetric space groups 1*P*222, 2*P*22, $3P_12$, $3P_22$. *Acta Crystallogr.* **9,** 635–651.

Main, P., Fiske, S. J., Hull, S. E., Lessinger, L., Germain, G., Declercq, J. P. and Woolfson, M.M. (1980). *MULTAN80: a System of Computer Programs for the Automatic Solution of Crystal Structures from X-ray Diffraction Data.* Universities of York and Louvain.

McCarthy, A. A., Baker, H. M., Shewry, S. C., Patchett, M. L. and Baker, E. N. (2001). Crystal structure of methylmalonyl-coenzyme A epimerase from *P. shermanii*: a novel enzymatic function on an ancient metal binding scaffold. *Structure* **9,** 637–646.

Miller, R., DeTitta, G. T., Jones, R., Langs, D. A., Weeks, C. M., and Hauptman, H. A. (1993). On the application of the minimal principle to solve unknown structures. *Science* **259,** 1430–1433.

Miller, R., Gallo, S. M., Khalak, H. G., and Weeks, C. M. (1994). *SnB*: crystal structure determination via *Shake-and-Bake*. *J. Appl. Cryst.* **27**, 613–621.

Morris, R. J., Perrakis, A., and Lamzin, V. S. (2003). *ARP/wARP* and automatic interpretation of protein electron density maps. *Method Enzymol.* **374,** 229–244.

Rajakannan, V., Selvanayagam, S., Yamane, T., Shirai, T., Kobayashi, T., Ito, S. and Velmurugan, D. (2004a). The use of *ACORN* in solving a 39.5 kDa macromolecule with 1.9Å resolution laboratory source data. *J. Synch. Rad.* **11,** 358–362.

Rajakannan, V., Yamane, T., Shirai, T., Kobayashi, T., Ito, S., and Velmurugan, D. (2004b). Applications of *ACORN* to data at 1.45Å resolution. *J. Synch. Rad.* **11,** 64–67.

Schenk, H. (1974). On the use of negative quartets. *Acta Crystallogr. A* **30**, 477–481.

Schneider, T. R. and Sheldrick, G. M. (2002). Substructure solution with *SHELXD*. *Acta Crystallogr. D* **58**, 1772–1779.

Sheldrick, G. M. (1990). Phase annealing in *SHELX*-90: direct methods for larger structures. *Acta Crystallogr. A* **46**, 467–473.

Sheldrick, G. M. (2002). Macromolecular phasing with *SHELXE*. *Z. Kristallogr.* **217,** 644–650.

Shmueli, U. and Wilson, A. J. C. (1996). Statistical properties of the weighted reciprocal lattice. In: *International Tables for Crystallography*, Shmueli, U., ed., Vol. B, pp. 184–200. Kluwer Academic Publishers, Dordrecht.

Smith, G. D. (2002). Matching selenium-atom peak positions with a different hand or origin. *J. Appl. Cryst.* **35,** 368–370.

Smith, G. D., Nagar, B., Rini, J. M., Hauptman, H. A., and Blessing, R. H. (1998). The use of *SnB* to determine an anomalous scattering substructure. *Acta Crystallogr. D* **54,** 799–804.

Terwilliger, T. C. (2003). *SOLVE* and *RESOLVE*: automated structure solution and density modification. *Method Enzymol.* **374,** 22–37.

Terwilliger, T. C. and Berendzen, J. (1999). Automated MAD and MIR structure solution. *Acta Crystallogr. D* **55,** 849–861.

von Delft, F., Inoue, T., Saldanha, S. A., Ottenhof, H. H., Schmitzberger, F., Birch, L. M., Dhanaraj, V., Witty, M., Smith, A. G., Blundell, T. L. and Abell, C. (2003). Structure of *E. coli* ketopantoate hydroxymethyl transferase complexed with ketopantoate and Mg2+, solved by locating 160 selenomethionine sites. *Structure,* **11,** 985–996.

Wang, B.-C. (1985). Solvent flattening. *Method Enzymol.* **115,** 90–112.

Weeks, C. M. and Miller, R. (1999). The design and implementation of *SnB* version 2.0. *J. Appl. Cryst.* **32,** 120–124.

Weeks, C. M., Blessing, R. H., Miller, R., Mungee, R., Potter, S. A., Rappleye, J., Smith, G. D., Xu, H. and Furey, W. (2002). Towards automated protein structure determination: *BnP*, the *SnB-PHASES* interface. *Z. Kristallogr.* **217,** 686–693.

Weeks, C. M., DeTitta, G. T., Hauptman, H. A., Thuman, P., and Miller, R. (1994). Structure solution by minimal-function phase refinement and Fourier filtering. II. Implementation and applications. *Acta Crystallogr. A* **50,** 210–220.

CHAPTER 10

# Phase refinement through density modification

Jan Pieter Abrahams, Jasper R. Plaisier, Steven Ness, and Navraj S. Pannu

## 10.1 Introduction

It is impossible to directly measure phases of diffracted X-rays. Since phases determine how the measured diffraction intensities are to be recombined into a three-dimensional electron density, phase information is required to calculate an electron density map of a crystal structure. In this chapter we discuss how prior knowledge of the statistical distribution of the electron density within a crystal can be used to extract phase information. The information can take various forms, for example:

*Solvent flatness.* On average, protein crystals contain about 50% solvent, which on an atomic scale usually adopts a random, non-periodic structure within the crystal and hence is featureless within the averaged unit cell. Therefore, if we know the location of the solvent regions within a macromolecular crystal, we already know a considerable part of the electron density (i.e. the part that is flat and featureless), and 'flattening' the electron density of the solvent region can improve the density of our macromolecule of interest.

*Non-crystallographic symmetry.* Many protein crystals contain multiple copies of one or more molecules within the asymmetric unit. Often the conformations of such chemically indistinguishable but crystallographically non-equivalent molecules are sufficiently alike to treat them as identical. In this case, we can improve the signal to noise ratio of the electron density of our molecule of interest by averaging the density of the multiple copies in the asymmetric unit.

*Electron density statistics.* At high resolution we know the shape of the electron density of an atom, in which case we only need to know its exact location to reconstruct the electron density in its immediate vicinity. At lower resolution we can impose an expected shape on the uni- or multivariate distributions of electron density within the protein region in a procedure that is known as histogram matching.

The problem is not so much in understanding the restrictions these types of prior knowledge impose on (suboptimal) electron density, but rather in using these restraints in reciprocal space. In practice an iterative procedure is followed. First the electron density of an initial model is calculated, which is then modified to satisfy the expected, previously determined restraints. From the modified map, the diffraction data are recalculated. The resulting phases are combined with the measured data and their associated phase probability distributions. On this basis the currently most probable phase set is calculated. The procedure is repeated until convergence. Below, we briefly describe the mathematical background of these procedures and discuss some of their essential aspects. We pay special attention to visual, geometric concepts, as we believe them important for developing an intuitive grasp for the process of density modification in phase refinement. To this end, we illustrate the concepts on a one-dimensional, centrosymmetric map, as this allows us to depict phases simply by their sign.

## 10.2 Fourier transforms and the phase problem

We want to know the electron density that is determined by the measured structure factor amplitudes and their phases. The electron density at point $x$ is calculated by a Fourier summation:

$$\rho(\mathbf{x}) = \frac{1}{V}\sum_{j=0}^{2N} |F(\mathbf{h}_j)| e^{i\phi(\mathbf{h}_j)-2i\pi\mathbf{x}\cdot\mathbf{h}_j} \quad (1)$$

In the above equation, $\rho(\mathbf{x})$ is the electron density at $\mathbf{x}$, while $V$ is the volume of the unit cell, $2N$ is the number of relevant structure factors, and $|\boldsymbol{F}(\mathbf{h}_j)|$ is the amplitude of the structure factor with Miller indices $\mathbf{h}_j = (h_j, k_j, l_j)$ and a phase of $\phi(\mathbf{h}_j)$. Note that $2N$ is determined by the size of the unit cell and the resolution.

Equation 1 is a discrete Fourier transform. It is discrete rather than continuous because the crystalline lattice allows us to sum over a limited set of indices, rather than integrate over structure factor space. The discrete Fourier transform is of fundamental importance in crystallography – it is the mathematical relationship that allows us to convert structure factors (i.e. amplitudes and phases) into the electron density of the crystal, and (through its inverse) to convert periodic electron density into a discrete set of structure factors.

Even though the Fourier transform in Eq. 1 is discrete, $\rho(\mathbf{x})$ is continuous, as it can be calculated for any grid point $\mathbf{x}$. Obviously one could calculate Eq. 1 on an arbitrary fine grid, but in that case the density at any grid point is correlated to that of its neighbours through interpolation. Since Fourier transforms neither create nor destroy information, the maximum number of uncorrelated, independent density grid points is limited to $2N$, the number of structure factors going into the summation of Eq. 1. To conclude, if there are $2N$ structure factors, the corresponding electron density map has $2N$ independent grid points and there are $2N$ independent equations of type 1 relating the former to the latter.

Intuitively, the correct application of restraints and constraints to electron density should improve the phases. To quantify this notion, it is useful, though unconventional, to proceed as if we are solving a system of non-linear equations. A solution of such a system requires at least as many independent equations as there are unknowns. Let us therefore count the number of unknowns: there are $2N$ unknown electron densities at the independent grid points and $2N$ unknown phases. These $4N$ unknowns are inter-related by $2N$ Fourier summations, so the system is underdetermined. Its solution clearly requires at least $2N$ additional equations.

However, in view of experimental errors, $2N$ additional equations are unlikely to be sufficient to solve the phase problem. In practice, we can only expect a statistically meaningful solution if we include many more equations and identify the solution that agrees most with all equations simultaneously. Furthermore, since Eq. 1 is non-linear in $\phi(\mathbf{h}_j)$, we cannot expect to find an analytic solution. Hence, we have to make initial guesses for the unknowns and improve from there.

Constraints effectively reduce the number of unknowns while restraints add to the number of equations in the system, without changing the number of unknowns. The effectiveness of a restraint is partially determined by the number of independent equations the restraint introduces in the minimization. Also the discriminating potential of the individual terms between right and wrong models affects the scale of the improvement. Clearly, a robust term with a sharp minimum contributes much more to phasing than a permissive term that hardly distinguishes a wrong model from a correct one. Below we summarize the general form of the additional information, constraints and restraints, which in practice leads to a system of equations that is no longer underdetermined.

## 10.3 Reciprocal space constraints

### 10.3.1 Friedel's law

In protein crystallography we assume that all electron density is real, and does not have an imaginary component. In reciprocal space this observation is known as Friedel's law, which states that a structure factor $F(\mathbf{h})$ and its Friedel mate $F(-\mathbf{h})$ have equal amplitudes, but opposite phases. The correspondence of these two assumptions follows straight from Fourier theory and, in consequence, explicitly constraining all electron density to be real is entirely equivalent to introducing $N$additional equalities of the following type:

$$\phi(\mathbf{h}_j) = -\phi(-\mathbf{h}_j) \tag{2}$$

Straight substitution of these equalities into Eq. 1 reduces the magnitude of the problem; rather than $4N$ unknowns, we are left with $3N$ unknowns.

### 10.3.2 Differences between corresponding $|F(\mathbf{h}_j)|$'s

In order to obtain initial phase estimates, crystallographers typically use either experimental phasing techniques or molecular replacement. Obtaining initial phase estimates and their associated phase probability distributions are treated in other chapters of this book, and for the remainder of this chapter we assume that the initial phases and the associated phase probabilities have been calculated. These phase probability distributions are conveniently described by Hendrickson–Lattman coefficients.

$$P(\phi(\mathbf{h}_j)) = K_j e^{A_j \cos(\phi(\mathbf{h}_j)) + B_j \sin(\phi(\mathbf{h}_j)) + C_j \cos(2\phi(\mathbf{h}_j)) + D_j \sin(2\phi(\mathbf{h}_j))} \tag{3}$$

Here, $P(\phi(\mathbf{h}_j))$ is the probability function of a phase $\phi(\mathbf{h}_j)$, whilst $A_j$, $B_j$, $C_j$, and $D_j$ are its Hendrickson–Lattman coefficients and $K_j$ is a normalizing constant. Clearly, $P(\phi(\mathbf{h}_j))$ cannot be inserted straight into Eq. 1. However, it does provide additional equations, one for each phase for which $A_j$, $B_j$, $C_j$, or $D_j$ are non-zero.

## 10.4 Real space restraints

### 10.4.1 Solvent flatness

In disordered solvent regions of the unit cell, the density is featureless and flat. In practice, the location and size of the solvent region are inferred from an initial electron density map and the molecular weight of the molecule. Since the average electron density of the solvent usually is very similar to that of the protein, automated procedures identify the solvent mask by determining the regions which have the smallest *variation* in electron density. If we know which regions of the unit cell are featureless, additional equations of the following type can be inferred:

$$\rho(\mathbf{x}_{\text{solvent}}) = \rho_{\text{solvent}} \tag{4}$$

Here, $\mathbf{x}_{\text{solvent}}$ is a real space coordinate within the solvent region and $\rho_{\text{solvent}}$ is the mean electron density of the solvent.

The number of additional terms based on Eq. 4 is determined by the solvent fraction of the crystal. If the solvent content is 50% (which is the average for protein crystals), $N$ independent, additional equations of type 4 are introduced. As these equations can be substituted into the Fourier summations of Eq. 1, they effectively reduce the number of unknowns by $2N$ times the solvent fraction – provided they accurately distinguish disordered solvent from protein.

### 10.4.2 Non-crystallographic symmetry (NCS)

If the electron density of one area of the asymmetric unit is sufficiently similar to that of another (after a translation and/or a rotation), additional equations of the following type can be inferred:

$$\rho(\mathbf{x}_{\text{unique}}) = \rho(\mathbf{T}\mathbf{x}_{\text{unique}}) \tag{5}$$

Here, $\mathbf{x}_{\text{unique}}$ is a real space coordinate within a region of density that is repeated elsewhere in the asymmetric unit after a rotation and/or translation defined by the transformation $\mathbf{T}$. Also these equations can be substituted in the Fourier summation of Eq. 1, effectively further reducing the number of unknowns in real space down to the number of independent grid points within the fraction of unique density.

### 10.4.3 Electron density statistics

The overall distribution of randomly phased electron density is Gaussian, whereas a correctly phased map is expected to have a non-Gaussian distribution at resolutions beyond about 2.5 Å. An electron density distribution is described by a histogram, in which for each density value the likelihood is plotted of finding such a value within the unit cell. The shape of this histogram is determined by the resolution of the map (at low resolution, extreme values are less likely) and the chemical composition (heavy atoms will cause more extreme histograms). Proteins share characteristic electron density histograms, provided they do not contain many heavy atoms or large, disordered volumes. This implies that for a given structure a good guess of the correct histogram can be obtained, resulting in the following equation:

$$H(\rho(\mathbf{x}_{\text{protein}})) = H^{\text{obs}}\left(\frac{1}{V}\sum_{j=0}^{2N}|F(\mathbf{h}_j)|e^{i\phi(\mathbf{h}_j)-2i\pi\mathbf{x}_{\text{protein}}\cdot\mathbf{h}_j}\right) \tag{6}$$

Here $\mathbf{x}_{\text{protein}}$ is a real space coordinate within the protein region, $H(\rho(\mathbf{x}))$ is the expected, non-Gaussian histogram of the electron density and $H^{\text{obs}}(\rho(\mathbf{x}))$ is the observed histogram of protein density which may or may not have phase errors.

Equation 6 cannot be substituted into Eq. 1 and therefore it does not further reduce the number of unknowns. However, it does provide additional equations, their number being determined by the number of independent grid points within the unique protein region. Its effectiveness is determined by the difference between the theoretical histogram of a protein at a given resolution, and that of randomly phased data.

## 10.5 The practice of phase refinement: Fourier cycling

In theory, density modification could produce perfect phases, if the Eqs 2 to 6 are sufficiently restrictive. Let us illustrate this by an example; assume a crystal with 50% solvent and three-fold non-crystallographic symmetry. There are $2N$ Fourier summations (Eq. 1) with $4N$ unknowns. After substitution with Friedel's Law (Eq. 2), only $N$ phases remain unknown, so now there are $3N$ unknowns in total. For all phases we have experimental information encoded in Hendrickson–Lattman coefficients (Eq. 3), so we can add $N$ equations to our set. As we know the location of the solvent region we can reduce the number of unknown densities at independent grid points from $2N$ to $N$ upon substituting with Eq. 4. Non-crystallographic symmetry further reduces the number of unknown densities to $N/3$ by substituting with Eq. 5. Histogram matching can further reduce the search space of solutions and improve convergence.

If there are more equations than unknowns, why then can we not determine phases accurately without building an atomic model? Well, the additional equations of type 3 and 6 are of a statistical nature and therefore may be less restrictive. Furthermore, inaccuracies in determining the solvent mask and the non-crystallographic symmetry operators and masks will compromise the procedure. Nevertheless, the additional information imposed often leads to substantial improvement.

Unfortunately, the relations between the electron density, the restraints we have discussed here, and the structure factors are non-linear. Thus, the only strategy we can adopt is to use the approximate phases we start out with and improve these iteratively. Even this is not straightforward, mainly because Eq. 1 is expensive to compute. However, there exists a powerful and straightforward procedure that is used in virtually all phase refinement programs: Fourier cycling.

In Fourier cycling, the approximate phases we have available at the beginning of the process of density modification are used to calculate an initial map. The real space restraints, solvent flatness, non-crystallographic symmetry averaging, and histogram matching, are imposed on this initial density map. After Fourier transformation, the structure factors obtained typically no longer obey the reciprocal space constraints such as the measured amplitude and the phase probability distribution. Therefore existing reciprocal space restraints are recombined with the phase probability distribution obtained after back transformation of the restrained electron density. These modified structure factors are used to calculate a new map. This new map may, in turn, no longer obey the real space restraints, so these are reimposed. The procedure is repeated until it converges on a density map satisfying the equations available as well as is possible. In Fig. 10.1, a flow chart of the process of Fourier cycling in shown.

Before we can understand why Fourier cycling works, we have to deepen our understanding of the Fourier transform. In particular, we need to understand the effects of modifying density on the structure factor amplitudes and phases. The mathematical tool that describes this is the convolution operator.

Convolution is a commonly used mathematical technique that takes as input two functions, say A(x) and B(x). To convolute A(x) with B(x), first take the function A(x) and place it at the origin of the second

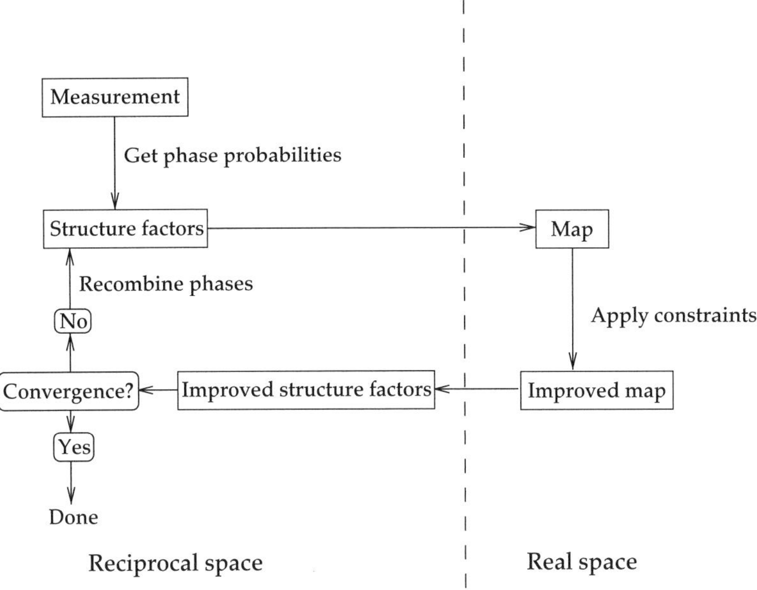


**Figure 10.1** A flow chart demonstrating phase refinement in practice with the iterative recycling between real and reciprocal space.

function, then multiply the two functions. Now, do this for each point in B, moving A to each point in B, multiplying the functions, and adding all the product functions. The result is the convolution of A with B. See Fig. 10.2 for an example of the convolution operator.

Mathematically, the convolution operator ⊗ is defined as follows:

$$C(\mathbf{y}) = (A \otimes B)(\mathbf{y}) = \int_{-\infty}^{\infty} A(\mathbf{y} - \mathbf{x})B(\mathbf{x})\, \delta\mathbf{x} \qquad (7)$$

A useful property of the Fourier transform and convolution is that the Fourier transform of the convolution of two functions is equal to the Fourier transform of the two functions multiplied together. Thus, since the convolution of two functions is typically a time consuming process, this property, together with the Fast Fourier Transform, is used to significantly speed up the process of convolving two functions.

In conclusion, when modifying the density in real space is equivalent to a multiplication with another map, in reciprocal space this results in the convolution of the Fourier transform of both maps (and *vice versa*).

## 10.6 Phase recombination in Fourier cycling

Clearly, for the procedure outlined in Fig. 10.1 to work, we need to take care of several critical steps; next to reasonable initial phase estimates required to formulate the initial restraints, we need a statistically valid procedure for the combination of the phases obtained by back transformation of the real space restrained map and the initial phase probability distribution. This recombination step is discussed below.

In general, estimates of high resolution phases will be less accurate than the ones at low resolution. The reason is that in the beginning of a structure determination, it is easier to establish low resolution contours than high resolution details. This is because contours are hardly affected by errors at high resolution. On the other hand, the contrast of high resolution features is severely affected by errors at low resolution. Hence, in phase refinement we generally see the improvement progressing from low to high resolution as we cycle through the procedure. It makes sense to weight down structure factors with erroneous phases. Therefore we need to introduce a weighting scheme that typically has a higher

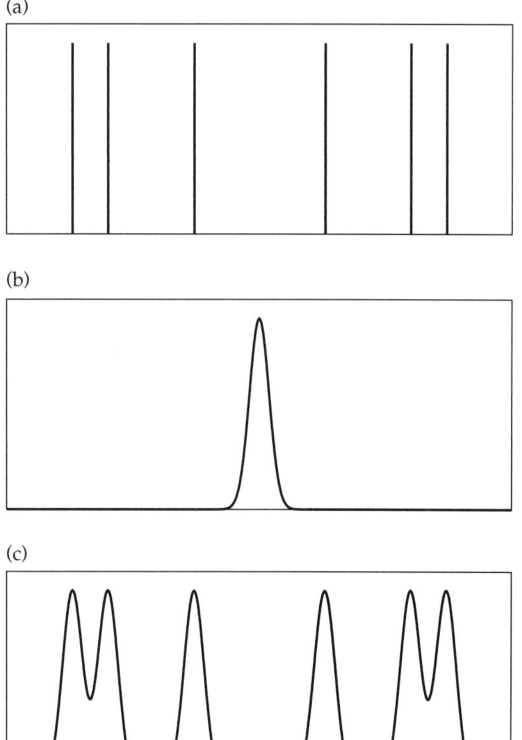


**Figure 10.2** The procedure of convolution, represented graphically. (a) A one-dimensional centrosymmetric structure. (b) A Gaussian distribution, which could potentially be an atomic shape function. (c) The convolution of the function in (a) and (b).

weight or lower fall-off as the phase refinement converges.

In practice, recombination of structure factors involves first weighting of the phases of the modified structure factors in a resolution dependent fashion, according to their estimated accuracy or probability. Every phase also has an experimental probability (determined by experimental phasing techniques and/or molecular replacement). The two distributions are combined by multiplication, and the new phase is calculated from this combined probability distribution. The measured associated structure factor amplitude is then scaled by the probability of the phase, and we have our set of recombined structure factors.

However, there is a problem with the phase recombination approach. Essentially we're combining probability distributions: (1) real space restraints give distributions for modified structure factors, while (2) the phasing experiments give partially independent phase probability distributions. Combining distributions is easy: we just multiply them, *provided we know they are independent*. However, here consecutive distributions are clearly not independent and treating them as independent would inevitably lead to an undesirable bias towards the very first map with which we started the Fourier cycling. How do we deal with this situation? We *separate out* the dependent component, and multiply the independent components. In order to explain how this is done in practice, we give a more quantitative explanation of the reason why Fourier cycling and phase recombination works, first for solvent flattening and subsequently for NCS and histogram matching.

## 10.7 Why does Fourier cycling improve phases in solvent flattening?

Before we can flatten the solvent, we need to know where it is. One of the implementations to obtain a good approximation of the solvent mask computes the variance of the electron density within a small sphere throughout the entire unit cell. Regions in the unit cell where a low variance is found then are considered to be solvent, whereas a high variance indicates protein. Most density modification programs use a binary solvent mask, with one value representing the protein region and the other value representing the solvent region. Some programs have reported good results by extending this and using real valued numbers between 0 and 1, where the value of the grid point indicates the probability of being in a protein region (Terwilliger, 2003).

Now return to Eq. 3, which describes the process of solvent flattening. As a restraint, it can be written down as follows:

$$\rho_{\text{mod}}(\mathbf{x}) = \rho_{\text{init}}(\mathbf{x})g(\mathbf{x}) + \rho_{\text{solvent}}\hat{g}(\mathbf{x}) \quad (8)$$

where:

$g(\mathbf{x})$ is a mask function which is equal to one in the protein region and is zero in the solvent region.
$\hat{g}(\mathbf{x})$ is a mask function that is zero in the protein region and one in the solvent region.

(a)

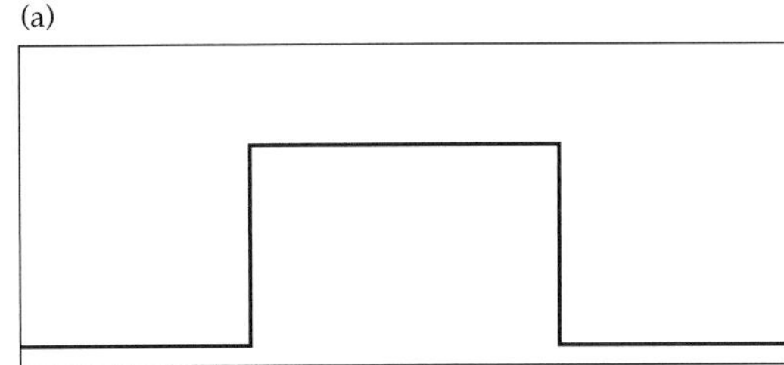

(b)

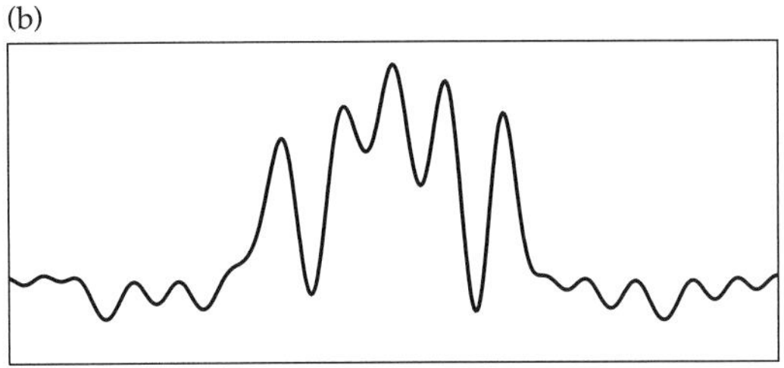

(c)

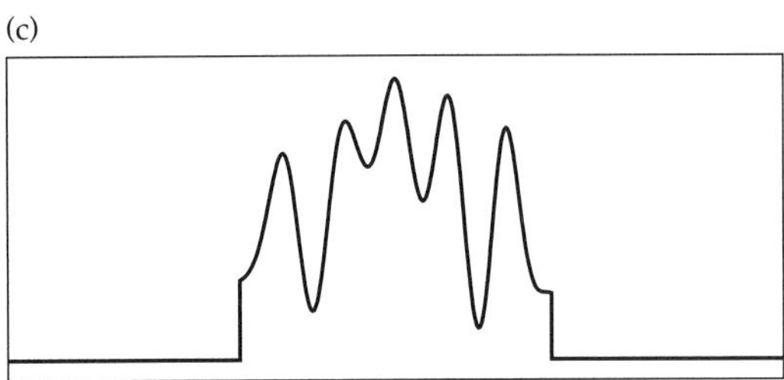

**Figure 10.3** A graphical representation of solvent flattening in real space. (a) represents a one- dimensional solvent mask, (b) is a one-dimensional, unflattened electron density, and (c) is the resulting flattened electron density map that imposes the solvent mask.

$\rho_{mod}(\mathbf{x})$ is the modified electron density.
$\rho_{init}(\mathbf{x})$ is the initial electron density.
$\rho_{solvent}$ is the mean density in the solvent region.

A graphical representation of solvent flattening in real space is shown in Fig. 10.3. In Eq. 8, we multiply the two functions $\rho_{init}(\mathbf{x})$ and $g(\mathbf{x})$ as we flatten the density within the solvent region. However, multiplication in real space is equivalent to a convolution in reciprocal space. Therefore, we can rewrite Eq. 8 as follows:

$$F_{mod}(\mathbf{h}) = (F_{init} \otimes G)(\mathbf{h}) + C(\mathbf{h}) \tag{9}$$

Where:

$F_{mod}(\mathbf{h})$ is the Fourier transform of the modified density map, $\rho_{mod}(\mathbf{x})$.
$F_{init}(\mathbf{h})$ is the Fourier transform of the unmodified density map, $\rho_{init}(\mathbf{x})$.
$G(\mathbf{h})$ is the Fourier transform of the protein mask, often referred to as an interference function.
$C(\mathbf{h})$ defines a small correction, mainly peaking at the origin.

Applying a mask function in real space is equivalent to combining many structure factors through a convolution in reciprocal space. This results in an improvement because the random error component of the structure factors will average out, whereas the true values of the structure factors will add up systematically. Fig. 10.4 gives a graphical example showing this phenomenon.

As is described in Abrahams (1997), plotting the radial distribution of the intensity of the interference function $G(\mathbf{h})$, most of the intensity is around the origin. Thus, when you convolute the structure factors F with the interference function G, each structure factor will mainly be recombined with structure factors that are close by.

However, this procedure is not entirely without problems. Importantly, there is a term in the convolution given in Eq. 9 which cannot be neglected: it is the value of the G(a) function at ($h = 0$). The magnitude of this term determines how much of the original (partially erroneous) map is unaffected by the convolution. The modified map is actually a scaled down version of the initial map, to which is added a new map containing new information. Now we have effectively identified the bias component: it is defined by the magnitude of $G(\mathbf{h})$ (or by the mean value of $g(\mathbf{x})$, as follows from Fourier theory). In order to do a proper phase recombination, we have to set G(0) to zero. Several ways of achieving this have been developed: the reflection omit method (Cowtan, 1996); the gamma-correction (Abrahams, 1997), which speeds up the procedure by an order of magnitude; and the perturbation gamma method (Cowtan, 1999), which generalizes the method to any type of bias determination.

## 10.8 Fourier cycling and NCS averaging

One of the earliest ways of doing density modification is non-crystallographic symmetry averaging. Sometimes, different molecules in the asymmetric

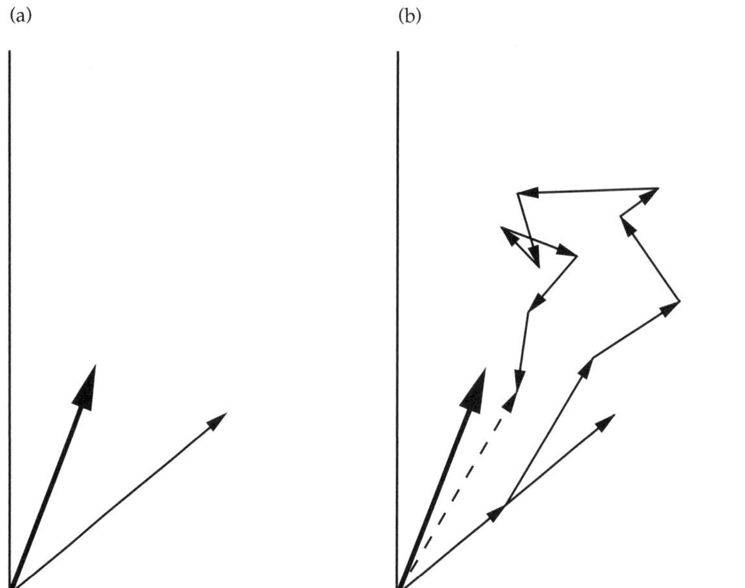


**Figure 10.4** Vector diagram of the effect of convolution of many structure factors in reciprocal space. In (a) is shown the true structure factor (thick arrow) and the initial, unmodified structure factor (thin arrow). In (b) the convolution operator applied to the initial structure factor (thin arrow) results in a closer estimate (dashed arrow) of the true structure factor (thick arrow).

unit are related by rotation and/or translation operators that do not belong to the crystal symmetry. Thus, when crystallographic symmetry operators are used, they superimpose the entire crystal lattice onto itself; non-crystallographic symmetry operators do not have this property.

Another way of looking at this is that non-crystallographic symmetry operators cannot be used to tile three-dimensional space, and thus are not of the class of crystallographic symmetry operators. Because these symmetry related molecules are not related by crystallographic symmetry, extra symmetry is introduced in reciprocal space over and above the symmetry of the Laue group. Bricogne (1974) gives mathematical relationships necessary to efficiently take advantage of this extra source of information.

In many ways, NCS averaging is the easiest density modification technique to intuitively understand, especially if considered in real space. In the process of NCS averaging, one simply takes all the different NCS related molecules in the asymmetric unit, superimposes them, and then replaces their density with the average density. Because these molecules are in a similar chemical environment, and are of similar shapes, when we superimpose them, regions of similar electron density reinforce each other. We therefore increase the signal from the protein, and as we overlay multiple proteins, their signal increases additively. Likewise, the noise decreases by $1/n^{1/2}$, where n is the number of non-crystallographic symmetry related molecules. This property of signal amplification and noise reduction in NCS averaging is illustrated graphically in Fig. 10.5.

In NCS-averaging, bias problems occur in Fourier cycling that are similar to the ones we mentioned in solvent flattening. For example, in two-fold averaging the result within the protein region is biased towards the initial map by 50%. Since we calculated the average of the two molecules at each grid point, half of the original density is retained. Therefore, similar treatment of the bias is required. Removing the bias results in swapping of the densities in two-fold averaging, and replacing the density of a molecule by the average of the others in high non-crystallographic symmetries.

(a)

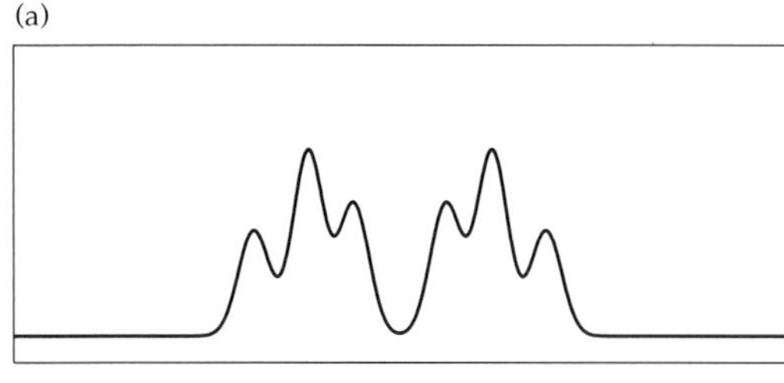

(b)

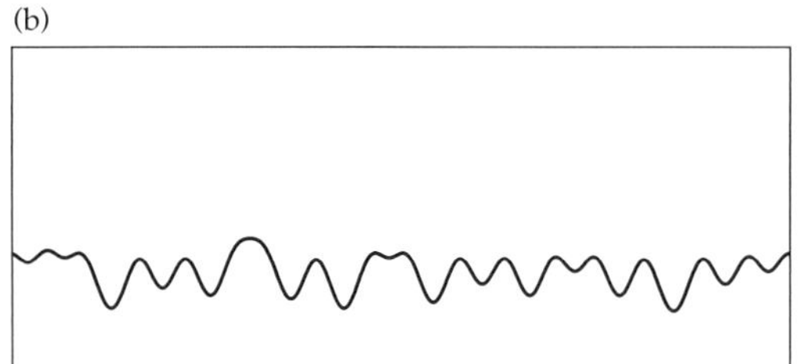

(c)

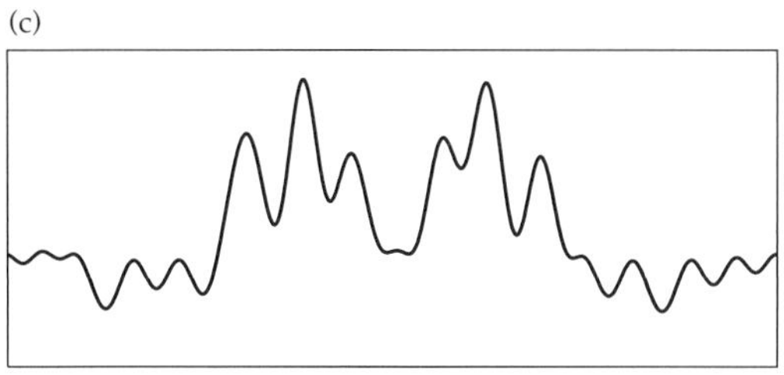

(d)

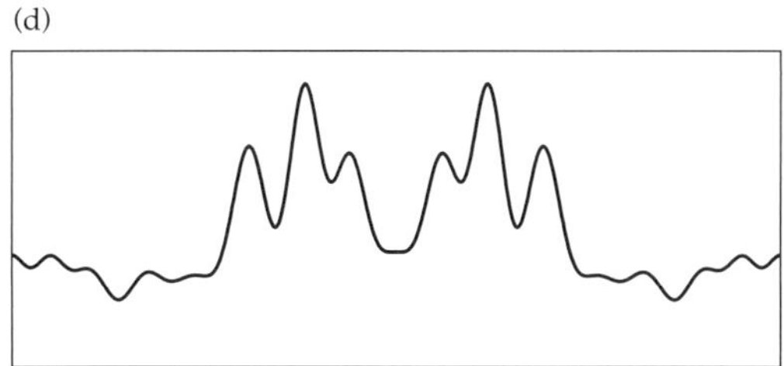

**Figure 10.5** Increase of the signal to noise ratio in non-crystallographic symmetry averaging. In (a) is shown a one-dimensional representation of the electron density of a macromolecule. In (b), a graph of the noise that results from the sources of errors in the crystallographic process, including experimental phasing and measurement errors. In (c), the observed density composed of the true electron density with the noise component. In (d), the effect of non-crystallographic symmetry improves the signal from the macromolecule while decreasing the noise level, the dotted lines shows the level of bias.

## 10.9 Histogram matching

Proteins often fold into a compact, globular form, composed of secondary elements such as beta sheets and alpha helices. This regular collection of building blocks leads to interesting similarities between the electron density across protein folds and families. These similarities can be exploited, and can provide a path to improve poor electron density.

Histogram matching is a technique that took its cue from the field of image processing. In various fields of study, one common problem is that of controlling contrast and brightness of an image. One such field is map making using aerial photography. Due to the wide variety of conditions and ground types, the aerial photographs are often badly corrected for contrast and brightness. One way of fixing this problem is that of histogram matching. In histogram matching, one does not look at the image *per se*, but instead, the histogram of the intensity values of each pixel, binning them into their appropriate histogram area.[1] After making a histogram of your model, comparing it to a histogram of a known good image and changing the model so that its histogram resembles the known image, the contrast and brightness of the starting image are dramatically improved.

As was mentioned before, solvent flattening helps to improve the low contrast solvent region, whereas histogram matching helps to improve the high contrast protein region. In addition, histogram matching may be used for phase extension, where by adding and phasing thin shells of data in reciprocal space, good phase estimates for structure factors of previously unphased structure factors can be obtained. One situation where this happens often is when solving a structure by MIR. Often, the diffraction pattern of a protein modified by isomorphous replacement will not diffract as far out into reciprocal space as those of the native crystal. By a judicious application of phase extension, phases that could not be obtained from the MIR experiment may be obtained by phase extension. In addition to the mainly textual description given above, a simple one-dimensional illustration is shown in Fig. 10.6.

In contrast to the situation in solvent flattening and non-crystallographic symmetry averaging, in histogram matching, the theoretical gamma correction performs less well than the perturbation gamma. Cowtan (1999) shows that, in the case of histogram matching, there is not a single,

(a)

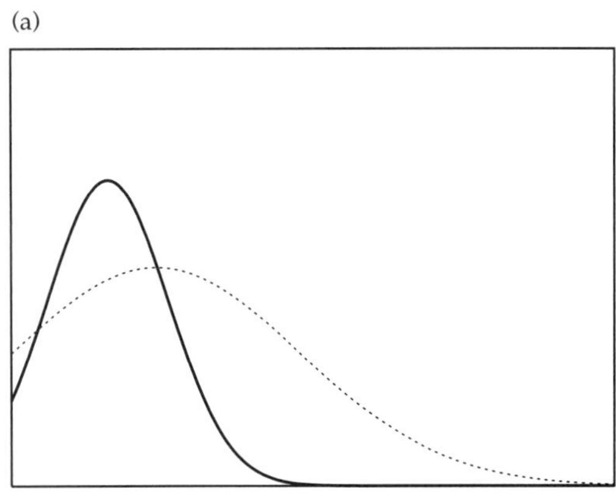

(b)

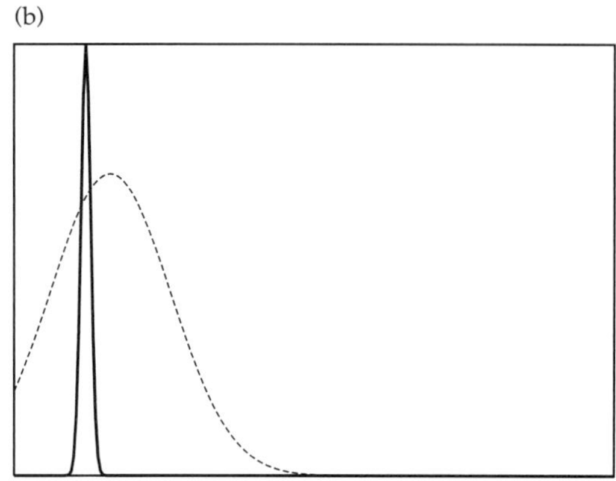

**Figure 10.6** Histogram matching. In (a) are shown a histogram from a first map in phase refinement (dashed line) and a theoretical protein histogram (solid line). In (b), the protein histogram (dashed line) and a sharp solvent histogram (solid line) are shown.

global gamma value because of non-linearities in the histogram matching method. The problem is underdetermined, leading to multiple solutions. The perturbation gamma algorithm runs a single histogram-matching cycle twice, once normally, and once with a map with a small amount of noise added to it. By comparing the correlation between these, one can determine a gamma correction for any subset of the data.

## Note

[1] A histogram plots frequency distributions of observed values. If observations within a certain interval occur frequently, a histogram will plot a high value for this interval, irrespective of where or when these observations occur. A histogram of the values cast by a single perfect die is flat, whereas a histogram for the total value cast by a pair of dice peaks at 7, where it is six times higher than at values of 2 or 12. For frequency distributions of correlated observations, multidimensional histograms may be useful.

## Technical notes

Some of the popular density modification programs implementing the real space restraints discussed above are Solomon (Abrahams, 1997), DM (Cowtan, 1999), RESOLVE (Terwilliger, 2003), Pirate (Cowtan, http://www.ysbl.york.ac.uk/~cowtan/pirate/pirate.html) and SHELXE (Sheldrick, 2002).

**Solomon.** This was the first density modification program to use solvent flipping, where density in the solvent region is inverted or 'flipped' to enhance density modification. Can be downloaded with the CCP4, but a superior script implementation is found in the SHARP suite.

Availability: http://www.ccp4.ac.uk
http://www.globalphasing.com

**DM.** One of the most popular density modification programs, DM comes bundled with the CCP4 suite. It incorporates many different ideas in density modification including histogram matching, NCS averaging, multi-resolution modification, Sayre's equation and skeletonization.

Availability: http://www.ccp4.ac.uk

**RESOLVE.** An easy to use density modification program that uses statistical density modification, which is an application of the general principles of maximum likelihood to density modification. One of its central algorithms is to iterate through every reflection in turn, examining all of the possible phases for the one that gives the most probable map. This procedure determines the most statistically valid hypothesis for every phase and is designed to help reduce bias. The RESOLVE program is also capable of performing model building and has many other advanced tools for protein structure solution.

Availability: http://solve.lanl.gov

**Pirate.** A new statistically based density modification program that uses sparseness/denseness and order/disorder in a statistical framework to model a new protein structure from ones that have been previously determined.

Availability: http://www.ysbl.york.ac.uk/~cowtan/pirate/pirate.html

http://www.ccp4.ac.uk

**SHELXE.** Part of the SHELX suite of programs, SHELXE uses a variety of novel algorithms to help perform density modification. One of it's primary algorithms is the 'sphere-of-influence' method, where a 2.42 Å sphere of 92 (or 272) points is systematically moved in real space through the starting electron density map, regions of high variance are assigned to be protein and regions of low variance are assigned to be solvent.

Availability: http://www.uni-ac.gwdg.de/SHELX

## 10.10 Conclusions

In this chapter we have presented the concept of density modification, where we take *a priori* information about the structure of the macromolecule we are studying to improve phase estimates. These concepts of density modification for the improvement of phases from crystallography have been implemented in many different programs in the past. Some of the popular density modification programs implementing the real space restraints discussed above are SOLOMON (Abrahams, 1997), DM (Cowtan, 1999), and RESOLVE (Terwilliger, 2003).

## References


Abrahams, J. P. (1997). Bias reduction in phase refinement by modified interference functions: introducing the correction. *Acta Crystallogr. D* **53**, 371–376.

Cowtan, K. (1999). Error estimation and bias correction in phase-improvement calculations. *Acta Crystallogr. D* **55**, 1555–1567.

Sheldrick, G. M. (2002). Macromolecular phasing with SHELXE. *Z. Kristallogr.* **217**, 644–650.

Terwilliger, T. C. (2003). Statistical density modification using local pattern matching. *Acta Crystallogr. D* **59**, 1688–1701.

CHAPTER 11

# Getting a macromolecular model: model building, refinement, and validation

R. J. Morris, A. Perrakis, and V. S. Lamzin

## 11.1 Introduction

Determination and analysis of three-dimensional (3D) structures is a cornerstone in modern molecular biology. Macromolecular structure plays an unchallenged role in the interpretation of biochemical data and constitutes a main key that can open the door to unravel mysteries of the function of biological macromolecules. The vast majority and a wide variety of chemical reactions and other processes in living organisms is carried out by proteins. They are typically specific, highly efficient and yet very versatile. The formation, development, and the sustenance of a living organism are governed by the correct function of proteins, which in a given environment is defined by their spatial structure. Therefore the investigation into macromolecular structures and the characteristics that determine their function has been of particular interest over the last decades, often giving rise to new and highly detailed biological insights at the atomic level.

A large number of methods have been established to address the gathering structural information. Some are based on the biophysical properties of the studied molecules, others on their chemical composition. A most important final step is to properly relate the structural knowledge to the inferred functionality. Currently, X-ray crystallography is the predominant technique for 3D structure determination, owing to its now practical simplicity, ease, and high level of automation as compared to the formerly tedious and time consuming measurements as well as computationally and labour-intense interpretation.

Especially during the last two decades, crystallography has developed into the primary tool for the investigation of biological macromolecules. Subject to the strength of the X-ray diffraction produced by fragile protein crystals and, therefore, the resolution of the crystallographic data, the method allows visualization of the electron densities down to the individual atoms and sometimes even further – approaching a resolution at which subtle, but biochemically most important, electronic differences may be studied. This obviously had important consequences in opening up an entire new wealth of possibilities for the structural investigation of proteins and the establishment of structure–function relationships at truly atomic level of detail.

Protein molecules are generally large and are composed of polypeptide chains of amino acids that provide amazing conformational flexibility. This is reflected not only in the way proteins are folded into an appropriately arranged globular body with a dedicated purpose (function) but also in the way they undergo conformational changes upon interactions with other proteins or small molecules (ligands). This conformational flexibility results in a challenge in obtaining sufficiently ordered protein crystals that, being placed into an X-ray beam, would be able to provide diffraction data to sufficient resolution. An additional feature of protein crystals that promotes this flexibility is that the molecules pack loosely into the crystalline lattice and are surrounded by layers of solvent. Indeed, protein crystals are approximately half liquid, with the fraction of solvent varying from 25 to 85%. In extreme cases, a protein crystal is not too dissimilar from a glass of good wine.

Recent years have witnessed further progress in the development of the underlying methodology for the determination of 3D macromolecular structures. A particular emphasis is given to high-throughput methodologies as an integral part and one of the specific goals of structural biology.

In this chapter, high-throughput automation efforts being developed to meet the needs of Structural Genomics initiatives will be cast in the framework of an optimization problem. A general overview will be given on optimization techniques with a bias specifically towards the problem of crystallographic refinement. This picture will be extended as we brush over model building, program flow control, decision-making, validation, and automation. Finer details of different approaches will be painted in a conclusive review of some popular software packages and pipelines.

## 11.2 Basics of model building and refinement

### 11.2.1 Introduction to optimization

Optimization is an important field of mathematics with applications covering virtually all areas of science, engineering, technology, transport, business, etc. It is hardly surprising that much effort has been invested in this area and that well-matured techniques and methods exist for solving many kinds of optimization problems. State-of-the-art optimization packages are highly complex and fine-tuned software masterpieces that often contain many ingenious ideas, robust heuristics, and decades of manpower on the underlying research and work. To run through the theoretical and algorithmic details of these tools is clearly beyond the scope of this chapter. Instead we walk through some basic ideas and considerations.

Optimization is concerned with finding extrema (minima and maxima) of functions (provided that they have them). The function $f(.)$ that should be optimized is called the *objective function* or *cost function*. In the general case, the objective function $f(.)$ will depend on several variables, $\mathbf{x}=(x_i,\ldots,x_N)$. The basic issue of an optimization problem is shown in Fig. 11.1. The plot depicts a one-dimensional function $f(\mathbf{x})$ and its dependency on the variable $\mathbf{x}=(x_1)$. Optimization theory aims to provide methods for determining the values of the variables $\mathbf{x}$ such that the objective function is either maximized or minimized. The variables that optimize $f(.)$ are known as *optimal values*. An important practical shortcut is to not necessarily obtain the optimal values of the variables but to approach them to a satisfactory accuracy (tolerance) within a reasonable amount of computational time.

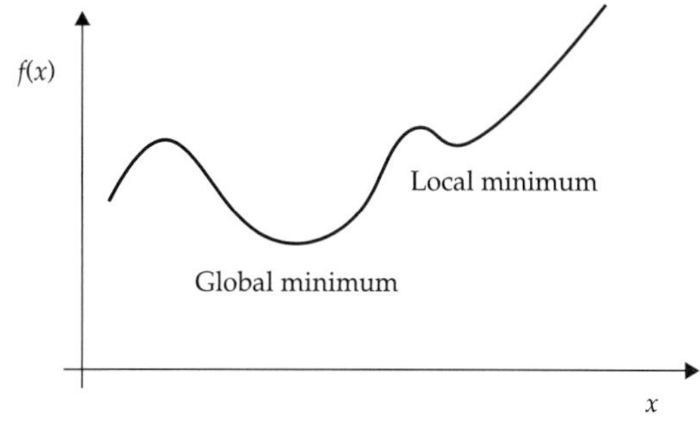


**Figure 11.1** The basic problem of an optimization problem with a one-dimensional function having more than one minimum. The slope (gradient) can give useful information concerning which direction to seek the minimum and which point to try next. Extrema are characterized by a gradient of zero, so methods that rely solely on this information will halt also in a local minimum.

Without loss of generality, one can formulate all optimization problems as minimization problems, with the maximum for the function $g(.)=-f(.)$ being the minimum for the function $f(.)$. By definition, $f(.)$ has a minimum at point $\mathbf{x}^0=(x_i^0,\ldots,x_N^0)$ if, and only if $f(\mathbf{x}^0) < f(\mathbf{x})$ for all $\mathbf{x}$ over which the function is defined. If the condition $f(\mathbf{x}^0) < f(\mathbf{x})$ is valid only within a small neighbourhood of $\mathbf{x}^0$, then $f(\mathbf{x})$ is said to have a local minimum in $\mathbf{x}^0$.

A crystallographic example of optimization would be the minimization of a least-squares or a negative log-likelihood residual as the objective function, using fractional or orthogonal atomic coordinates as the variables. The values of the variables that optimize this objective function constitute the final crystallographic model. However, due to the complexity of the crystallographic objective function a macromolecular model never truly optimizes the function, that is the global minimum is never reached. Instead, the model – although typically very good – is an approximation to what Nature has gathered within the macromolecular sample. This point should always be kept in mind and would indeed still be valid even if the model did represent the global minimum of the target function. This should become clearer further down the text.

Optimization problems in crystallographic structure refinement are seldom convex, that is very rarely characterized by a *unimodal* function $f(\mathbf{x})$. Regularization of a two-atom model is an example of such a unimodal function, Fig. 11.2a. In contrast, Fig. 11.2b shows a profile of a function for modelling an amino acid side chain – the peaks correspond to the possible rotamers. In this case, the shape of the function $f(\mathbf{x})$ is called *multimodal*. Such functions arise naturally in structural macromolecular optimization problems and possess a highly complex multiminima energy landscape that does not lend itself favourably to standard robust optimization techniques.

Typically one has a function at hand that may be described by

$$f(\mathbf{x}) : A \rightarrow \Re$$

in which the domain $A$, the *search space*, is a subset of Euclidean space $\Re^n$, often specified by a set of constraints in the form of equalities or inequalities that these solutions must satisfy thereby reducing the effective dimensionality of the search space. Solutions that satisfy all constraints are known as *feasible solutions* in the sense that they are plausible under given boundary conditions. Examples for feasible solutions in macromolecular crystallography are cell parameters that obey the space group restraints, or bond distances that agree with known stereochemistry, or torsion angles that fall within an allowed region in the Ramachandran plot (Ramakrishnan and Ramachandran, 1965).

In many crystallographic problems, the choice of the variables $\mathbf{x}$ is subject to constraints (boundary conditions represented by equations). The problem is then known as a *constrained optimization problem*. An example would be the refinement of a macromolecule with constraints on the bond lengths and angles, or TLS (translation-libration-screw) or NCS (non-crystallography symmetry) refinement (e.g. Tronrud, 2004), where a protein model is split into a number of rigid bodies within which the atomic positions and displacement parameters are constrained. Problems with no limitations on the freedom of the variables are called *unconstrained optimization problems*.

A special but very important topic is the tightness of the boundary conditions. A (holonomic) *constraint* is an absolutely tight condition (an equality – a precise setting or sharp probability density function of values), which in effect reduces the number of parameters to be refined. We may, however, wish to formulate the problem where the condition is relatively soft (an inequality – a range of plausible values and a potentially broader probability function). For example, within the above mentioned constrained refinement case we may want to allow bond lengths to vary slightly within reasonable chemical limits, or for the values of *atomic displacement parameters* (also known as *temperature factors*) not to jump too sharply from one atom to the next, or to give a preference to one of the rotamers in Fig. 11.2c if it falls into electron density. The degree to which we would like in the latter example to take the density height into account defines the tightness of this additional condition. Such conditions are called *restraints*. Mathematically, this does not reduce the number of refined parameters but instead increases the number of observations. It now becomes apparent that a restrained optimization problem may deal with observations, which have different physical origin (e.g. measured X-ray intensities combined with stereochemistry), different metrics, and different degrees of variability. Thus the term 'total number of observations' becomes ambiguous and, if quoted, should be taken with a pinch of salt, unless precisely defined.

Optimization problems and the computational techniques to tackle them are often classified further depending on the properties of these constraints, the objective function, and the domain itself. *Linear Programming* deals with cases in which the objective function $f(\mathbf{x})$ is linear and the set $A$ is specified through linear equalities and inequalities. If the variables $\mathbf{x}$ can only acquire integer values,

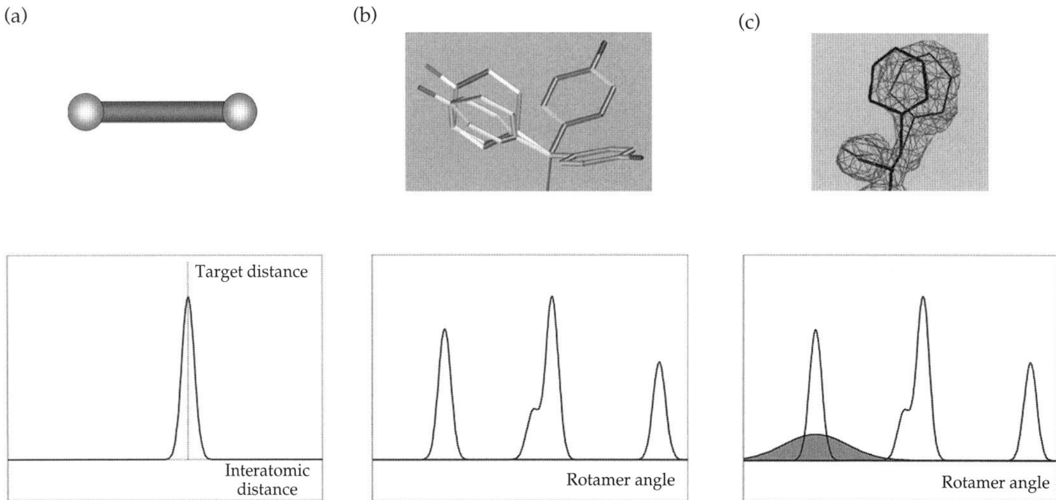


**Figure 11.2** This figure shows the probability functions for two typical crystallographic examples. In (a) two atoms show a preference for a certain bond distance, which is inversely related to an energy function of minimization target. This function is unimodal and the corresponding optimization problem is easy to solve. In (b) four different rotamer positions are depicted. The probability distributions are derived from knowledge-based potential of this side chain. This function is multimodal and requires a sufficiently wide sampling to solve as an optimization problem. Even so, the 'best' result may not be correct. In (c) the electron density, shown in darker colour, is also taken into account that, when correctly weighted, points clearly to the right solution. It would be misleading to state that this is a standard situation but is does show the significance of taking all available data into account and weighting them appropriately.

such studies of linear problems are called *Integer Programming*. Further generalization is covered by *Quadratic Programming* that handles objective functions with quadratic terms, but still with linear constraints. If there is no requirement for linearity and the variables are still integer, the problem is considered to be one of *Combinatorial Optimization*. *Non-linear Programming* is an even more general topic and allows the objective function or the constraints or both to contain non-linear parts. *Dynamic Programming* cuts the optimization problem into smaller subproblems from which the global solution is constructed. Many more classifications exist and dedicated handbooks or software manuals should be consulted for further information. The book on statistical learning by Hastie *et al.* (2001) provides a good basic introduction to many of these topics.

Most optimization procedures work in an iterative manner. An initial guess is made at where the minimum may lie and various techniques are applied to improve that estimate. This is generally achieved by random to very sophisticated sampling methods. If the objective function $f(\mathbf{x})$ is differentiable then its gradient

$$\frac{\partial f(\mathbf{x})}{\partial \mathbf{x}}$$

(the first derivative vector of the function with respect to the variables – *slope*) can be used effectively to determine where to sample next. A simple method that uses gradient information to find a local minimum is *gradient descent* or the *method of steepest descent*. If the second derivative of the function exists, the Hessian (the second-order derivative matrix of all combinations of variables) can be employed to help the search and to classify stationary points. The use of Hessian information significantly improves the search and reduces the number of required steps to reach the minimum. Here each step in the iterative procedure is proportional to the gradient at the current point. Given an appropriate step size $\alpha$ (which can be problematic to determine) this approach works very well, also in high-dimensional space:

$$\mathbf{x}_{\text{new}} = \mathbf{x}_{\text{old}} - \alpha \mathbf{H}^{-1} \left( f(\mathbf{x}) \frac{\partial f(\mathbf{x})}{\partial \mathbf{x}} \right)$$

A better alternative is often the *conjugate gradient* method. *Newton's* method uses gradient and Hessian information to compute the next point, in effect approximating the objective function at each point by a quadratic function and finding the minimum of this function. A drawback is that the Hessian may not be easy to compute and even more so the inversion thereof which is needed for the step computation. The matrix inversion can be an unstable and computationally very expensive operation. In this context *quasi-Newton*, *Gauss–Newton* and *Levenberg–Marquardt* methods and especially the Broyden–Fletcher–Goldfarb–Shanno method should be mentioned. The interested reader is advised to consult specialized literature for further details, or Tronrud (2004) for a crystallographic overview. In particular the Numerical Recipes books (Press *et al.*, 2002) provide a good introduction to many important approaches with implementation details.

There are methods that deliberately avoid the use of gradient and Hessian information. Such approaches typically require many more iterations but can nevertheless save overall on computation. Some popular ones are the *Simplex Method*, *Genetic Algorithms*, *Simulated Annealing*, *Particle Swarm* and *Ant Colony Optimization*, and variants thereof.

Many open-source, freely available and commercial packages exist for handling many kinds of problems; especially the statistics package *R* (Dalgaard, 2002) is recommended as a good starting point. It is important to choose the best method for the problem at hand, that is to match the optimization approach to the nature of the function, the variables, and the boundary conditions. There are many features to consider in choosing the method. Is the function linear, or not? Are the variables real or integer (or binary)? Are derivatives available? Is it a global or local optimization? What accuracy is required? Etc.

In macromolecular crystallography the optimization problems are too complex so that hardly any existing package could be used as a black box without considerable development and specification. Below we outline how some basic crystallographic problems in refinement and modelling are addressed starting from a foundation of *pattern recognition*.

### 11.2.2 A brief introduction to pattern recognition

Many tasks in the crystallographic structure solution process may be seen as pattern recognition problems, especially the areas of model building and validation do little more than just pattern matching. The objective of pattern recognition is to assign *objects*, which are described by a collection of numerical *features* (a *feature vector*), to the *class* to which they correspond. For example, a feature vector that assigns a quality 'wine' to different quality classes may consist of features such as grape variety, year, location, colour, alcohol content, nose, balance, acidity, sediment, container type (bottle or tetrapak), etc. Some of these features can be very useful in *discriminating* between the classes (such as year and location in this example) while others may not have such a high *discriminative power* (such as alcohol content perhaps). Often features can be combined to provide new features that contain a similar power of discrimination (in this example, all these features could be approximated by the price tag).

Pattern recognition can be formulated as an optimization problem that minimizes the risk of making a wrong class assignment. Feature selection can be formulated as an optimization problem that aims at maximizing the discriminative power, often subject to further constraints, such as minimizing the dimensionality and therefore casting most of a population's variance on as fewer features as possible. Such combinations of features for the task of classification are particularly well explored by newly emerging computational techniques, such as *neural networks* (e.g. Bishop, 2002). They are generally based on the principle of a multilayer perceptron. Networks with already as few as three layers and sigmoidal activation functions can approximate any smooth mapping to an arbitrary accuracy (Lapedes and Farber, 1988). An important feature of neural networks is that they do not necessarily contain feedback loops and that their outputs can be calculated as explicit functions of the inputs.

In order to achieve the goal of classification, a good *classifier* is needed. *Supervised learning* is a popular approach that may be used to *train* the classifier to discriminate between the classes using the features from a *training set*. A training set is a set of objects with already assigned classes. Supervised learning therefore implies the availability of prior knowledge for construction of the *training set*. It is explicitly known what the result of the classification should be. Concurrently, a *test set* is kept apart to validate the classifier, while developing it, against data that have not been used for this training. A *validation set* is kept apart for the final testing of the developed algorithms. In macromolecular crystallography having available a large set of structures and their experimental data, will provide an invaluable resource for supervised learning but also for general algorithm development. There is a wide collection of *classifiers* (or *classifier functions*) that can be used for supervised learning, depending on the nature of the features: neural networks (i.e. a multilayer perceptron); likelihood-based methods (including linear and quadratic discriminators); nearest neighbour methods (i.e. a linear vector quantizer); and rule-based classifiers (such as binary trees and support vector machines). These methods are easily accessible for testing through publicly available software (e.g. Lippmann *et al.*, 1993). Any ensemble of classifiers can be linked to form a *committee*, where each classifier has a 'vote' and, if enough classifiers reach the same conclusion, this result is taken. In many practical implementations it is preferred that each classifier is given a smooth weight (rather than a binary 'voting') so that the 'votes' of all *committee* 'members' are used to the extent of their reliability.

As in the above example in which the 'wine quality' is a somewhat ill-defined concept subject to individual taste, many classification schemes are often heavily biased by the viewpoint of the researcher (and this can influence the performance). *Unsupervised learning* (e.g. Ritter *et al.*, 1992) largely avoids this bias but at the cost of often less powerful methods and the missing interpretation of the arising classes, where often it is not obvious what these classes represent.

Similar to optimization, the field of pattern recognition is well developed with many high quality textbooks (e.g. Theodoridis and Koutroumbas, 2006; Duda *et al.*, 2000) and some excellent and user-friendly software packages.

### 11.2.3 Crystallographic model building

The aim of crystallographic model building is to construct a model that explains the experimental data under the condition that it should make physical and chemical sense. To build a crystallographic macromolecular model the most crucial information is the reconstructed *electron density map*. Interpretation of a map consists of examining topological features of the density and using knowledge of the chemical nature of macromolecules to determine the atomic positions and their connectivity. Before the development and availability of automated software approaches this was a cumbersome task that could easily take many months and required experience in protein structure and chemistry, and often a vivid imagination.

In order to construct an electron density map, into which the macromolecular model should be built, both the *amplitudes* and the *phases* of the diffracted X-rays need to be known. Since only the amplitudes are directly attainable in a diffraction experiment, obtaining phases (solving the so-called phase problem) plays a central role in crystallography. The initial phase estimates, provided either by the use of a homologous model – *molecular replacement* (Turkenburg and Dodson, 1996) or by experimental techniques involving the use of heavy atoms and synchrotron radiation, (M/S)IR(AS), (S/M)AD (Ogata 1998), are often of rather poor quality and result in inadequate electron density maps. The model, which can be built into such maps may be incomplete and even, in parts, incorrect. It may therefore require rounds of extensive refinement combined with model rebuilding. Visual examination of the electron density and manual adjustment of the current model is a tedious, time demanding, and subjective heavily step that relies on user experience. It has been recently eased by automated methods, and these and other related developments are discussed below.

### 11.2.4 Crystallographic refinement

Macromolecular crystallographic refinement is an example of a restrained optimization problem. Standard refinement programs adjust the atomic positions and, typically, also their atomic displacement parameters of a given model with the goal of maximizing the agreement between the measured experimental diffraction data and predicted structure factors from the model. At the same time, they optimize the agreement to a variety of *a priori* available information, most commonly the stereochemistry of the model. The objective function for the refinement is a high-dimensional function. Its dimensionality is equal to the number of refined independent parameters and thus typically is in the order of many thousands. The reader may be referred to Tronrud (2004) for a discussion on crystallographic model parametrization. The landscape of this objective function in its multidimensional space depends primarily on the atomic coordinates of the model and is characterized by many minima. An additional complexity arises from the crystallographic phase problem and the often initially poor phase quality, which affect the completeness of the model, where even challenging techniques, such as conjugate-gradient, do not necessarily lead to the global minimum.

Macromolecular diffraction data are rarely of sufficient quality and quantity to allow construction of atomic models that would obey basic stereochemistry just based on this optimization of parameters to data. The observation-to-parameter ratio is a key factor in optimization procedures. For the optimization of a crystallographic model, experimental diffraction data extending to at least atomic resolution (about 1.2 Å) are necessary to provide a sufficient number of reflections and thus a high enough observation-to-parameter ratio to justify an atomic model. The dependence of the observation-to-parameter ratio is exemplified in Fig. 11.3. The number of diffracted X-ray reflections for a complete data in P1 lattice can be computed as

$$N_{\text{refl}} = \frac{2\pi}{3d^3} V$$

where $V$ is the volume of the asymmetric unit and $d$ is the resolution. Assuming the asymmetric unit to contain only one 'average' atom, nitrogen, with a molecular mass of 14 and the xyzB refinement (four parameters per atom) it is straightforward to compute the observation-to-parameter ratio for a given value of the solvent content. From Fig. 11.3 it becomes evident that at atomic resolution the optimization problem is well overdetermined and one can easily afford even anisotropic treatment of atomic displacement parameters, while at a resolution lower than about 2.7 Å the number of observed data becomes smaller than the number of parameters. This problem of lack of data is (to an extent) overcome by restraining (or less usually constraining) the model parameters to values determined from small molecule structure solutions. Most of the

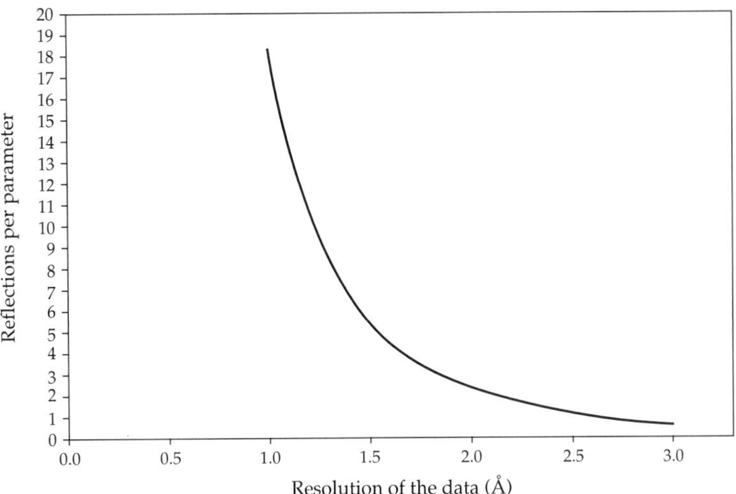


**Figure 11.3** The ratio of the number of reflections to the number of parameters in the XYZB crystallographic refinement of a protein model as a function of the resolution of the data. The data are assumed to be complete and the solvent content to be 50%.

refinement programs use stereochemical parameters derived from the Engh and Huber (1991) dictionary.

Refinement is thus the optimization of the model parameters to simultaneously fit both the experimental diffraction observations and a set of *a priori* known stereochemical information. It should become clear from the formula above, that solvent content crucially affects the observation-to-parameter ratio. Since crystals that diffract to lower resolution typically have higher solvent content, refinement can often be carried out decently well even at nominally low resolution.

The crystallographic $R$ factor, defined as

$$R = \frac{\sum_{\mathrm{hkl}} \left| F_{\mathrm{hkl}}^{\mathrm{obs}} - F_{\mathrm{hkl}}^{\mathrm{calc}} \right|}{\sum_{\mathrm{hkl}} F_{\mathrm{hkl}}^{\mathrm{obs}}}$$

is used to measure the fit between the model and the observed data and to monitor the refinement progress. In addition to the standard crystallographic $R$ factor, the so-called free $R$ value can be used as a cross-validating indicator to monitor the overall progress and to avoid fitting to noise (Brunger, 1992). The free $R$ factor is, in principle, the same as the standard $R$ value, but only for a small subset of the data that are never used throughout the refinement process and can therefore be taken as an independent evaluation of quality of fit as the model has not been influenced by this set of intensities; a technique that falls in the general category of bootstrapping methods for validation. Full cross-validation by bootstrapping would require that all data in turn are tested, that is a model is refined first with, for example, 5% of data excluded for validation, the refinement repeated several times from the beginning using different subsets of 'free' data. In practice, however, this would not be efficient and require huge CPU costs for thorough validation; a singe 'free' set is in general sufficient for crystallographic refinement and is typically employed in structure determination.

The use of cross-validation should be widely recommended for crystallographic model refinement. Its simplicity to understand for the non-expert and its power in discriminating models that are consistent with the experimental data make it indispensable. However, in modern crystallography model refinement is often falsely seen as the task to minimize the value of $R$ free, instead of correctly obtaining the model that best describes the data. One may decide to run several model refinement protocols in parallel and choose the one with the lowest $R$ free factor. However, it is important to understand that from this point onwards the free factor becomes biased to the chosen protocol.

A misconception, which still occurs in the literature, is that a cross-validation could serve as a local indicator rather than a global guide for the whole structure determination protocol. Cross-validation cannot be sufficiently sensitive to indicate as to whether a particular side chain could be built into one or another conformation.

In using cross-validation it is essential to avoid, or at least minimize, bias to the free $R$ factor itself. In the era of emerging automated procedures for modelling and refinement a frequent mistake is to set aside a fraction of reflections for minimization of the residual in reciprocal space and, at the same time, to use all data for computation of electron density and model rebuilding. Since local adjustment of the model in real space is equivalent to global phase adjustment in reciprocal space, the free reflection set becomes biased towards the current model and loses its validation credibility.

From a totally different perspective, omitting the free set of reflections from the experimental observations effectively reduces the observation-to-parameter ratio and can adversely affect the refinement. Clearly, an optimization problem crucially depends on the number of observations that are used to find optimal values for the model parameters. If done by a crystallographer, the result may depend on the human creativity and skills combined with wishful thinking. The danger of over-interpretation in normally not a caveat of automated model building algorithms and cross-validation may not be necessary since the growing polypeptide chains of traced residues can themselves be seen as an independent cross-validation criterion. Based on one of the authors' subjective experience (i.e. not fully cross-validated) it can be advised that when an objective model building software package is used in parallel with maximum likelihood refinement, it may be of advantage to not set aside the free set for validation but instead use all the data. This can diminish the 'rippling' effect caused by Fourier transforms with incomplete datasets, on top of depriving the optimizer algorithm of the refinement program from valuable diffraction data.

### 11.2.5 Model building and refinement are both phase improvement procedures

ARP/wARP challenged the common separation of model building and refinement by taking a more general view of the underlying phase optimization through coordinate manipulation and by allowing the extension of the macromolecular model as part of the whole process (Perrakis *et al.*, 1999). This placed structure solution on a more unified foundation and encouraged automation in linking the entire procedure.

Model building is an interpretation of the currently available electron density. Refinement is the adjustment of the built model to fit better to the experimental data. A crucial point here is that a density map computed from the refined model is generally better than the map obtained from the same model before the refinement. This then allows for an even better model to be built. Thus, refinement is needed to improve the outcome of model building by generating a better electron density map and model building is needed to provide a model in the first place and to provide stereochemical restraints for the subsequent refinement to proceed smoothly. This viewpoint merges these two steps into one model optimization process.

The ARP/wARP procedure is largely based on the concept of describing the electron density with a collection of *free atoms*, originally introduced for phase extension by Isaacs and Agarwal (1985). The main *a priori* knowledge in this case is the fact that proteins consist of atoms, although this information alone is hardly sufficient. ARP/wARP, however, takes this further – before linking the atoms, one has to position them correctly, not necessarily into peaks of the density but preserving at least the first order stereochemical information – the interatomic distances. Free atoms themselves account for the electron density features without the need to apply additional chemical knowledge or distance restraints between them. The coordinates of the free atoms are refined, updated where needed, and used by the ARP/wARP autobuilding module for construction of a polypeptide chain that is in agreement with the electron density. In the regions where it is not possible to make sense of the electron density features in terms of a protein model, free atoms are kept. This results in what is called a *hybrid model*, a mixture of free atoms and fragments of the protein structure. Much in the same way as in the 'traditional' X-ray structure solution, the hybrid model undergoes a refinement procedure with the highly successful maximum likelihood based refinement program REFMAC5 (Murshudov *et al.*, 1997) in which the parameters of the model are adjusted to best fit the experimental data and stereochemical expectations. If the quality of the model is sufficiently high, the phases improve overall and result in an enhanced electron density into which a more accurate and more complete model may be built at the next cycle. ARP/wARP, like human crystallographers, links model building and refinement together into a unified process that iteratively proceeds towards a more complete macromolecular model.

## 11.3 Software packages

### 11.3.1 An overview of refinement programs

Refinement is an optimization problem. Therefore we choose to outline the current programs based on how the optimization problem is formulated (including the variables and the objective function) and the method of choice for locating the minima (the solution method). The parameterization of a macromolecular model is a crucial step and should reflect the amount of available data that can be used to optimize the model parameters. This fluid parameterization is typically replaced by standard atomic coordinates (xyz), a displacement parameter (B), and occupancy (occ) parameterization in which the amount of data is accounted for in the weights (tightness) of the stereochemical restraints. Other parameterization approaches use torsion angle, rigid body, and NCS constrained/restrained (hard/soft) parameterization. Additional parameters include those used for modelling the bulk solvent. Within the free atoms framework, the atom type itself enters as a parameter since such atoms are initially simply diffracting balls of density without necessarily having a meaningful chemical identity or connectivity.

One of the most popular refinement programs is the state-of-the-art package Refmac (Murshudov *et al.*, 1997). Refmac uses atomic parameters (xyz, B, occ) but also offers optimization of TLS and anisotropic displacement parameters. The objective function is a maximum likelihood derived residual that is available for structure factor amplitudes but can also include experimental phase information. Refmac boasts a sparse-matrix approximation to the normal matrix and also full matrix calculation. The program is extremely fast, very robust, and is capable of delivering excellent results over a wide range of resolutions.

Another highly popular package is CNS (Brunger *et al.*, 1998). It offers a wide range of capabilities common to other packages and also includes torsion angle refinement, which is particularly powerful and useful in the low resolution regime. CNS also uses likelihood targets and the restraints are formulated via energy functions and force fields. CNS employs a conjugate-gradient based optimizer combined with simulated annealing. XPLOR (Brunger, 1993) and its commercial reincarnations are based on similar principles. The PHENIX (Adams *et al.*, 2002) package, which encompasses a dedicated refinement module, follows some of the CNS ideas but uses other, very advanced minimizers.

BUSTER/TNT (Bricogne and Irvin, 1996) is another likelihood based refinement package that excels especially in cases in which the model is still severely incomplete (Blanc *et al.*, 2004; Tronrud *et al.*, 1987). It uses atomic parameters but also has a novel solvent and missing model envelope function. The optimization method is a preconditioned conjugate gradient as implemented in the TNT package (Tronrud *et al.*, 1987) that had a faithful audience in the pre-likelihood era.

SHELXL (Sheldrick and Schneider, 1997) is often viewed as a refinement program for high-resolution data only. Although it undoubtedly offers features needed for that resolution regime (optimization of anisotropic temperature factors, occupancy refinement, full matrix least squares to obtain standard deviations from the inverse Hessian matrix, flexible definitions for NCS, easiness to describe partially overlapping occupancies, etc.), its power for lower resolution should not be underestimated.

### 11.3.2 Automation in model building

To review all existing model building concepts and programs is beyond the scope of this chapter. Instead, a small, selected number of key ideas and methods will be mentioned. The current software packages have anyway much in common so programs that are not described here should nevertheless be comprehensible with minimal extrapolation. Intuitively, the approaches generally try to mimic what an experienced crystallographer would do. Therefore, many of the methods follow the algorithms originally employed in molecular-graphics packages such as O (Jones *et al.*, 1991) or QUANTA (Accelrys Inc.) to aid the manual model building process. All techniques that address automation in macromolecular model building are, to a larger or smaller extent, based on the pattern-recognition aspect in the interpretation of crystallographic electron density map.

A popular method to represent an electron-density distribution in a way that captures the connectivity of the map was *skeletonization*, proposed back in 1974 (Greer, 1974) but put into practice by Jones and coworkers in the late 1980s (Jones and Thirup, 1986). It is a simple and elegant way to reduce an often cluttered 3D electron density map to a set of lines that capture the essential chemistry (connectivity) of the map. The method is an iterative procedure that removes points from a grid representation of the electron density map as long as this does not break their connectivity. A small set of grid points remains that can be used to produce a skeleton of the original density. Despite more sophisticated developments, skeletonization remains a powerful and commonly employed method and is perhaps still the most widely employed technique for computer-assisted manual macromolecular model building in experimental electron density maps. Related approaches that make use of electron density extremes include the core-tracing algorithm (Swanson, 1994) and molecular scene analysis (Fortier *et al.*, 1997).

The idea of ESSENS (Kleywegt and Jones, 1997) is to recognize secondary structural templates around each point in the map by an exhaustive search. This was one of the first successful approaches to automate construction of new protein structures from previously seen fragments, following the syntactic concept of pattern recognition. FFFear (Cowtan, 1998) presented an elegant implementation of the ESSENS idea in a fast Fourier framework that allowed not only to speed-up program execution but also the use of structure templates in a probabilistic manner.

The software packages ARP/wARP (Lamzin and Wilson, 1993; Perrakis *et al.*, 1997, 1999, 2001; Lamzin *et al.*, 2001; Morris *et al.*, 2002, 2003, 2004; Cohen *et al.*, 2004, Zwart *et al.*, 2004), TEXTAL/CAPRA (Ioerger *et al.*, 1999; Holton *et al.*, 2000; Ioerger and Sacchettini, 2002, 2003), MAID (Levitt, 2001), and RESOLVE (Terwilliger, 2001, 2003a, 2003b, 2003c, 2004) aim to deliver a complete macromolecular model starting from an electron density map. RESOLVE is perhaps one of the most advanced programs in terms of implementing a proper statistical pattern-recognition system. A promising development of RESOLVE involves the use of template matching of a number of commonly observed density distributions to perform density modification. FFT-based methods are employed to calculate derivatives of a log-likelihood map, and structure factor statistics are combined with prior knowledge of density distributions to provide scores, which are used to match templates to grid points and their surrounding region to perform density modification. TEXTAL uses a high-dimensional feature space consisting of rotationally invariant properties derived from the electron density. These include the average density and higher moments, the variance over grid points, the moments of inertia and spokes of density values (the three bonding tubes of density that are to be expected around a C$\alpha$ atom). TEXTAL computes these feature vectors and attempts to recognize the closest density pattern from a large database of precomputed patterns. After the density features have been matched, the coordinates from the database template are used to place atoms into density. The program MAID relies on skeletonization of a given electron density map to determine the path of the main chain. The skeleton grid points then serve as features in which secondary structural elements patterns are sought, very much like the computer graphics model building steps with a program such as O. A stereochemically accurate atomic model is then built in these secondary structure regions whilst taking care not to over-interpret them. Secondary structural main-chain fragments of sufficient length (15–20 residues) are slid along the sequence and the best match to the side-chain density is used to assign the residue type. Loops are built by extending the built main-chain fragments by sampling Ramachandran space.

The ARP/wARP package follows the full path of syntactic construction of a model starting from free atoms, through candidate C$\alpha$ atoms, peptide and dipeptide units, chain fragments to sequence docking, and side-chain building. It was the first software package to combine most steps of model building with refinement in an iterative manner. The power of ARP/wARP lies precisely in this iterative nature (Perrakis *et al.*, 1999). Although iteration is offered in other packages it is *de facto* less important and less powerful due to the lack of the use of the free atoms concept and the focus on a 'snapshot' interpretation of the current density map. The concept of ARP/wARP is different. When atoms are placed in electron density only the fragments that could be interpreted with confidence are built and refined as part of the restrained atomic model. The parts that were less certain are still allowed to move as free atoms. Atoms that do not refine well can be removed and density that is yet unaccounted for can be filled with atoms at any point. In this way, not only the atomic parameters but the atoms themselves and the connectivity between them are made part of the optimization that so encompasses both model building and refinement.

For a detailed comparison of some popular automated model building packages, see Badger (2003). For an analysis of structural features and their relationship to model building strategies, see Morris (2004). Further details should be sought in the original articles.

### 11.3.3 Automation in crystallographic structure solution: decision-making systems

An emerging goal in the field of macromolecular crystallography methods development is the overall automation of the time-consuming steps in structure solution. To a large extent, two software packages exemplified that concept at the turn of the previous decade: SOLVE (Terwilliger and Berendzen, 1999) for experimental phasing of the diffraction data and ARP/wARP (Perrakis *et al.*, 1999) for model building and refinement. Over the last few years, a drive from Structural Genomics initiatives has boosted a variety of outstanding activities. Most notable is the PHENIX project (Adams *et al.*, 2002) that has recently been released. Other automation attempts include a series of the so-called software pipelines, for example autoSHARP (Bricogne *et al.*, 2002; de La Fortelle and Bricogne, 1997), BnP (Weeks *et al.*, 2005), Elves (Holton and Alber, 2004), ACrS (Brunzelle *et al.*, 2003), CRANK (Ness *et al.*, 2004), and AutoRickshaw (Panjikar *et al.*, 2005).

Commonly known in the Structural Genomics jargon as *software pipelines*, these are systems that combine a number of macromolecular crystallographic computer programs with several decision-making steps linking the computational modules. Pipelines can be a simple chain-type sequence of steps or can involve internal loops at several levels of complexity, exemplifying the inadequacy of the term 'pipeline' to describe them. In the latter case there could be many different paths through which the structure determination evolves and these paths can either be predefined or modified on the fly, making that software rather complicated decisions systems with significant amounts of pre-existing scientific knowledge and new ideas incorporated in them. Some of these automated decisions systems rely mainly on one software package while others are more comprehensive.

Below we briefly describe the crystallographic software pipelines using AutoRickshaw as an example, with its flexibility and the ability to decide on the path to be taken dependent on the outcome of a previous step. On one hand, AutoRickshaw has features and general steps, which are also shared by many other pipelines. On the other hand, AutoRickshaw is perhaps the first software pipeline which aims not at the delivery of a fully built, refined, and validated model but rather at fast evaluation of the quality of the X-ray data in terms of interpretability of the obtained electron density map.

AutoRickshaw considers crystal structure determination as a multistep process in which each step in structure solution, from substructure determination to model building and validation, requires certain decisions to be made. These decisions involve choice of the crystallographic program most suitable for each task, together with optimal input parameters for each of these programs. The important parameters include the space group of the crystal, the number of molecules in the asymmetric unit, the type of the heavy atom derivative, the extent of derivatization, the diffraction limit of both the native and the derivatized crystal and the quality of the collected diffraction data. If the data collected in the first experiment during, for example, MAD measurements can be successfully interpreted, further data collection can be halted (Dauter, 2002). The choice of model building step within AutoRickshaw depends on the maximum resolution of the X-ray data. If it is lower than 2.6 Å, ARP/wARP Version 6.1 or *ESSENS* are invoked to identify helices in the electron density map. For higher resolution ARP/wARP is used for tracing polypeptide chains.

An important aspect in all pipelines is the evolution of the decision-making. As more data become available, the structure determination paths can be scrutinized thoroughly in order to increase the efficiency of the overall workflow.

## 11.4 Model building and refinement cookbook

Coming up with robust recipes for model building and refinement is a challenging task. Automated pipelines attempt to capture as many successful approaches as possible but still fail in some cases. If automation fails, the user has to understand the fundamental logic behind the software and grasp, at least qualitatively, the underlying mathematics, and successfully identify potential problems before making informed decisions. Seeking the advice of an experienced expert crystallographer can rarely be avoided. In this section we give some general guidelines for situations that an average crystallographer can encounter. It must be emphasized that each 'case study' should be treated critically, especially the resolution margins that characterize each case and the reader may be advised to keep in mind Fig. 11.3 and adjust the conclusions accordingly. It should be noted, that in quoting the resolution limits we will refer to the highest resolution of at least one dataset, which typically (but not necessarily) is the 'native' dataset. Phase extension procedures and cross-crystal averaging should allow to extend phases to the highest available resolution in all but some extreme cases. We start with emphasizing a few important issues:

• Checking the data quality is strongly recommended: inspection of the Wilson plot and data reduction statistics is very useful in judging the extent of the resolution to which the data can realistically be used. Pathologically bad data, for example those from a split crystal, twinned data, systematically incomplete data, low resolution overloads will always make model building and refinement hard if not impossible.
• If experimental phases are available, getting the best out of them before embarking on model building is worth the effort and will definitely pay off. Using software like SHARP (de La Fortelle and Bricogne, 1997) which puts particular emphasis on detail and proper statistical treatment can be highly recommended.
• Automated procedures typically underperform compared to a reasonably experienced crystallographer. If an automated package did not provide a solution, thinking about the problem and adjusting the strategy to the particularities of the case may lead to a success. Even if an automated structure solution package provided good initial phases but automated model building failed, careful inspection and improvement of the phasing step can help subsequent automated model building to work better.

### 11.4.1 Model building and refinement starting from experimental phases

In the following text we list very general case studies for native data extending to different resolutions. The recommendations apply to software versions that were available in mid-2006; it is likely that rapid progress in software development will make automated model building considerably more powerful, particularly at lower resolution. The suggestions about refinement are sufficiently general but, again, the features of software are likely to change in the near future. The software packages referred to are those which the authors have experience with; other software exists or may exist that offers similar or more advanced features.

*Very high resolution* (above 1.4 Å): Automated model building with ARP/wARP should work even with relatively poor starting phases since the free-atoms concept is particularly well suited for high resolution maps. Due to the large number of diffraction data at this resolution, the observation-to-parameter ratio is sufficient to allow detailed model parameterization and, together with stereo-chemical restraints, is high enough to correctly drive the optimizers. Refinement should generally be carried out with anisotropic ADPs using REFMAC or SHELXL. The use of strong NCS restraints may only harm. Double conformations of side chains and even double conformations of main chain should be detectable in the density and can be modelled. Solvent networks (except overlapping networks with partial occupancy) can very well be modelled automatically.

*High resolution* (between 1.4 and 2.0 Å): Automated model building with ARP/wARP should work with most phase sets. RESOLVE, which uses a template-based rather than atom-based approach, should also perform well but may be computationally more consuming. Refinement can best be carried out with REFMAC or PHENIX using isotropic ADPs since the amount of data is no longer sufficient for an anisotropic description of atomic displacement parameters. The use of TLS (Winn *et al.*, 2003) is highly recommended. A use of NCS restraints should be critically evaluated and in most cases the refinement can proceed without them. Double conformations of side chains should be visible and modelled. Ordered solvent can be modelled automatically.

*Medium-to-high resolution* (between 2.0 and 2.4 Å): Automated model building with ARP/wARP or RESOLVE should work for good phase sets that produce clear maps easily interpretable by a human. Such maps typically have average FOMs after phasing well above 0.6, which after density modification may rise up to 0.9 or higher. Refinement should best be carried out with REFMAC or PHENIX using isotropic ADPs. The use of TLS can still be very advantageous. NCS restraints in this resolution range are still not absolutely necessary but may be useful if indeed all copies of each molecule are identical. Some double conformations of side chains can still be modelled if clearly visible. Solvent atoms can be modelled automatically but may require visual inspection.

*Medium resolution* (between 2.4 and 2.7 Å): Automated model building with RESOLVE or ARP/wARP can work for good phase sets and in many cases RESOLVE can provide more complete models. However, feeding the RESOLVE model as input to ARP/wARP can sometimes produce a model, which is better that a model produced by any of the two programs on their own. Procedures that incorporate directly anomalous data to iterative model building and refinement, as described by, for example, Skubak *et al.* (2004) can be used when available. Refinement can still best be carried out with REFMAC or PHENIX using isotropic ADPs. BUSTER can offer significant improvements, especially with models that have missing regions. Phase restraints can be useful in providing additional 'observations' and helping the optimizer find a chemically plausible minimum. The use of TLS can still be recommended and NCS restraints are very likely to be useful. Solvent can still be modelled automatically but may require critical inspection.

*Medium-to-low resolution* (between 2.7 and 3.3 Å): Automated model building with RESOLVE is likely to give some partial models. BUCANNEER — a novel software from the CCP4 suite — can also be tried. In reality, most of the model will have to be built manually using interactive graphics software. Refinement with REFMAC, PHENIX, or BUSTER using isotropic ADPs should perform well in most cases. However, refinement in CNS (which should be also available in PHENIX by the time this text reaches the press) using a torsion angle parameterization (Rice and Brunger, 1994) should be tried. This way of describing the freedom of the molecular model requires a much smaller number of parameters and should therefore improve the performance of most optimizers. Phase restraints are very likely to be useful and should definitely be exploited, regardless of which parameterization has been chosen. The use of TLS can still be helpful. Macromolecules are viewed as collections of rigid bodies subject to anisotropic motions and TLS describes this with very little expense in terms of additional parameters. From the optimizational point of view a single TLS body is as 'expensive' for the minimizer as one alanine residue. NCS restraints must be used in this resolution range if available, unless the NCS related copies are clearly different. A few well-ordered solvent molecules could be modelled and might even be better visible than some side chains.

*Low resolution* (below 3.3 Å): Automated model building is unlikely to produce anything useful, unless the experimental maps are of outstanding quality. Refinement with CNS using a torsion angle parameterization would be a first choice. Phase restraints must be exploited. NCS restraints must be used if available and, in fact, they will offer the only realistic chance for the successful refinement at low resolution. Although one should always try hard to get the best possible model, inability to reduce the free R factor below 35% will not necessarily imply a failure, especially if the model is strongly supported by high quality experimental electron density and excellent stereochemistry as judged by the Ramachandran plot.

### 11.4.2 Model building and refinement starting from an available model

The discussion here is limited to model building; advice on refinement can be found in the corresponding paragraphs of the preceding section. A more detailed overview is given in Perrakis *et al.* (2001) and Carson (2006).

The rule of thumb is relatively simple: the higher the resolution of the X-ray data, the poorer (less similar, less complete) the starting model can be for the model building to still succeed. With a resolution of around 2.0 Åor higher ARP/wARP probably offers the best chances for success. Initial models having only 20% sequence identity (Weichenrieder *et al.*, 2004) or encompassing only 60% of the scattering mass (Pichler *et al.*, 2005) have proved sufficient to provide an adequate starting point. When the resolution is lower than 2.0 Åor the model is less than about two-thirds complete, automated model building should better be preceded by either statistical density modification as implemented in RESOLVE, 'standard' density modification techniques (e.g. Abrahams, 1997) or new approaches (e.g. PIRATE from the CCP4 suite). RESOLVE and ARP/wARP both have a good chance of bootstrapping a correct model after density modification.

With poor models or lower resolution is may be advantageous to compute model phases and then treat them as 'experimental' ones. Guidelines described in the preceding section would then be applicable.

## 11.5 Concluding remarks

While the automation of initial model building has largely been alleviated (at least for good quality electron density maps and sufficient resolution of the X-ray data), an increasing number of non-expert crystallographer users do need a substantial amount of time to complete and finalize the model. This step that a few years back seemed to be of secondary importance is now becoming a bottleneck and will be one of the targets of the new developments. Inherent components of this task are building up poorly ordered regions, modelling alternate conformational networks of atoms, and the automated construction of ligands and nucleic acids bound to the macromolecule. This can only be efficiently realized with the use of a sophisticated decision-making system that will judge the model completeness and learn from the accumulated history. Although great advances have been made in the field of crystallographic model building and refinement, further effort must be invested to turn this technique into truly accessible and easy-to-use tool for biologists. The coming years will no doubt see a greater involvement of researchers with different expertise in achieving this goal.

A major challenge will be the refinement based on multiple diffraction data sets. For example, a protein with two models available at high resolution and a complex structure determined at lower resolution, one could envisage the simultaneous refinement of all three structures. Areas that are unaffected by the formation of the complex and short-range features (i.e. bonds and angles that are likely to be invariant) would benefit from the high resolution data, while long-range features (relative placement of helices for example) could be determined from the contribution of the low resolution data. Such developments still await theoretical underpinning and implementations. Similar ideas can be used for the refinement of the same protein in differently liganded states.

With the continuously increasing usage of X-ray crystallography, the number of macromolecular structures deposited in the PDB, Fig. 11.4, their size, and complexity are rapidly growing. Particularly stunning is the increase in the volume (and the content) of the crystallographic asymmetric unit.

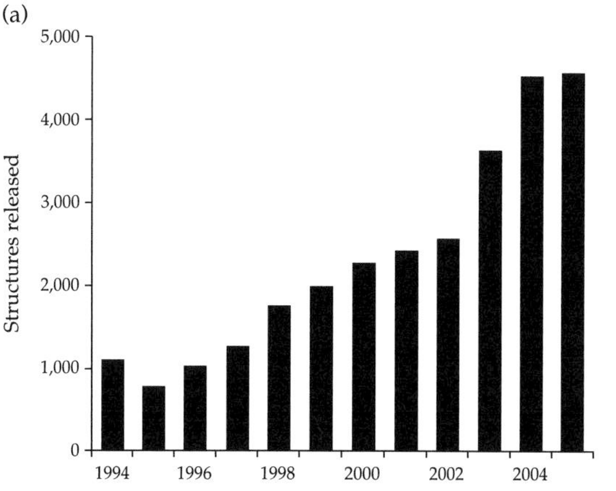


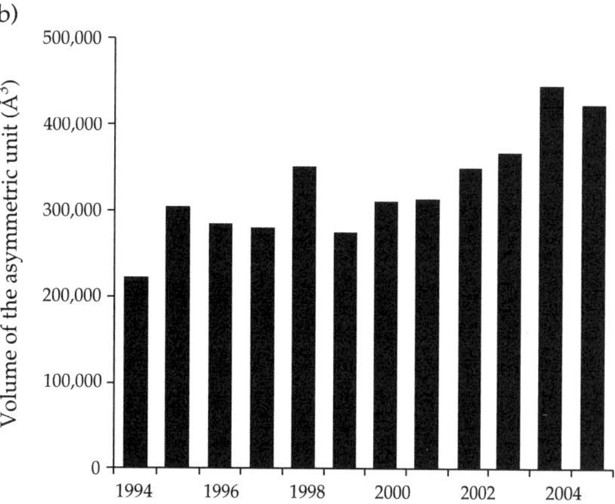


**Figure 11.4** (a) The boost in the number of crystal structures released by the PDB. (b) The growing complexity of the structures is indicated by the increase in the average volume of the crystallographic asymmetric unit.

Its average value, recently jumped over 400,000 Å$^3$ (1500 amino acid residues for the 50% solvent content) may be affected by fewer giant structures such as the one for avian birnavirus (Coulibaly *et al.*, 2005) with 38 MDa content in the asymmetric unit. Although the unrefined model of this structure consists of approximately positioned C$\alpha$ atoms into a 7 Å density map, it does serve as an excellent example of the potential of macromolecular crystallography for challenging projects. It is important, that the data for these and even more complicated structures are obtained using state-of-the-art technologies and that the derived models present an as accurate and complete interpretation as possible.

## References


Abrahams, J.P. (1997) Bias reduction in phase refinement by modified interference functions: introducing the gamma correction. *Acta Crystallogr.* **D53**, 371–376.

Adams, P. D., Grosse-Kunstleve, R. W., Hung, L.-W., Ioerger, T. R., McCoy, A. J., Moriarty, N. W., Read, R. J., Sacchettini, J. C., Sauter, N. K. and Terwilliger, T. C. (2002). PHENIX: building new software for automated crystallographic structure determination. *Acta Crystallogr. D* **58**, 1948–1954.

Badger, J. (2003). An evaluation of automated model building procedures for protein crystallography. *Acta Crystallogr. D* **59**, 823–827.

Bishop, C. M. (2002). *Neural Networks for Pattern Recognition*. Oxford University Press, New York.

Blanc, E., Roversi, P., Vonrhein, C., Flensburg, C., Lea, S. M. and Bricogne, G. (2004). Refinement of severely incomplete structures with maximum likelihood in BUSTER-TNT. *Acta Crystallogr. D* **60**, 2210–2221.

Bricogne, G. and Irwin, J. J. (1996). *Proceedings of the CCP4 Study Weekend. Macromolecular Refinement*, Dodson, E., Moore, M., Ralph A. and Bailey, S., eds, pp. 85–92. Warrington: Daresbury Laboratory.

Bricogne, G., Vonrhein, C., Paciorek, W., Flensburg, C., Schiltz, M., Blanc, E., Roversi, P., Morris, R. and G. Evans, G. (2002). Enhancements in autoSHARP and SHARP, with applications to difficult phasing problems. *Acta Crystallogr. A* **58** (Suppl.), C239.

Brunger, A. T. (1992). The free R-value: a novel statistical quantity for assessing the accuracy of crystal structures. *Nature* **355**, 472–474.

Brunger, A. T. (1993). *X-PLOR version 3.1 Manual*. Yale University Press, New Haven, CT, USA.

Brunger, A. T., Adams, P. D., Clore, G. M., DeLano, W. L., Gros, P., Grosse-Kunstleve, R. W., Jiang, J.-S., Kuszewski, J., Nilges, M., Pannu, N. S., Read, R. J., Rice, L. M., Simonson, T. and Warren, G. L. (1998). Crystallography and NMR System: A new software suite for macromolecular structure determination. *Acta Crystallogr. D* **54**, 905–921.

Brunzelle, J. S., Shafaee, P., Yang, X., Weigand, S., Ren, Z. and Anderson, W. F. (2003). Automated crystallographic system for high-throughput protein structure determination. *Acta Crystallogr.* **D 59**, 1138–1144.

Cohen, S. X., Morris, R. J., Fernandez, F. J., Jelloul, M. B., Kakaris, M., Parthasarathy, V., Lamzin, V. S., Kleywegt, G. J. and Perrakis, A. (2004). Towards complete validated models in the next generation of ARP/wARP. *Acta Crystallogr. D* **60**, 2222–2229.

Coulibaly, F., Chevalier, C., Gutsche, I., Pous, J., Navaza, J., Bressanelli, S., Bernard Delmas, B. and Rey, F. A. (2005). The birnavirus crystal structure reveals structural relationships among icosahedral viruses. *Cell* **120**, 761–772.

Cowtan, K. (1998). Modified phased translation functions and their application to molecular-fragment location. *Acta Crystallogr. D* **54**, 750–756.

Dalgaard, P. (2002). *Introductory Statistics with R. Springer*.

Dauter, Z. (2002). One-and-a-half wavelength approach. *Acta Crystallogr. D* **58**, 1958–1967.

de La Fortelle, E. and Bricogne, G. (1997). Maximum-likelihood heavy-atom parameter refinement for multiple isomorphous replacement and multiwavelength anomalous diffraction methods. *Method Enzymol.* **276**, 590–620.

Duda, R. O., Hart, P. E. and Stork, D. G. (2000). *Pattern Classification*. Wiley-Interscience.

Engh, R. A. and Huber R. (1991). Accurate bond and angle parameters for X-ray protein structure refinement. *Acta Crystallogr. A* **47**, 392–400.

Fortier, S., Chiverton, A., Glasgow, J. and Leherte, L. (1997). Critical-point analysis in protein electron-density map interpretation. *Method Enzymol.* **277**, 131–157.

Greer, J. (1974). Three-dimensional pattern recognition: an approach to automated interpretation of electron density maps of proteins. *J. Mol. Biol.* **82**, 279–301.

Hastie, T., Tibshirani, R. and Friedman, J. H. (2001). *The Elements of Statistical Learning. Data Mining, Inference and Prediction*. Springer, New York.

Holton, J. and Alber, T. (2004). Automated protein crystal structure determination using ELVES. *Proc. Natl. Acad. Sci. USA* **101**, 1537–1542.

Holton, T., Ioerger, T. R., Christopher, J. A. and Sacchettini, J. C. (2000). Determining protein structure from electron-density maps using pattern matching. *Acta Crystallogr. D* **56**, 722–734.

Ioerger, T. R. and Sacchettini, J. C. (2002). Automatic modeling of protein backbones in electron-density maps via prediction of Calpha coordinates. *Acta Crystallogr. D* **58**, 2043–2054.

Ioerger, T. R. and Sacchettini, J. C. (2003). TEXTAL system: artificial intelligence techniques for automated protein model building. *Methods Enzymol.* **374**, 244–270.

Ioerger, T. R., Holton, T., Christopher, J. A. and Sacchettini, J. C. (1999). TEXTAL: a pattern recognition system for interpreting electron density maps. *Proc. Int. Conf. Intell. Syst. Mol. Biol.* 130–137.

Isaacs, N. W. and Agarwal, R. C. (1985). Free atom insertion and refinement as a means of extending and refining phases. *Methods Enzymol.* **115**, 112–117.

Jones, T. A. and Thirup, S. (1986). Using known substructures in protein model building and crystallography *EMBO J.* **5**, 819–822.

Jones, T. A., Zou, J.-Y., Cowan, S. W. and Kjeldgaard, M. (1991). Improved methods for building protein models in electron density maps and the location of errors in these models. *Acta Crystallogr. A* **47**, 110–119.

Kleywegt, G. J. and Jones, T. A. (1997). Template convolution to enhance or detect structural features in macromolecular electron-density maps. *Acta Crystallogr. D* **53**, 179–185.

Lamzin, V. S. and Wilson, K. S. (1993). Automated refinement of protein models. *Acta Crystallogr. D* **49**, 129–147.

Lamzin, V. S., Perrakis, A. and Wilson, K. S. (2001). The ARP/wARP suite for automated construction and refinement of protein models. In: *International Tables for Crystallography. Volume F: Crystallography of biological macromolecules*, Rossmann, M. G. and Arnold, E. eds., pp. 720–722. Dordrecht, Kluwer Academic Publishers, The Netherlands.

Lapedes, A. and Farber, R. (1988). How neural nets work. In: *Neural Information Processing Systems*, Anderson, D. Z., ed. American Institute of Physics, New York.

Levitt, D. G. (2001). A new software routine that automates the fitting of protein X-ray crystallographic electron-density maps. *Acta Crystallogr. D* **57**, 1013–1019.

Lippmann, R. P., Kukolich, L. C. and Singer, E. (1993). LNKnet: neural network, machine learning, and statistical software for pattern classification. *Lincoln Laboratory J.* **6**, 249–268.

Morris, R. J. (2004). Statistical pattern recognition for macromolecular crystallographers. *Acta Crystallogr. D* **60**, 2133–2143.

Morris, R. J., Perrakis, A. and Lamzin, V. S. (2002). ARP/wARP's model-building algorithms. I. The main chain. *Acta Crystallogr. D* **58**, 968–975.

Morris, R. J., Perrakis, A. and Lamzin, V. S. (2003). ARP/wARP and automatic interpretation of protein electron density maps. *Method Enzymol.* **374**, 229–244.

Morris, R. J., Zwart, P. H., Cohen, S., Fernandez, F. J., Kakaris, M., Kirillova, O., Vonrhein, C., Perrakis, A. and Lamzin, V. S. (2004). Breaking good resolutions with ARP/wARP. *J. Synchr. Rad.* **11**, 56–59.

Murshudov, G. N., Vagin, A. A. and Dodson, E. J. (1997). Refinement of macromolecular structures by the maximum-likelihood method. *Acta Crystallogr. D* **53**, 240–255.

Ness, S. R., de Graaff, R. A., Abrahams, J. P. and Pannu, N. S. (2004). CRANK: new methods for automated macromolecular crystal structure solution. *Structure* **12**, 1753–1761.

Ogata, C. M. (1998). MAD phasing grows up. *Nat. Struct. Biol.* **5**, 638–640.

Panjikar, S., Parthasarathy, V., Lamzin, V. S., Weiss, M. S. and Tucker, P. A. (2005). Auto-Rickshaw: an automated crystal structure determination platform as an efficient tool for the validation of an X-ray diffraction experiment. *Acta Crystallogr. D* **61**, 449–457.

Perrakis, A., Harkiolaki, M., Wilson, K. S. and Lamzin, V. S. (2001). ARP/wARP and molecular replacement. *Acta Crystallorg. D* **57**, 1445–1450.

Perrakis, A., Morris, R. and Lamzin, V. S. (1999). Automated protein model building combined with iterative structure refinement. *Nat. Struct. Biol.* **6**, 458–63.

Perrakis, A., Sixma, T. K., Wilson, K. S. and Lamzin, V. S. (1997). wARP: improvement and extension of crystallographic phases by weighted averaging of multiple refined dummy atomic models. *Acta Crystallogr. D* **53**, 448–455.

Pichler, A., Knipscheer, P., Oberhofer, E., van Dijk, W.J., Korner, R., Olsen, J.V., Jentsch, S., Melchior, F. and Sixma, T.K. (2005) SUMO modification of the ubiquitin-conjugating enzyme E2-25K. *Nat. Struct. Mol. Biol.* **12**, 264–269.

Press, W. H., Flannery, B. P., Teukolsky, S. A. and Vetterling, W. T. (2002). *Numerical Recipes in C: The Art of Scientific Computing*. Cambridge University Press, New York, USA.

Ramakrishnan, C. and Ramachandran, G. N. (1965). Stereochemical criteria for polypeptide and protein chain conformations. *Biophys. J.* **5**, 909–903.

Rice, L.M. and Brunger, A.T. (1994) Torsion angle dynamics: reduced variable conformational sampling enhances crystallographic structure refinement. *Proteins* **19**, 277–290.

Ritter, H., Martinetz, T. and Schulten, K. (1992). *Neural Computations and Self-Organized Maps: An Introduction.* Addison-Wesley.

Skubak, P., Murshudov, G.N. and Pannu, N.S. (2004) Direct incorporation of experimental phase information in model refinement. *Acta Crystallogr.* **D60**, 2196–2201.

Sheldrick, G. M. and Schneider, T. R. (1997). SHELXL: High-resolution refinement. *Method Enzymol.* **277**, 319–343.

Swanson, S. M. (1994). Core tracing: depicting connections between features in electron density. *Acta Crystallogr. D* **50**, 695–708.

Terwilliger, T. (2004). SOLVE and RESOLVE: automated structure solution, density modification and model building. *J. Synchr. Rad.* **11**, 49–52.

Terwilliger, T. C. (2001). Map-likelihood phasing. *Acta Crystallogr. D* **57**, 1763–1775.

Terwilliger, T. C. (2003a). Automated main-chain model building by template matching and iterative fragment extension. *Acta Crystallogr. D* **59**, 38–44.

Terwilliger, T. C. (2003b). Automated side-chain model building and sequence assignment by template matching. *Acta Crystallogr. D* **59**, 45–49.

Terwilliger, T. C. (2003c). Improving macromolecular atomic models at moderate resolution by automated iterative model building, statistical density modification and refinement. *Acta Crystallogr. D* **59**, 1174–1182.

Terwilliger, T. C. and Berendzen, J. (1999). Automated MAD and MIR structure solution. *Acta Crystallogr. D* **55**, 849–61.

Theodoridis, S. and Koutroumbas, K. (2006). *Pattern Recognition.* Academic Press, London.

Tronrud, D. E. (2004). Introduction to macromolecular refinement. *Acta Crystallogr. D* **60**, 2156–2168.

Tronrud, D. E., Ten Eyck, L. F. and Matthews, B. W. (1987). An efficient general-purpose least-squares refinement program for macromolecular structures. *Acta Crystallogr. A* **43**, 489–501.

Turkenburg, J. P. and Dodson, E. J. (1996). Modern developments in molecular replacement. *Curr. Opin. Struct. Biol.* **6**, 604–610.

Weeks, C. M., Shah, N., Green, M. L., Miller, R. and Furey, W. (2005). Automated web- and grid-based protein phasing with BnP. *Acta Crystallogr. A* **61** (Suppl.), **C152**.

Weichenrieder, O., Repanas, K. and Perrakis, A. (2004) Crystal structure of the targeting endonuclease of the human LINE-1 retrotransposon. *Structure* **12**, 975–986.

Winn, M.D., Murshudov, G.N. and Papiz, M.Z. (2003) Macromolecular TLS refinement in REFMAC at moderate resolutions. *Methods Enzymol.* **374**, 300–321.

Zwart, P. H., Langer, G. G. and Lamzin, V. S. (2004). Modelling bound ligands in protein crystal structures. *Acta Crystallogr. D* **60**, 2230–2239.

CHAPTER 12

# High-throughput crystallographic data collection at synchrotrons

**Stephen R. Wasserman, David W. Smith, Kevin L. D'Amico, John W. Koss, Laura L. Morisco, and Stephen K. Burley**

## 12.1 Introduction

The discussion in Chapter 5 on in-house data collection methods highlighted the effect that advances in current technology have had on laboratory protein crystallography. There have been parallel developments in macromolecular crystallography utilizing modern synchrotron X-ray sources. The success of these efforts, combined with the recent scientific emphasis on genomics and proteomics, has yielded a thriving structural biology community that routinely uses these tunable, high-intensity X-ray sources for crystallographic experiments.

The development of X-ray synchrotrons over the last four decades and their use for X-ray analyses have been detailed elsewhere (Mills, 2002). Synchrotron sources offer several advantages for the acquisition of diffraction data from protein crystals. They provide extremely intense beams, with photon fluxes that are many orders of magnitude greater than those available from rotating anode sources. Current, third-generation synchrotrons make use of insertion devices, that is devices inserted into the main synchrotron ring, to produce very small, highly directional X-ray beams. Winick has provided a qualitative description of the insertion devices, called undulators and wigglers, which are used to generate these X-ray beams (Winick, 1987). Particularly when undulators are used, the X-ray beam is highly collimated, delivering all of the photons into a small area.

Having such bright, intense X-ray beams enables examination of very small protein crystals. Samples between 30 and 100 microns in their longest dimension are routinely used to determine the desired protein structure. Given the often prodigious effort required to grow larger crystals, synchrotron beamlines have proven cost effective when compared with in-house sources, despite their initial construction cost of US$5 to $10 million per beamline.

Unlike home X-ray sources, which are limited to a few specific wavelengths corresponding to the $K_\alpha$ radiation from elements such as copper, molybdenum, and chromium, synchrotrons provide access to a continuous range of X-ray energies. During the 1980s, it was recognized that matching the energy of the X-ray photon to the absorption edge of an atom within the protein crystal offers the possibility of extracting, from a single protein crystal, all the phase information required for the structure determination (Hendrickson, 1991). Successful replacement of methionine residues within a protein with selenomethionine results in a crystal that is structurally isomorphous to the crystal from the native protein and contains an element, selenium, ideal for direct determination of experimental X-ray phases. A discussion of anomalous dispersion methods can be found in Chapters 8 and 9.

These technical developments stimulated a dramatic growth in the number of beamlines available for protein crystallography. There are currently at least 22 synchrotrons worldwide supporting studies of the crystalline forms of proteins, with a further three under construction. In 2006, the number of beamlines used for protein crystallography exceeded 110, with at least 45 in the United States alone. This number has grown more than five-fold since 1990 (Helliwell, 1990).

Rapid collection of diffraction data depends on access to such powerful X-ray sources. This chapter describes how high-quality, high-throughput data collection can be achieved. We use SGX-CAT, the SGX Collaborative Access Team beamline, located at the Advanced Photon Source of Argonne National Laboratory, as an example to illustrate the concepts behind the design of, and the hardware used at, synchrotron beamlines. Many of these features are found, individually or in combination, at other beamlines. Data collection at synchrotron sources produces enormous quantities of data. We, therefore, also discuss the information technology infrastructure and software that is necessary for effective data management.

High-throughput data collection requires seamless interoperation of various hardware components. In addition, user-supplied descriptions of protein crystals must be directly linked with the diffraction data. Such linkages can be achieved efficiently with computer databases. A database that tracks production of the protein samples, crystallization, and diffraction from the resultant crystals serves as the glue that holds the entire gene-to-structure process together. In this chapter, we first discuss data collection processes and hardware. We then illustrate how a well-constructed database ensures information flow through the steps of data acquisition. With such a database, synchrotron beamline measurements can be directly and efficiently integrated into the process of protein crystallographic structure determination.

The approaches to data acquisition summarized in this chapter apply to both the *de novo* determination of protein structures and, as is routine in drug discovery, examination of protein–ligand cocrystals. Each type of experiment benefits from the advantages of modern synchrotron X-ray sources. Recently in pharmaceutical development there has been an emphasis on discovering lead compounds through structure-guided molecular elaboration that begins with small chemical fragments (Jhoti, 2004). Because of the large number of cocrystals to be examined in this approach, often exceeding several hundred, rapid access to high-quality diffraction data is required. This requirement and those of the current efforts in structural proteomics place a premium on highly automated evaluation of crystals, data collection, and data processing.

## 12.2 Access to synchrotrons

Crystallographers have traditionally travelled to synchrotrons for data collection. However, this approach is not generally compatible with high-throughput data collection. For example, significant time is often expended merely in transporting technical staff to the synchrotron facility.

Two related alternatives for 'mail-in' use of synchrotrons have arisen. Several facilities permit users to send crystals to the synchrotron beamline, where local staff load them onto a robot that places the crystals into the X-ray beam (see Section 12.3.3). The user then operates the beamline remotely, during a specific period allocated for their experiments. Considerable effort has been expended by various synchrotron facilities to provide robust, secure internet connections between the beamline and the end user.

Even greater efficiencies can be achieved when the entire process of data collection is delegated to the staff of the beamline and the automated facilities they operate. This latter approach is used exclusively at SGX-CAT. All decisions on data collection protocols are made by the local staff aided by facility software systems. Such reliance on professional experts for data collection was anticipated nearly a decade ago (Sweet, 1998).

## 12.3 Beamline hardware

### 12.3.1 X-ray optics

Synchrotron X-ray sources include both bending magnets and insertion devices. For protein crystallography, an undulator insertion device is preferred because it provides greater intensity at a specific wavelength and has lower beam divergence. This latter property results in smaller X-ray reflections.

The configuration of X-ray optical elements at the SGX-CAT undulator beamline is shown in Fig. 12.1. Generally, beamline components are housed in lead-walled enclosures or hutches. In Fig. 12.1, the first

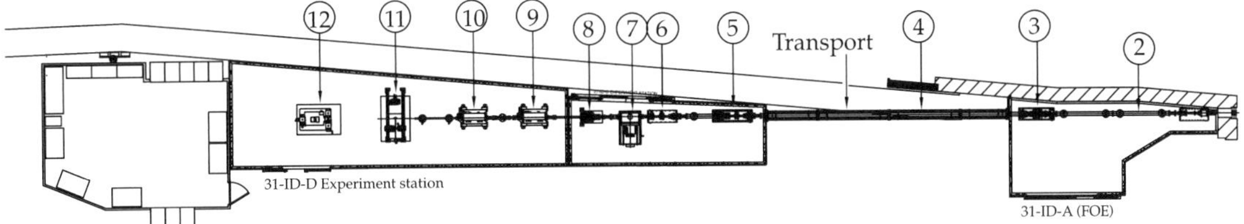


**Figure 12.1** SGX-CAT beamline schematic. The components of the beamline include: (1(not shown), 8) photon shutters; (2, 4) beam transport tubes; (3, 5) collimators and vacuum pumps; (6) beam-defining slits; (7) monochromator; (9, 10) focusing and harmonic rejection mirrors; and (12) CCD detector, supporting base, and sample robot.

enclosure contains optics for selection of the desired X-ray energy. Usually this enclosure includes the components to focus the beam into a small area to maximize brightness. However, at SGX-CAT these components are in the second hutch, which also houses the equipment necessary to mount crystals and perform the actual X-ray diffraction experiment. Shutters control access of the X-ray photons to each enclosure.

The selection of the X-ray wavelength or energy is achieved with a monochromator. At SGX-CAT, this component is a double-crystal device from Kohzu Precision Co. that contains two small diamond crystals (111, 2d = ~4.1188 Å). Many protein crystallography beamlines use Si crystals (111, 2d ~6.2712 Å). Silicon crystals provide greater photon fluxes than diamond crystals at a given energy. However, they do so at the expense of energy resolution and do not provide access to the shorter X-ray wavelengths available from diamond-based devices. Because of the high levels of heat deposited into monochromator crystals by modern undulator beamlines, they must be cooled. Otherwise the crystals tend to expand or contract, thereby subtly changing the wavelength of the transmitted X-rays. In order to minimize the effects of monochromator crystal heating, silicon must be maintained at liquid nitrogen temperatures. For diamond, a low-maintenance water-cooling system that keeps the crystals at ~25°C suffices.

X-ray energies are selected by rotating the crystal pair relative to the incoming beam. The wavelength delivered by a given crystal at a specific reflection angle (2θ) is given by Bragg's Law (Eq. 1).

$$n\lambda = 2d \sin\theta \qquad (1)$$

For high-throughput data collection, control of the X-ray wavelength must be rapid and reproducible. Current state-of-the-art monochromators have mechanical precisions on the order of 0.0001 angular degrees.

Undulator insertion devices generate multiple harmonics of their fundamental X-ray wavelength at intensities comparable to that of the fundamental. The unwanted wavelengths are usually removed or 'rejected' using X-ray mirrors. A shallow angle between the incoming beam and the mirror is chosen so that the desired X-ray wavelength is efficiently reflected, while most of the harmonics are absorbed. In order to preserve the intensity of the reflected X-ray beam, such mirrors are polished to near atomic smoothness. Typically, X-ray mirrors have more than one surface coating, so that harmonic control can be exercised over the widest possible range of X-ray wavelengths. If only one surface type were available, harmonic rejection at higher energies could involve a significant deflection of the beam and require a reconfiguration of the downstream components of the beamline. The problem of harmonics is greatest with insertion devices at the most modern synchrotron sources. With earlier first and second-generation synchrotrons, the operating energies of the rings were too low to generate significant harmonic content in the X-ray beam. Any harmonics could usually be eliminated by setting one of the crystals in the monochromator slightly non-parallel to the other.

At SGX-CAT, the inherent size of the X-ray beam at the position of the sample, if left unfocussed, would be approximately 1500 μm × 700 μm. A beam of this size would greatly exceed the cross-sections of typical crystalline protein samples, thereby wasting most of the photons emerging from the undulator X-ray source. Therefore, the X-ray beam is focused to match the size of the crystal. On undulator beamlines, focusing is usually achieved either through sagittal focusing, using specially designed monochromator crystals, or with the aid of a pair of bendable mirrors. In a typical Kirkpatrick–Baez focusing system (Kirkpatrick and Baez, 1948) two mirrors are used to compress the beam, one in the vertical dimension and the other in the horizontal. At SGX-CAT, such mirrors yield an X-ray beam that is approximately 70 $\mu$m × 80 $\mu$m at the sample.

### 12.3.2 Detectors

Chapter 5 described the types of detectors available for in-house data collection. In-house diffraction experiments generally rely on image-plate or CCD detectors, whereas virtually all modern protein crystallography synchrotron beamlines use only CCD detectors (Fig. 12.2). Although image-plates provide a greater dynamic range, the time required for data read out is too long relative to the synchrotron exposure time. For example, readout for a Rigaku/MSC RAXIS IV$^{++}$ is 50 to 100 sec, with a minimum total cycle time of 3 min, including reset of the image plate (Fujisawa *et al.*, 2003). The fastest CCD detectors have a readout time of less than 2 sec, which is comparable to typical exposure times used at third-generation synchrotrons (0.4 to 12 sec). The speed of these detectors, combined with the short exposures available at the synchrotron, results in very rapid data collection. With current technology, a dataset containing 180 diffraction images can be recorded easily in less than 9 min.

CCD detectors use a phosphor to convert the incoming X-ray to visible light, which is in turn detected by the CCD chip. During this conversion process, the apparent size of an X-ray reflection increases, a phenomenon known as the point-spread factor. Typically, this change in reflection size does not present a significant challenge. In extreme cases (i.e. large unit cell dimensions or a very short sample to detector distance), however, it can lead to overlaps between adjacent reflections. When overlaps do occur, diffraction data can be recorded using smaller sample oscillations/rotations at the expense of recording more diffraction images over a greater period of time. We have found that this approach may not always eliminate the problem of overlaps. In addition, such fine slicing introduces a non-trivial complication to subsequent data processing. Use of smaller oscillation/rotation ranges reduces the number of reflections that are fully recorded on a single diffraction image, which makes it more difficult to scale together data from different images. At SGX-CAT, it has proven more effective to maintain as large an oscillation range as possible and increase the sample-to-detector distance. This approach increases the distance between reflections on the face of the CCD detector. The sizes of the reflections remains essentially unchanged because of the collimated X-ray beam produced by the undulator insertion device.

A new generation of pixel-array detectors is currently under development for use in protein crystallography (Brönnimann *et al.*, 2004, 2006). These detectors exhibit minimal point-spread factors. In addition, their extremely rapid readout, <10 msec, permits diffraction data to be acquired on a near continuous basis.

### 12.3.3 Crystal handling

Acquisition of the diffraction images is only one part of the data collection process. Cryogenically preserved crystals have to be placed in the X-ray beam, while being maintained at liquid nitrogen temperatures. There has been a considerable effort in the development of robotic systems to transfer crystals from a storage dewar to the goniostat for data collection. Custom systems have been developed at various synchrotrons for this purpose (Cohen *et al.*, 2002; Snell *et al.*, 2004; Jacquamet *et al.*, 2004; Pohl *et al.*, 2004; Karain *et al.*, 2002; Ueno *et al.*, 2004). This variety of robots reflects the fact that each beamline tends to be of unique design. As automated sample changing becomes more established, it is likely that beamline robotics will reflect a greater degree of standardization. Three commercial systems are now available from Rigaku (Abad-Zapatero, 2005) (based on an earlier design from Abbott Laboratories (Muchmore *et al.*, 2000)), Bruker, and MAR Research.

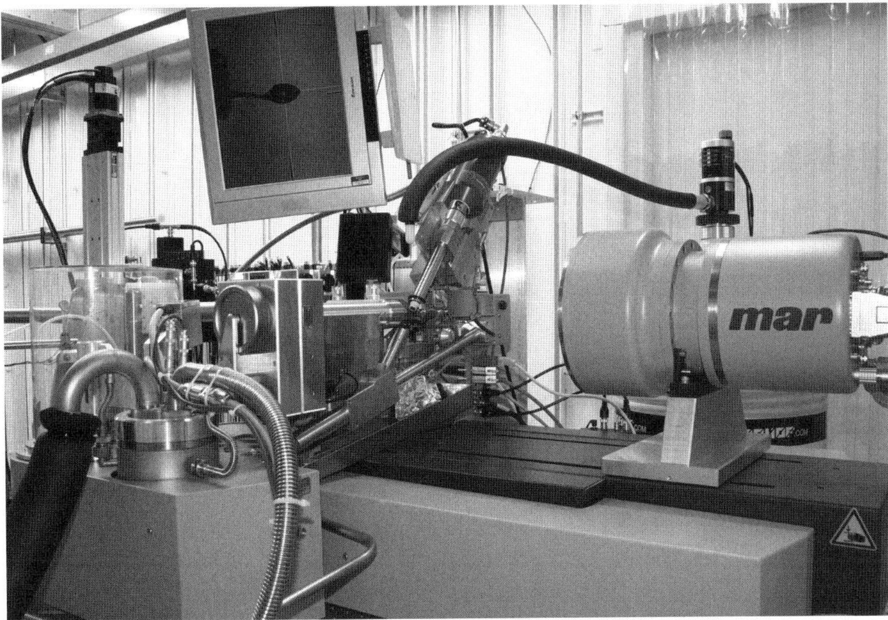


**Figure 12.2** SGX-CAT data acquisition station. The CCD detector and cryogenic sample handling robot appear in the foreground. The pump to deliver the liquid nitrogen that removes ice from the sample is located at the rear right.

Sample changing robots (Fig. 12.2) consist of two parts. The first is a mechanism for moving the crystal from the storage dewar to the data collection position. In most systems standard, multiaxis industrial robots are used, the MAR CryoSample Changer (MARCSC) being a notable exception. The MAR device is also unique in that the samples are kept within liquid nitrogen during the transition from storage to sample position. The second component of the sample changer is the liquid nitrogen storage dewar itself. The capacity of this dewar determines the maximum number of samples that can be examined automatically. Current capacities range from 19 (MARCSC) to 288 (Stanford Automounting System) samples. Auxiliary robots can be used to increase sample capacity. For example, SGX-CAT is currently developing a system that increases the capacity of the MAR CSC by six-fold.

Automated positioning of samples relies on compatibility between the robot and the hardware used for crystal mounting. In 2004, the SPINE (Structural Proteomics in Europe) consortium (www.spineurope.org) established a standard specification for sample holders. This specification is compatible with all three commercial sample changers and with most beamline- and facility-specific systems. The SPINE protocols, which are based on a mounting system originated by MAR Research, include use of a two-dimensional (2D) bar code within the mounting hardware (Fig. 12.3). When barcode readers are included within the automation hardware, barcodes can be used to identify the crystal and unambiguously link the physical sample to its record in the database (see Section 12.9).

### 12.3.4 Crystal mounting and positioning

The SPINE protocols also specify the interaction between the robotic hardware and the sample hardware. Specifically, the sample must be located 22 mm above the base that attaches the crystal to the robot. Precisely how samples are mounted on the base itself, however, is at the discretion of the user.

The usual method for mounting crystals is to suspend them, surrounded by a drop of cryoprotectant, within a loop of nylon attached to a pin. The pin is, in turn, attached to a magnetic base

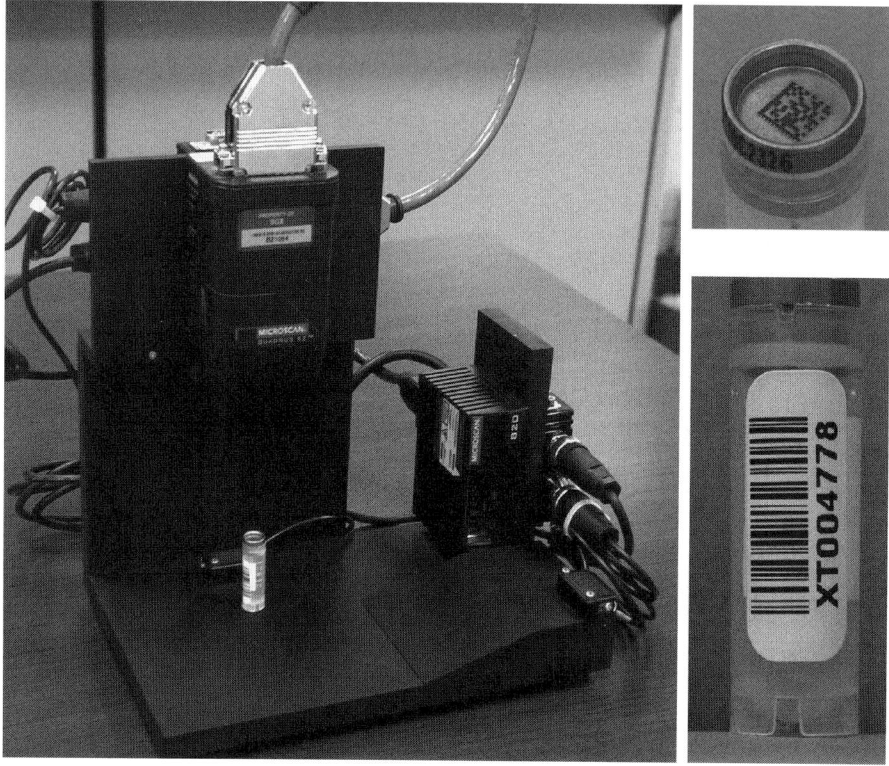


**Figure 12.3** SGX-CAT barcode reader station. The 2D barcodes on the base of each crystal mount are read on the left. The scanner on the right reads the unique 1D barcodes on the vials used to ship the crystals. (NB: During operations, crystals are maintained in liquid nitrogen throughout this process.)

bearing the 2D barcode. An alternative method has been developed recently, which uses microfabricated mounts created from polyimides such as Kapton (Thorne *et al.*, 2003). These mounts have greater rigidity than that of nylon loops, facilitating precise placement of the sample across a perforation in the crystal mount. As with nylon loops, different microfabricated mounts are available to match the hole in the mount with the size of the crystal. Unlike nylon loops, the manufacturing process for these new mounts guarantees uniformity of dimensions, which can be important during automatic positioning of protein crystals in the X-ray beam (see below). These polyimide mounts include registration or fiducial 'cross hairs', which visual recognition systems can use to automatically place crystals precisely in the X-ray beam.

Despite this obvious advantage, micromounts have the drawback of requiring that the amount of cryoprotectant be minimized to ensure reliable recognition of the fiducial marks. While this requirement presents little problem when preparing a small number of crystals, it can necessitate many extra hours of labour when dealing with large numbers of samples. Therefore, because of the high volume of crystals generated at SGX headquarters in San Diego, we continue to use nylon loops for mounting of crystals.

Once a crystal has been mounted, it must be placed in the X-ray beam. Most modern goniostats are motorized, permitting this process to be executed remotely or, for high-throughput data collection, automatically. At third-generation synchrotron sources, highly focused X-ray beams and small crystals (i.e. <50 μm in diameter) make sample positioning akin to threading a needle. For larger sized beams, such as those found with home sources or at older synchrotrons, or crystals with greater dimensions, this process can be considerably easier.

With appropriate crystal lighting and video-based visualization optics, it is usually possible for an end user to locate the crystal within the nylon loop and position it in the X-ray beam. At present, automation of this process does entail some compromise. With current technology, it is considerably easier to identify the centre of a quasicircular nylon loop than to recognize a protein crystal, particularly because individual crystals can vary greatly in both shape and size. Centring based on detection of the loop, however, will place the crystalline sample in the X-ray beam only if the diameter of the loop is commensurate with that of the crystal or the beam. This centring method, therefore, requires upstream control of how crystals are mounted, ensuring that the loop and the crystal are matched in size and that the crystal is placed at or near the centre of the loop. Illumination of the loop and visualization optics have a significant impact on automated recognition of the crystal mount. The lighting strategy must provide sufficient contrast between the loop and the background of the video image and minimize glare from the surfaces of cryogenically frozen samples. Suboptimal lighting can obscure the boundaries of the loop and compromise the effectiveness of automated detection algorithms.

For high-throughput data collection, sample centring via loop detection is currently the method of choice for placing crystals in the X-ray beam. Alternative methods rely on the direct detection of the crystal itself by monitoring the intensity of either X-ray diffraction from the crystal or X-ray fluorescence from an element in the crystal that is not present in the cryoprotectant (Pohl *et al.*, 2004). Both of these approaches have drawbacks. First, some of the limited X-ray lifetime of the crystal (Section 12.6) must be committed to detection instead of data collection. Second, the loop itself has to be placed close to the X-ray beam prior to crystal detection, followed by step-wise translation of the crystal through the beam, thereby increasing the time required to centre the sample.

Non-X-ray signatures from crystals can also be used for detection/centring. Ultraviolet fluorescence from the aromatic residues in the polypeptide chain can indicate the position of the crystal within the mount. Recent work has utilized high intensity UV lasers for fluorescence excitation (Vernede *et al.*, 2006). Differential transmission and reflection of infrared radiation from the crystal, relative to that of the loop and the surrounding cryoprotectant, has also been used to detect crystals within nylon loops (Snell *et al.*, 2005). The success of this latter technique at a synchrotron beamline is not yet at the level achieved in the developmental laboratory. One of these methods may ultimately become a standard means for detecting protein crystals directly. Currently, however, control of the location of the crystal within a commensurate loop prior to cryogenic preservation in liquid nitrogen, followed by positioning of the loop, represents the most reliable method for rapid, accurate placement of protein crystals in X-ray beams at synchrotron sources.

### 12.3.5 Removal of surface ice

It is common for liquid nitrogen frozen protein crystals to acquire a patina of ice on the surface of the cryoprotectant. Diffraction of X-rays from even small ice crystals can mask reflections from the protein crystal. In addition, the presence of excessive amounts of ice can obscure the true position of the nylon loop, thereby resulting in the failure to place the crystal in the X-ray beam. It is, therefore, essential to remove ice crystals prior to diffraction analysis.

Traditionally, crystallographers have manually removed ice by either pouring a small amount of liquid nitrogen over the crystal or by gently abrading the surface with an implement such as a single paint brush bristle. The Stanford Auto-mounting System provides for removal of a sample from the goniostat and rapid 'washing' within a bath of liquid nitrogen to eliminate ice (Cohen *et al.*, 2005).

SGX-CAT uses a unique system based on a small liquid nitrogen pump for a similar purpose. The pump controls delivery of a gentle stream of liquid nitrogen that is directed onto the surface of the sample after it has been positioned within the gaseous cryostream by the sample-changing robot. A phase separator is used to ensure that only cold liquid strikes the sample. Washing with liquid nitrogen is used routinely at SGX-CAT during automated data collection, because the presence or absence of ice cannot be known in advance. The process is performed in parallel with other manipulations of the sample, so that the time added to the mounting and positioning of a protein crystal is less than 30 sec.

Constant use of this liquid nitrogen delivery system requires control of the environment within the X-ray enclosure to eliminate condensation of water on the phase separator located above the crystal sample. If this water were to reach the sample, the crystal would experience a temporary increase in temperature that can result in loss of crystallinity. Dehumidification of the hutch eliminates this potential problem. Removal of ambient humidity has the added benefit of reducing the formation of ice on all components that use liquid nitrogen, particularly the dewars used to store crystals before and after X-ray analysis.

## 12.4 Evaluation of crystal quality

A fully automated protein crystallography beamline at a third-generation synchrotron source can screen several hundred crystals daily. Automatic evaluation of the diffraction images to ascertain crystal quality is, therefore, a critical step for high-throughput data acquisition. Evaluation of each image requires software that mimics the traditional visual assessment of crystal quality.

Automatic diffraction analysis is available through the Diffraction Image Screening Tool and Library (DISTL) (Zhang *et al.*, 2006). DISTL incorporates automatic indexing and identification of extraneous features such as ice rings, and provides an estimate of the resolution of the diffraction data. Neural networks have also been tested for the evaluation of crystal quality (Berntson *et al.*, 2003).

Given the nascent nature of these software tools, SGX developed its own method to evaluate or score the diffraction quality. The SGX system is based on two established software programs, d*TREK (Pflugrath, 1999) and Mosflm (Leslie, 1992). These programs index diffraction images to determine the appropriate Laue group. In addition, they provide an analysis of the properties of the observed reflections, including the intensity of spots ($I/\sigma$), their shape, the percentage of the detected spots that satisfy the indexing process and, once indexed, how well the spots refine within the chosen point/space group. Such analyses yields a numerical score that permits direct comparison between crystals within a set. The estimated diffraction resolution of the crystal is also determined, from which the optimal sample-to-detector distance is calculated.

Software at SGX-CAT readily distinguishes crystals of good quality from those that diffract poorly or display artefacts which compromise the data. Somewhat surprisingly, we have found that the use of automatic scoring also improves the selection of crystals for data collection. Figure 12.4a presents an example of a crystal that most members of the structural biology community would reject for data collection. However, our evaluation system suggested that this crystal was likely to yield a useable dataset. The electron density map of the active site, automatically generated from diffraction data recorded from the same crystal, is shown in Fig. 12.4b. Utilization of the scoring system has significantly improved data collection efficiency at SGX, because the entire process can be conducted without staff oversight. In addition, crystals that previously would have been rejected upon visual inspection are routinely used for structural studies of protein–ligand complexes. This benefit of our system has the added advantage of reducing the amount of time devoted to crystallization, thereby improving throughput in structure-based drug discovery. Our scoring system also monitors the number of overloaded reflections, which permits identification of samples for which X-ray beam attenuation and/or exposure time reduction may be required.

Within our crystal quality evaluation process, data collection strategy is determined. Because of the extended times for data collection with in-house and earlier-generation synchrotron sources, optimization of data acquisition remains a significant consideration (Chapter 5). At third-generation synchrotron sources, experience has shown that brute force data collection (180 or 360 degrees, depending on the Laue group) with standardized, short exposure times usually suffices. In most cases, sample integrity can be maintained throughout the

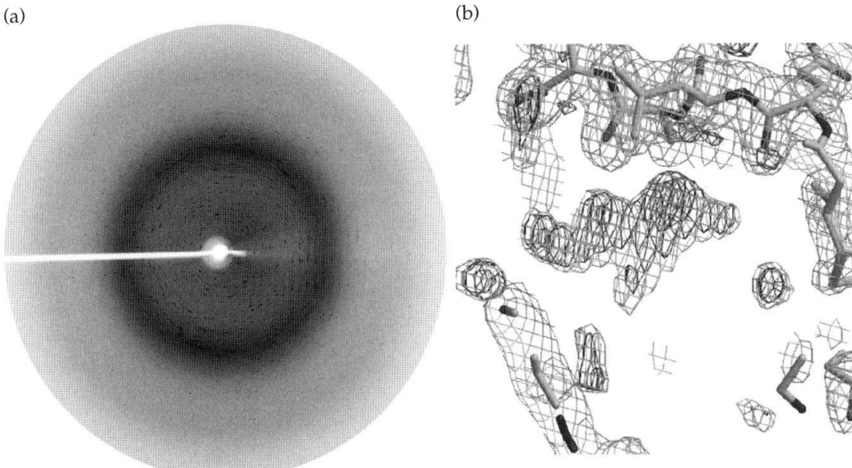


**Figure 12.4** SGX-CAT automated scoring. (a) Diffraction from a crystal of a protein kinase/ligand complex. Automated crystal scoring suggested that a dataset should be acquired from this sample. Standard visual inspection would almost certainly reject such a crystal. (b) Experimental electron density map of the same crystal generated automatically from diffraction data without human intervention. The presence of a ligand in the active site is readily apparent. The resulting structure was used by the SGX structure-based drug discovery team to guide further optimization of ligand affinity.

collection. In rare instances of unacceptable sample degradation, manually optimized data collection is invoked.

Automated crystal quality evaluation permits rapid, objective assessment of each crystal, determining whether or not it is worthy of data collection. In addition, quantitative comparisons among duplicate crystals improves throughput by focusing data collection on only the best crystals. To automate our selection process, a link between replicate crystals must exist within the software control system. At SGX, this link is made via the database (Section 12.9). This combination of software linkage and automatic scoring has had a significant impact on data collection throughput. Currently, approximately 70% of the samples examined at SGX-CAT are evaluated without manual examination of the diffraction images, either during screening or data collection. We expect this proportion to increase as further software enhancements are made.

## 12.5 Selection of X-ray wavelength

A key advantage of synchrotron sources is their ability to provide X-rays over a wide range of energies or wavelengths. When using synchrotron radiation, unlike a home source, the energy that best serves the diffraction experiment must be selected.

If the diffraction experiment does not require a particular wavelength, such as for structure determinations with molecular replacement (Chapter 7), convenient choices for the wavelength are 1 Å or 0.9793 Å, the wavelength corresponding to the K absorption edge of selenium. Such X-ray wavelengths are readily available at most synchrotron beamlines.

Measurements that rely on anomalous scattering, however, are usually most successful when the X-ray energy matches that of the absorption edge of the anomalous element. Because of concerns that the chemical environment of the anomalous atoms affects the precise location of the absorption edge (Smith and Thompson, 1998), it has been customary to measure the absorption spectrum of the anomalous element within the protein crystal itself. In doing so, the crystal is exposed to X-rays prior to the diffraction experiment, which can damage the crystal. However, many of the effects of atomic environment and oxidation state on the precise location of an absorption edge are well-understood (Koningsberger and Prins, 1988). Knowledge from the field of X-ray Absorption Near

Edge Spectroscopy (XANES) can be used to estimate the expected absorption energy of the anomalous atoms. For example, SGX has determined the energy that corresponds to the maximum absorption of X-rays by pure selenomethionine. This value is used for all experiments that utilize selenomethionine to determine the required phases. With this approach, the limited X-ray lifetime of protein crystals can be devoted to determination of diffraction. In addition to preserving crystals, this approach maximizes the time devoted to data-collection in high-throughput mode.

It should be noted that the absolute energy calibrations of beamlines vary. Determining the peak position of an element within a protein relative to a known standard (such as the metallic form of that element) can save time in future structural experiments. Accurate determination of absorption edge energies for various metals has been reported by Kraft *et al.* (1996).

## 12.6 Selection of exposure duration

The exposure time used for each diffraction image represents a compromise between maximizing the intensity of the diffraction signal and minimizing crystal degradation during data collection. Nave and Garman have provided a brief review of the current investigations on the sources of radiation damage (Nave and Garman, 2005). Henderson estimated that the limiting X-ray dose for a protein crystal is approximately $2 \times 10^7$ Grays (Gray = J/kg) (Henderson, 1990). Recent results suggest that a dose 50% higher than Henderson's estimate can be tolerated by most protein crystals (Owen *et al.*, 2006). The program RADDOSE can be used to estimate the maximum time that a crystal should remain in the X-ray beam (Murray *et al.*, 2004, 2005). Such calculations rely on knowledge of both the characteristics of the beamline and the estimated composition of the unit cell. In addition, recent studies suggest that there are significant variations in the X-ray dose that can be tolerated by the selenium in selenomethionine when determining diffraction phases through anomalous dispersion (Holton, 2007).

While it is possible to optimize exposure conditions for crystals individually, high-throughput efforts require an alternative approach. At SGX-CAT, which uses an undulator as its X-ray source, standard exposure times are used for most samples. At the selenium K edge, we typically use ~1 sec for each 1 degree oscillation. At the bromine K edge, where the X-ray intensity from the first harmonic of the undulator is approximately 40% of that at the Se edge, the exposure time is doubled. When smaller or larger oscillations are used, exposure times are adjusted proportionately. These values are appropriate for datasets that include data from an angular wedge up to 180 degrees. For anomalous dispersion experiments, the amount of data collected generally increases by a factor of two. In these cases, exposure times are typically reduced by a third, to preserve sample integrity throughout the duration of data collection. Experience at SGX-CAT has shown that this approach minimizes radiation damage while ensuring adequate diffraction intensity.

Crystals with very strong diffraction may have an unacceptably high number of saturated reflections. As noted above, such crystals are identified automatically during the crystal evaluation process, based on the average number of overloaded reflections on the screening images. For such crystals, the exposure time is reduced. If a further decrease in the illumination of the sample is required, the intensity of the incoming beam is adjusted.

## 12.7 Sample tracking

When large numbers of crystals are examined at synchrotron beamlines, unambiguous identification of the samples is critical. The SPINE standard includes a unique barcode on the magnetic base for that purpose. However, the SPINE protocol also assumes that the bases will be reused. Hence, these 2D barcodes alone do not permit unique identification of a given crystal.

SGX uses two bar codes to track samples. The first identifier is the 2D bar code on the pin bases, which are used multiple times to mount individual crystals. The second identifier, a 1D bar code, is placed on the vials in which crystals are shipped to the beamline. This barcode is unique to the crystal and, unlike the base, is not reused. The device used to read the barcodes is shown in Fig. 12.3. This system requires that the 1D barcode, which links the physical crystal with its description in the database, constitutes part of the database record for that crystal. At the time that crystals are loaded into the sample-changing robot carousel, the 2D barcode is also read into the database. This two bar code system provides redundancy in sample identification. And, since the 2D bar code is part of the mounting hardware and its value is stored in the diffraction image file, this identifier is always associated with the crystal.

## 12.8 Diffraction data processing

Any high-throughput system for protein crystallography requires efficient processes for converting measured diffraction images into experimental electron density maps and structures. Chapters 6 to 11 and 13 cover the various approaches to data analysis in considerable depth.

Several systems have been developed for automatic protein structure determination via X-ray crystallography, including software packages such as *PHENIX* (Adams *et al.*, 2002), *ELVES* (Holton and Alber, 2004), *Auto-Rickshaw* (Panjikar *et al.*, 2005), and *ACrS* (Brunzelle *et al.*, 2003). These packages utilize existing individual programs to process the diffraction images and data. Additional programs and scripts analyse interim results, and shepherd the output from one program to another. The modularity of these systems permits rapid incorporation of enhancements in crystallographic software.

Data processing occurs in two stages. Initially, diffraction images are reduced to a tabulation of reflection indices and intensities or, after truncation, structure factors. The second stage involves conversion of the observed structure factors into an experimental electron density map. The choice of how to execute the latter step depends on the method used to determine the phase for each reflection. The software packages enumerated above generally focus on exploitation of anomalous signals to overcome the phase problem for a protein of unknown structure.

Structure-based drug discovery involves determination of many cocrystal structures with different ligands bound to the same target protein. In such cases, where the structure of the protein is well known, automated structure determination procedures rely on molecular replacement to supply the necessary phases, followed by fitting of the ligand into the appropriate electron density feature within the map. The Astex AutoSolve® program and the SGX target to lead platform have both adopted this approach. When anomalous signals are used, such as in the SGX FAST (Fragments of Active Structures) fragment-based lead discovery process, they help to differentiate between ligands and to provide information on the orientation of the ligand within the target protein. A summary of the SGX automated data processing system is provided in the high-throughput example in Section 12.11.

## 12.9 Information technology infrastructure

Modern high-throughput data collection permits examination of a large number of samples and generates enormous quantities of diffraction data. With a CCD detector, data collection rates are typically 1 to 5 MB/sec, although higher rates, up to 24 MB/sec, are currently possible at third generation synchrotrons (Bakaikoa, 2006). The pixel array detectors currently available provide data at rates of 2 to 12 MB/sec. (Hülsen *et al.*, 2006).

Because of these collection rates, high-throughput synchrotron operations are as much an issue of sample and data management as of data collection. For this reason, SGX-CAT operations were included in the information management systems for the structural biology platform at SGX at the time of beamline commissioning. These data systems directly link beamline operations to SGX efforts in drug discovery and structural proteomics.

The SGX Laboratory Information Management System (LIMS) is based on an Oracle database platform. As an enterprise level relational database, Oracle is robust and can be expanded to meet any future needs. The expense to implement and maintain the database and to create the tools to retrieve the data stored within, including an administrator dedicated to managing the system, can be significant. At SGX, use of the LIMS to monitor all of our scientific activities reduces the incremental cost to the beamline to an acceptable level.

The SGX database tracks every aspect of the preparation of crystalline samples, including gene cloning, gene expression/solubility testing, protein purification, crystallization, cocrystallization with small molecule ligands, cryopreservation, crystal evaluation, diffraction data collection, data reduction, and structure determination and validation. Since samples are shipped to SGX-CAT from SGX headquarters in San Diego, locations of the samples on canes and within the shipping dewars are also stored within the database. Other information, such as the assignment of barcodes to a particular project, is also readily retrievable. As of January 2007, the volume of data maintained within our database was approximately 60 GB, of which approximately two-thirds were graphics images, such as spectra or electron density maps. The SGX database currently consists of over 300 tables, approximately 2000 individual columns (fields), and more than 200 triggers and procedures to update related sections of the database when required.

SGX-CAT maintains a direct T1 network connection from the Advanced Photon Source in Illinois to SGX San Diego. Database inquiries are handled over this link. Interactions with the database occur in three ways. An extensive web-based system is used for data entry and retrieval. For crystals generated by external users of the beamline, upload of an electronic spreadsheet transfers the required crystal data attributes to the SGX LIMS. For automated operations, such as crystal screening and data collection, custom scripts place the computed results directly into the database.

Since the SGX LIMS contains virtually all information on the provenance and results from a particular sample, the data therein are available to track the progress of individual crystals at the beamline. The interface supporting this functionality is shown in Fig. 12.5. This web page is automatically updated every 15 min, providing near real-time assessment of sample progression at SGX-CAT. The display monitors the status of up to 380 crystals, providing access to more than 40 fields of data and the diffraction screening images for each crystal. This LIMS tool uses the results of automated crystal evaluation to select the best crystal within a group of replicate samples for data collection and monitors whether both data collection and data processing activities have completed successfully. As this interface handles most of the routine decisions that would normally be made by a crystallographer, we have greatly enhanced throughput in data collection operations at SGX-CAT.

## 12.10 Beamline monitoring and maintenance

Synchrotron beamlines are a complex hybrid of hardware and software. Although current designs have achieved a level of robustness inconceivable a decade ago, tight process control is essential. For example, at SGX-CAT the position of the beam is controlled to within 0.5 $\mu$radians (0.000028°). This tolerance corresponds to keeping the X-ray beam centroid within a 25 $\mu$m diameter circle at a location 50 m from the undulator source. This performance, reflecting the combined capabilities of the synchrotron and the beamline, is impressive to say the least.

At SGX-CAT, beam alignment is accomplished automatically. A two-stage process first maximizes the intensity of the beam exiting the monochromator. Combined motions of the robot and the monochromator are then used to place the X-ray beam directly at the sample position. In addition, critical components, including the operational parameters of the cryostream, the sample robot, and the X-ray intensity monitor are continually assessed. If any drift out of their allowed tolerance is detected, a beamline staff member is automatically notified through a paging system. Necessary adjustments can then be performed, in many cases via remote internet access to the beamline control system.

Continual vigilance and routine maintenance is required to ensure this level of performance. While ongoing maintenance is not difficult, it is a critical to ensuring optimal beamline performance. Thus, data collection and beamline control by a dedicated, professional staff appears decidedly preferable to that by occasional users.

## 12.11 High-throughput data collection: an example

We have in this chapter summarized various aspects of data collection at a synchrotron beamline. In this section, the operations of SGX-CAT are presented to illustrate these concepts in action.

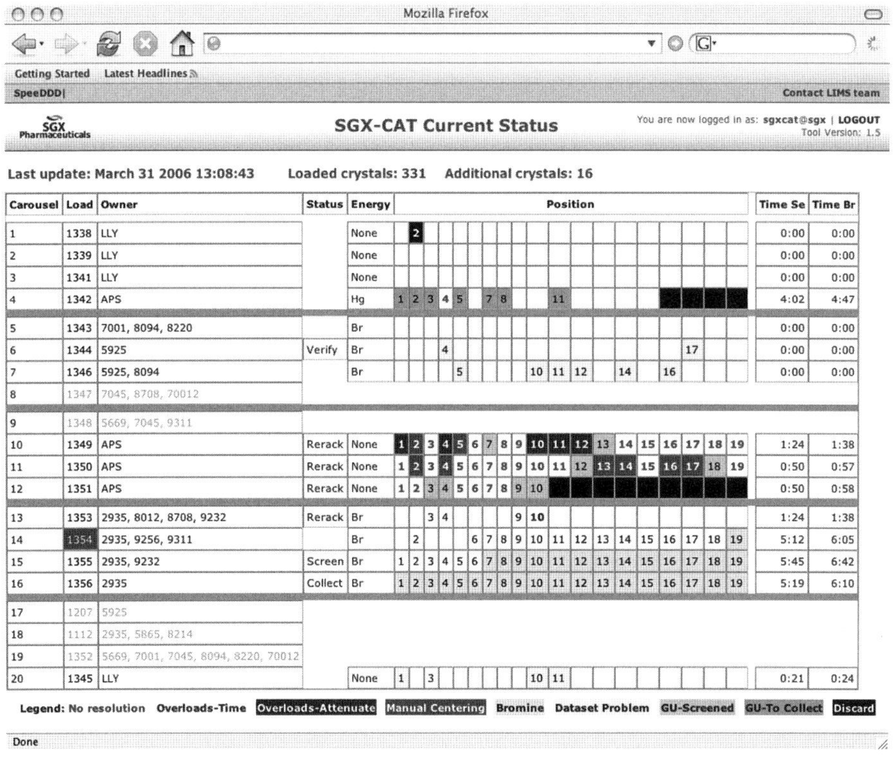


**Figure 12.5** SGX-CAT crystal status display. The webpage summarizes the status of up to 380 crystals using information extracted from the SGX LIMS. Each row represents a carousel of 19 crystals, each square in the row corresponding to a specific sample. Every square provides both an overlay and a dynamic link to further information on each crystal, including the screening images. The web page is updated every 15 min.

*Arrival:* Samples arrive at SGX-CAT in dry shipping dewars, which are filled with liquid nitrogen upon delivery. Descriptions of the crystals have been entered into the SGX LIMS prior to arrival.

*Loading of crystals:* Crystal samples are arranged into carousel loads that are tracked in the database. Each load can have up to 19 samples, the capacity of the MAR sample changer. Samples from different sources can be grouped together on a single carousel, because the LIMS, rather than physical proximity, tracks related crystals. The crystals comprising each load are placed in a carousel. Loading at the beamline, rather than at the home laboratory, ensures identification of defective samples early in the process. A beamline staff member uses the barcode reading station (Fig. 12.3) to associate the 1D barcode located on the outside of the vial and the 2D barcode from the base with sample information already resident in our LIMS. Validation is performed on the barcodes to verify that sample properties are indeed present in the database. After validation, records describing the load represented by this particular carousel are created within the database. Each load has a unique identification number. The database also tracks the identity of the physical carousel in use. Finally, the carousel is placed in a liquid nitrogen dewar for storage until screening or data collection.

*Screening:* The carousel containing the crystals to be screened is removed from the storage dewar and placed into the dewar of the MarCSC sample-changing robot. A protocol file that contains the screening parameters for each crystal is generated automatically by extracting the necessary information from the database. These data are then provided to the program, marccd, which operates the sample changing robot and the detector system. This data file also includes protocol commands for turning the liquid nitrogen wash on and off, archiving of data, and initiation of the crystal quality evaluation system. The entire carousel load is then screened automatically. For each crystal, the nylon mounting loop is automatically centred, after which four screening diffraction images are recorded at orientations with $\phi = 0°$, $45°$, $90°$, and $135°$. Once these images have been acquired, the crystal scoring system then analyses the images, determining diffraction quality and establishing whether or not that crystal diffracts to the desired resolution limit. All of the parameters generated from this evaluation process are stored permanently in the database for future use. Once quality scoring has been completed for all of crystals within a given set of replicate samples, the best crystal is marked automatically for collection on the status web page.

*Collection*: Collection of datasets is executed in much the same way as screening. In order to maximize the use of beam time while minimizing manual effort, new carousels of crystals are created that contain only crystals deemed suitable for data collection. At this stage, the crystals are identified through the 2D barcodes located on the base of the pin. Sample mounting, liquid nitrogen washing, and automated centring are performed during data collection as during screening. Optimal sample-to-detector distance for each sample is calculated from the estimated resolution limit and the X-ray wavelength. Protocol files for data collection also include the oscillation size and number of diffraction images specified in the database, both of which are determined automatically from knowledge of the sample Laue point group.

*Data processing*: Immediately upon completion of the acquisition of a diffraction dataset, data processing is launched as part of the process for changing to the next sample. At SGX-CAT, data are indexed and integrated using either Mosflm or d*TREK. Scaling of the integrated data is performed using Scala from CCP4 (Collaborative Computational Project, 1994). Scaled intensities are converted to structure factors by Truncate (also from CCP4). Processing for each dataset is assigned to one of nine servers located at the beamline. Reduced data are sent over the T1 line to SGX San Diego, where further processing occurs on a Linux cluster containing 296 processors.

Most of the data collected at SGX-CAT represent cocrystal of ligands with proteins of known structure. Cocrystal structures are determined via molecular replacement. EPMR (Kissinger *et al.*, 1999), MOLREP (Vagin and Teplyakov, 1997), or Phaser (Read, 2001) are used to find the best molecular replacement solution. Refinement of the structure and docking of bound ligands is accomplished using RefMac5 (Murshudov *et al.*, 1997) or CNX (Brünger, 1992).

Alternative scripts are available for *de novo* structure determination. Initial processing usually uses the Laue class determined during crystal evaluation. The phasing method depends on the type of experiment used to generate the data. In the case of a MAD/SAD dataset, ShakeNBake (Hauptman, 1997) or SHELXD (Schneider and Sheldrick, 2002) is used to determine the locations of anomalous scatterers. MLPHARE (Otwinowski, 1991) (heavy atom parameter refinement and phasing) and SOLOMON (Abrahams and Leslie, 1996) or DM (Cowtan, 1994) (phase improvement using solvent flattening/flipping) are used to generate and improve the phases based on the anomalous signal present in the data. Once the phases have been refined to convergence, ARP/wARP (Lamzin and Wilson, 1993) is used to build the model, followed by visual inspection and rebuilding with Xfit (McRee, 1999) or COOT (Emsley and Cowtan, 2004), and further refinement with RefMac5. When only relatively low resolution data (2.5–3.5 Å) are available, MAID (Levitt, 2001) is used instead of ARP/wARP to build the initial atomic model.

*Validation*: Once a structure has been determined, it is validated using a custom structure validation system (Badger and Hendle, 2002) to detect local errors. The system is based on PROCHECK (Morris *et al.*, 1992), WHATCHECK (Vriend, 1990), SFCHECK (Vaguine *et al.*, 1999), and PHISTATS and OVERLAPMAP (from CCP4).

## 12.12 Conclusion

High throughput data collection at third-generation synchrotron sources using automated systems produces enormous amounts of high-quality data. In 2006, ~9900 crystals were examined at SGX-CAT and ~4400 diffraction datasets were recorded. The previous year of operations saw similar productivity. Automation has reduced the average time required to screen a crystal from 13 min to less than 4 min, including all of the manipulations required to position the sample within the X-ray beam. This productivity is possible because of the close integration of the beamline activities with the SGX LIMS, which permits expert systems to manage most data collection without human intervention. In addition to serving the needs of SGX internal structure-based drug discovery programs, SGX-CAT provides 'mail-in' beamline access for several major pharmaceutical and biotechnology companies. The SGX-CAT paradigm also permits full-service support of a General User Program (www.sgxcat.com) that serves the needs of not-for-profit users, including one of the large scale production centres for the NIH-funded Protein Structure Initiative (www.nysgxrc.org).

SGX-CAT represents a state-of-the-art example of the concepts in high-throughput data collection presented in this chapter. Because of the high level of automation, the beamline functions exclusively as a 'mail-in' protein crystallographic facility, its activities supported by a dedicated professional staff of minimal size. The overarching goal of SGX-CAT operations is maintenance of the highest quality data while maximizing sample throughput. Critical enhancements to our data collection processes, in particular exploitation of the powerful SGX LIMS system and automatic crystal scoring and processing, have contributed greatly to the achievement of this goal.

## Acknowledgement

Design and construction of SGX-CAT and implementation of high-throughput operations would not have been possible without the cooperation of the entire SGX organization. We thank all of our colleagues, present and past, for their manifold contributions to the design, commissioning, and operation of SGX-CAT. Use of the Advanced Photon Source of Argonne National Laboratory is supported by the U. S. Department of Energy, Office of Science, Office of Basic Energy Sciences, under Contract No. DE-AC02–06CH11357.

## References

Abad-Zapatero, C. (2005). Notes of a protein crystallographer: my nights with ACTOR. *Acta Crystallogr. D* **61**, 1432–1435.

Abrahams, J. P. and Leslie, A. G. W. (1996). Methods used in the structure determination of bovine mitochondrial $F_1$ ATPase. *Acta Crystallogr. D* **52**, 30–42.

Adams, P. A., Grosse-Kunstleve, R. W., Hung, L.-W., Ioerger, T. R., McCoy, A. J., Moriarty, N. W., Read, R. J., Sacchettini, J. C., Sauter, N. K. and Terwilliger, T. C. (2002). *PHENIX*: building new software for automated crystallographic structure determination. *Acta Crystallogr. D* **58**, 1948–1954.

Badger, J. and Hendle, J. (2002). Reliable quality-control methods for protein crystal structures. *Acta Crystallogr. D* **58**, 284–291.

Bakaikoa, V. R. (2006). Control Software at ESRF beamlines. APS 2006 Users Meeting, Workshop 11, Beamline Controls at the APS. www.aps.anl.gov/aod/bcda/meetings/2006-05-04-workshop/rey-2006-05-04-wkshp11.pdf.

Berntson, A., Stojanoff, V. and Takai, H. (2003). Application of a neural network in high-throughput protein crystallography. *J. Synchrotron Rad.* **10**, 445–449.

Brönnimann, Ch., Bühler, Ch., Eikenberry, E. R., Horisberger, R., Hülsen, G., Schmitt, B., Schulze-Briese, C., Suzuki, M., Tomizaki, T., Toyokawa, H. and Wagner, A. (2004). Protein crystallography with the PILATUS 1M detector at the Swiss light source. *Synchrotron Radiation News* **17**, 23–30.

Broennimann, Ch., Eikenberry, E. F., Henrich, B., Horisberger, R., Huelsen, G., Pohl, E., Schmitt, B., Schulze-Briese, C., Suzuki, M., Tomizaki, T., Toyokawa, H., and Wagner, A. (2006). The PILATUS 1M detector. *J. Synchrotron Rad.* **13**, 120–130.

Brünger, A.T. (1992). *X-PLOR Version 3.1: A System for X-ray Crystallography and NMR*. Yale University Press, New Haven.

Brunzelle, J. S., Shafaee, P., Yang, X., Weigand, S., Ren, Z. and Anderson, W. F. (2003). Automated crystallographic system for high-throughput protein structure determination. *Acta Crystallogr. D* **59**, 1138–1144.

Cohen, A. E., Ellis, P. J., Miller, M. D., Deacon, A. M. and Phizackerley, R. P. (2002). An automated system to mount cryo-cooled protein crystals on a synchrotron beamline, using compact sample cassettes and a small-scale robot. *J. App. Crystallogr.* **35**, 720–726.

Cohen, A. E., McPhillips, S. E., Song, J. and Miller, M. D. (2005). Automation of high-throughput protein crystal screening at SSRL. *Synchrotron Radiation News* **18**, 28–35.

Collaborative Computational Project, Number 4. (1994). The *CCP4* Suite: Programs for Protein Crystallography. *Acta Crystallogr. D* **50**, 760–763.

Cowtan, K. (1994). 'dm': An automated procedure for phase improvement by density modification. *Joint CCP4 and ESF-EACBM Newsletter on Protein Crystallography* **31**, 34–38.

Emsley, P. and Cowtan, K. (2004). *Coot*: model building tools for molecular graphics. *Acta Crystallogr. D* **60**, 2126–2132.

Fujisawa, T., Nishikawa, Y., Yamazaki, H., and Inoko, Y. (2003). Evaluation and improvements of the Rigaku imaging plate reader (R-Axis IV++) for the use in synchrotron X-ray solution scattering. *J. Appl. Crystallogr.* **36**, 535–539.

Hauptman, H. A. (1997). Shake-and-bake: an algorithm for automatic solution *ab initio* of crystal structures. *Method Enzymol.* **277**, 3–13.

Helliwell, J. I. (1990). *Macromolecular Crystallography with Synchrotron Radiation*. Cambridge University Press, Cambridge.

Henderson, R. (1990). Cryo-protection of protein crystals against radiation damage in election and X-ray diffraction. *Proc. R. Soc. Lond. B* **241**, 6–8.

Hendrickson, W. A. (1991). Determination of macromolecular structures from anomalous diffraction of synchrotron radiation. *Science*, **254**, 51–58.

Holton, J. M. (2007). XANES measurements of the rate of radiation damage to selenomethionine side chains. *J. Synchrotron Rad.* **14**, 51–72.

Holton, J. and Alber, T. (2004). Automated protein crystal structure determination using ELVES. *PNAS* **101**, 1537–1542.

Hülsen, G., Broennimann, C., Eikenberry, E. F., and Wagner, A. (2006). Protein crystallography with a novel large-area pixel detector. *J. Appl. Crystallogr.* **39**, 550–557.

Jacquamet, L., Ohana, J., Joly, J., Legrand, R., Kahn, R., Borel, F., Pirocchi, M., Charrault, P., Carpentier, P. and Ferrer, J.-L. (2004). A new highly integrated sample environment for protein crystallography. *Acta Crystallogr. D* **60**, 888–894.

Jhoti, H. (2004). High-throughput crystallography. In: *Protein Crystallography in Drug Discovery*, Babine, R. E. and Abdel-Meguid, S.S., eds. Wiley-VCH Verlag, Weinheim.

Karain, W. I., Bourenkov, G. P., Blume, H. and Bartunik, H. D. (2002). Automated mounting, centering and screening of crystals for high-throughput protein crystallography. *Acta Crystallogr. D* **58**, 1519–1522.

Kirkpatrick, P. and Baez, A. V. (1948). Formation of optical images by X-rays. *J. Opt. Soc. Am.* **38**, 766–774.

Kissinger, C. R., Gehlhaar, D. K. and Fogel, D. B. (1999). Rapid automated molecular replacement by evolutionary search. *Acta Crystallogr. D* **55**, 484–491.

Koningsberger, D. C. and Prins, R., ed. (1988). *X-ray Absorption: Principles, Applications, Techniques of EXAFS, SEXAFS and XANES*. John Wiley and Sons, New York.

Kraft, S., Stümpel, J., Becker, P. and Kuetgens, U. (1996). High resolution x-ray absorption spectroscopy with absolute energy calibration for the determination of absorption edge energies. *Rev. Sci. Instrum.* **67**, 681–687.

Lamzin, V. S. and Wilson, K. S. (1993). Automated refinement of protein models. *Acta Crystallogr. D* **49**, 129–147.

Leslie, A. G. W. (1992). Recent changes to the MOSFLM package for processing film and image plate data. *Joint CCP4 + ESF-EAMCB Newsletter on Protein Crystallography*, **26**.

Levitt, D. G. (2001). A new software routine that automates the fitting of protein X-ray crystallographic electron-density maps. *Acta Crystallogr. D* **57**, 1013–1019.

McRee, D. E. (1999). *Practical Protein Crystallography, 2nd edn*. Academic Press, San Diego.

Mills, D. M., ed. (2002). *Third-Generation Hard X-ray Synchrotron Radiation Sources: Source Properties, Optics, and Experimental Techniques*. John Wiley and Sons, New York.

Morris, A. L., MacArthur, M.W., Hutchinson, E. G. and Thornton, J. M. (1992). Stereochemical quality of protein structure coordinates. *Proteins* **12**, 345–364.

Muchmore, S. W., Olson, J., Jones, R., Pan, J., Blum, M., Greer, J., Merrick, S. M., Magdalinos, P. and Nienaber, V. L. (2000). Automated crystal mounting and data collection for protein crystallography. *Structure* **8**, R243–R246.

Murray, J. W., Garman, E. and Ravelli, R. (2004). X-ray absorption by macromolecular crystals: the effects of wavelength and crystal composition on absorbed dose. *J. Appl. Crystallogr.* **37**, 513–522.

Murray, J. W., Rudiño-Piñera, E., Owen, R. L., Grininger, M., Ravellid, R. B. G. and Garman, E. F. (2005). Parameters affecting the X-ray dose absorbed by macromolecular crystals. *J. Synchrotron Rad.* **12**, 268–275.

Murshudov, G. N., Vagin A.A. and Dodson, E. J. (1997). Refinement of macromolecular structures by the maximum-likelihood method. *Acta Crystallogr. D* **53**, 240–255.

Nave, C. and Garman, E. F. (2005). Towards an understanding of radiation damage in cryocooled macromolecular crystals. *J. Synchrotron Rad.* **12**, 257–260.

Otwinowski, Z. (1991). Maximum likelihood refinement of heavy atom parameters. *Daresbury Study Weekend Proceedings* DL/SCI/R32, 80-86.

Owen, R. L., Rudiño-Piñera, E. and Garman, E. F. (2006). Experimental determination of the radiation dose limit for cryocooled protein crystals. *PNAS* **103**, 4912–4917.

Panjikar, S., Parthasarathy, V., Lamzin, V. S., Weiss, M. S. and Tucker, P. A. (2005). *Auto-Rickshaw*: an automated crystal structure determination platform as an efficient tool for the validation of an X-ray diffraction experiment. *Acta Crystallogr. D* **61**, 449–457.

Pflugrath, J. W. (1999). The finer things in X-ray diffraction data collection. *Acta Crystallogr. D* **55**, 1718–1725.

Pohl, E., Ristau, U., Gehrmann, T., Jahn, D., Robrahn, B., Malthan, D., Dobler, H. and Hermes, C. (2004). Automation of the EMBL Hamburg protein crystallography beamline BW7B. *J. Synchrotron Rad.* **11**, 372–377.

Read, R.J. (2001). Pushing the boundaries of molecular replacement with maximum likelihood. *Acta Crystallogr. D* **57**, 1373–1382.

Schneider, T. R. and Sheldrick, G. M. (2002). Substructure solution with *SHELXD*. *Acta Crystallogr. D* **58**, 1772–1779.

Smith, J. L. and Thompson, A. (1998). Reactivity of selenomethionine – dents in the magic bullet? *Structure* **6**, 815–819.

Snell, E. H., van der Woerd, M. J., Miller, M. D. and Deacon, A. M. (2005). Finding a cold needle in a warm haystack: infrared imaging applied to locating cryocooled crystals in loops. *J. Appl. Crystallogr.* **38**, 69–77.

Snell, G., Cork, C., Nordmeyer, R., Cornell, E., Meigs, G., Yegian, D., Jaklevic, J., Jin, J., Stevens, R. C. and Earnest, T. (2004). Automated Sample Mounting and Alignment System for Biological Crystallography at a Synchrotron Source. *Structure* **12**, 537–545.

Sweet, R. M. (1998). The technology that enables synchrotron structural biology. *Nature Struct. Biol.* (synchrotron suppl.) **5**, 654–656.

Thorne, R. W., Stum, Z., Kmetko, J., O'Neill, K. and Gillilan, R. (2003). Microfabricated mounts for high-throughput macromolecular cryocrystallography. *J. Appl. Crystallogr.* **36**, 1455–1460.

Ueno, G., Hirose, R., Ida, K., Kumasaka, T. and Yamamoto, M. (2004). Sample management system for a vast amount of frozen crystals at SPring-8. *J. Appl. Crystallogr.* **37**, 867–873.

Vagin, A. A. and Teplyakov, A. (1997). *MOLREP*: an automated program for molecular replacement. *J. Appl. Crystallogr.* **30**, 1022–1025.

Vaguine, A. A., Richelle, J. and Wodak, S. J. (1999). *SFCHECK*: a unified set of procedures for evaluating the quality of macromolecular structure-factor data and their agreement with the atomic model. *Acta Crystallogr. D* **55**, 191–205.

Vernede, X., Lavault, B., Ohana, J., Nurizzo, D., Joly, J., Jacquamet, L., Felisaz, F., Cipriani, F. and Bourgeois, D. (2006). UV laser-excited fluorescence as a tool for the visualization of protein crystals mounted in loops. *Acta Crystallogr. D* **62**, 253–261.

Vriend, G. (1990). WHAT IF: A molecular modelling and drug design program. *J. Mol. Graph.* **8**, 52–56.

Winick, H. (1987). Synchrotron radiation. *Scientific American* **257**, 88–99.

Zhang, Z., Sauter, N. K., van den Bedem, H., Snell, G. and Deacon, A. M. (2006). Automated diffraction image analysis and spot searching for high-throughput crystal screening. *J. Appl. Crystallogr.* **39**, 112–119.

CHAPTER 13

# Electron density fitting and structure validation

**Mike Carson**

## 13.1 Introduction

The Human Genome Project went three-dimensional in late 2000. 'Structural genomics' efforts will determine the structures of thousands of new proteins over the next decade. These initiatives seek to streamline and automate every experimental and computational aspect of the structural determination pipeline, with most of the steps involved covered in previous chapters of this volume. At the end of the pipeline, an atomic model is built and iteratively refined to best fit the observed data. The final atomic model, after careful analysis, is deposited in the Protein Data Bank, or PDB (Berman *et al.*, 2000). About 25,000 unique protein sequences are currently in the PDB. High-throughput and conventional methods will dramatically increase this number and it is crucial that these new structures be of the highest quality (Chandonia and Brenner, 2006).

This chapter will address software systems to interactively fit molecular models to electron density maps and to analyse the resulting models. This chapter is heavily biased toward proteins, but the programs can also build nucleic acid models. First a brief review of molecular modelling and graphics is presented. Next, the best current and freely available programs are discussed with respect to their performance on common tasks. Finally, some views on the future of such software are given.

## 13.2 Initial molecular models

Small molecule crystal structures solved through direct methods yield very accurate atomic positions at much higher resolution than is typical for proteins. Currently about 200,000 error-free organic compounds with conventional R factors less than 0.05 are available through the Cambridge Structural Database (CSD) (Allen *et al.*, 1979). These data have been mined for conformational analysis, hydrogen bonding directionality, non-bonded packing interaction, and more, as recently reviewed (Allen and Motherwell, 2002). The CSD provides an invaluable source of coordinate geometry for inhibitors and cofactors, which should be trusted more than the energy minimized output of any modelling program.

A common feature of modelling and refinement programs is a dictionary of ideal residues derived from the results of small-molecule crystallography. Ideal bond lengths and angles for the amino acid and nucleic acid building blocks of macromolecules have been gathered from the CSD (Engh and Huber, 1991). The atomic bond and angle parameters are tightly constrained for macromolecular refinement and may be regarded as fixed, with the only degrees of freedom coming from torsional rotation about single bonds.

The favoured dihedral angles for protein main chains were derived from energy considerations of steric clashes in peptides giving the well known Ramachandran plot (Ramachandran and Sasisekharan, 1968). These phi/psi combinations characterize the elements of secondary structure. Accurate main chain models can be constructed from 'spare parts', that is short pieces of helices, sheets, turns, and random coils taken from highly refined structures, provided a series of C-alpha positions can be established from the electron density map

(Jones and Thirup, 1986). Rotamers, the favoured dihedral angles for protein side chains, were first observed from an analysis of well-refined protein structures with no dihedral restraints (Ponder and Richards, 1987).

## 13.3 Initial fitting to density

In their classic paper (Brändén and Jones, 1990) the late Carl-Ivar Brändén and Alwyn Jones observed that fitting the linear protein sequence into an imperfect electron density map was a process 'between objectivity and subjectivity'. They observed the conventional R factor was not as reliable an indicator as previously thought as several published structures were incorrect. Three validation techniques have practically eliminated grossly incorrect structures, where the sequence is misfolded into the density: the definition of the real-space R factor by Jones' group (Jones *et al.*, 1991) with a per residue comparison between observed and calculated density; the threading and three-dimensional (3D) sequence profiling methods developed by Eisenberg's group (Bowie *et al.*, 1991; Lüthy *et al.*, 1992); and the use of statistical cross-validation with the development of the free R value by Brünger (Brünger, 1992) to prevent over-fitting of the data.

A nearly complete initial structure roughly fit to the electron density may be available through homology modelling and molecular replacement, as discussed by Delarue in Chapter 7 of this volume. A relatively new technique is the automatic interpretation of crystallographic electron density maps with construction and refinement of preliminary models. The ARP/wARP software suite is recommended (www.embl-hamburg.de/ARP) (Perrakis *et al.*, 1999). A main chain model or a more complete model with side chains may be output provided data around 2 Å is available. Attempts to use artificial intelligence have periodically been applied to automate map fitting (Feigenbaum *et al.*, 1977; Glasgow *et al.*, 1993). A promising new method using AI techniques has been developed to work optimally with maps around 2.8 Å resolution (Ioerger and Sacchettini, 2003). These methods are covered by Morris, Perrakis, and Lamzin in Chapter 11 of this volume.

## 13.4 Map fitting and refinement

One of the most difficult steps in X-ray crystallography is map interpretation. A major problem is correctly connecting secondary structure elements. This topic has been discussed in detail by several authors (Richardson and Richardson, 1985; McRee, 1993; Kleywegt and Jones, 1997). Most crystallographers are fortunate enough to learn map fitting from an experienced mentor. It is critical to have the most accurate data and phase information possible. The original experimental map should always be kept for references. A major concern in the later stages of refinement is model bias which is usually addressed by calculating omit maps (Bhat and Cohen, 1984) or simulated annealing omit maps (Hodel *et al.*, 1992).

One of the first protein refinement programs rotated torsion angles against real-space density gradients (Diamond, 1971). This is the idea used by most automatic and interactive building programs. The current CNS version performs reciprocal-space torsion angle refinement (Rice and Brünger, 1994). There is debate over the best refinement program – many favour REFMAC (Murshudov *et al.*, 1997). Refinement is discussed in Chapter 11 of this volume.

## 13.5 Validation of structures

The process of structure solution is iterative: model building, refinement, and analysis until one is satisfied with the model. Each lab. or researcher uses their favourite programs and protocols and may arrive at a model via different paths. Grossly incorrect structures are detected as discussed in the previous section. The validation criteria of atomic models from crystallography is fairly standardized, as reviewed by Gerard Kleywegt (Kleywegt, 2000). The PDB's automated deposition system uses Janet Thornton's PROCHECK (Laskowski *et al.*, 1993) as the validation module, checking all bonds, angles, dihedrals, and close contacts and providing a summary report. The SFCHECK (Vaguine *et al.*, 1999) program validates the structure factors.

Dihedrals are not generally constrained to expected values in refinement programs. Any outlier is highly suspect, even though the density fit is decent. For example, a valine side chain may require 'flipping' by a 180 degree rotation. WHATCHECK (Hooft *et al.*, 1996) (www.cmbi.kun.nl/gv/whatcheck), the validation module of Gert Vriend's molecular modelling program WHATIF (Vriend, 1990), is preferred locally to carry out conformational analyses. Its flagging of questionable rotamers may not be as good as PROCHECK, but it offers many important additional checks. Stretches of residues are analysed for packing and via a database comparisons. This can identify bad stretches that actually have reasonable dihedral angles, or carbonyls that need to be flipped. The hydrogen bonding pattern analysis reveals any unsatisfied main-chain hydrogen bonds, and Asn, Gln, and His side chains that should be flipped. An example is shown in Fig. 13.1.

The problem of validation has been of long-term interest to the author. *Ribbons* (Carson, 1997) was presented as a visual 'sanity check' of a structure, mapping properties of crystallographic interest to the ribbon drawing (Carson and Bugg, 1988). Residues were colour-coded by main-chain and side-chain dihedral sensibility, agreement with the electron density map, and potential energy. The real-space R-factor (RSR) (Jones *et al.*, 1991) is a particularly powerful tool for detecting stretches of residues that are out of register. The temperature factor, or B-factor, can be a sink for absorbing errors. Any residues with values over 40 may be closer to fantasy than reality. In a study designed to find errors in a molecular replacement model (Carson *et al.*, 1994), RSR, B-factors, convergence of refinement, dihedral fit to the database, and geometric strain were found to predict error. In perhaps a tautology, the most important factors were the RSR and B-factors. The contribution of the X-ray data is the most important – the structure cannot be judged by geometry alone.

The emerging field of structural genomics will generate thousands of new protein models through crystallography, each of which must be carefully checked computationally and visually. The handling and curation of this atomic structural data represents a challenging problem in bioinformatics. The human pattern recognition ability and human intuition is indispensable in validating atomic models and

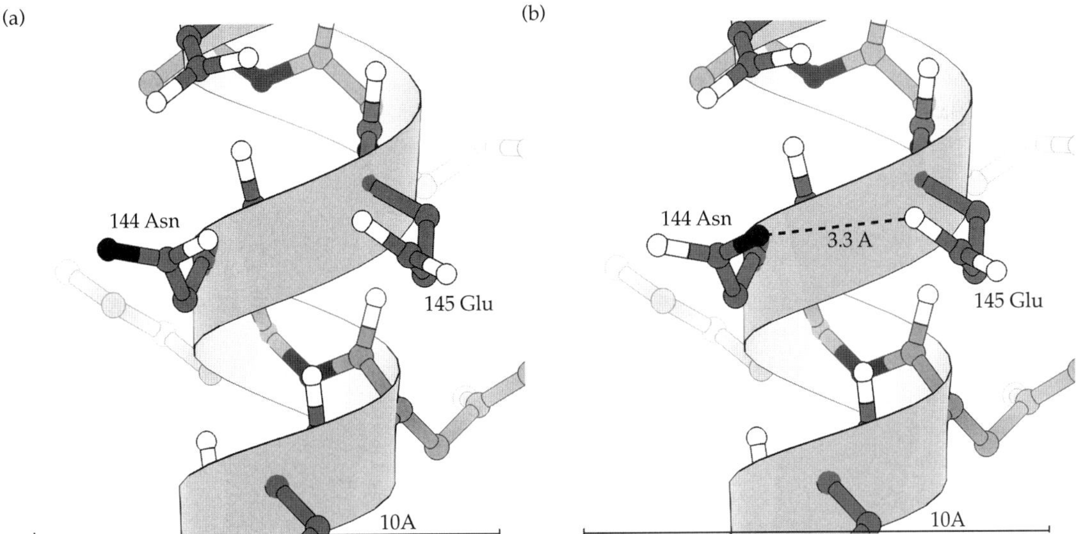


**Figure 13.1** Correction of a structure by side-chain flipping. The original structure after CNS refinement is shown in (a) and after the recommended correction of WHATCHECK in (b). Carbon atoms are grey, oxygens white, and nitrogens black. Asn 144 and Glu 145 are the labelled adjacent residues in a helix, shown as a ribbon. After a 180 degree rotation about the chi-2 dihedral of Asn 144, the side chain amide nitrogen is now in position to hydrogen bond, shown as the labelled dashed line, to the carboxylate of Glu 145. Image rendered by *ribbons*.

despite technological advances, covered in Chapter 11, methods are still very labour intensive.

## 13.6 Model building and molecular graphics

Structural genomics has been made possible by many concurrent advances in technology besides those in molecular biology. In particular, advances in computing power and algorithms go hand-in-hand with advances in crystallography. Software to perform most crystallographic tasks, in particular the CCP4 suite of programs (Collaborative Computational Project, 1994) and the CNS package (Brünger *et al.*, 1998), is freely available from academic groups. A very important aspect of crystallographic computing is the role of computer graphics.

### 13.6.1 A brief history of molecular graphics

Physical models have enjoyed a long and successful history in chemistry. Watson and Crick used cardboard cutouts of the DNA bases and observed the similarity in shape between an AT and a GC base pair. This insight led them eventually to the double helix structure of DNA and a molecular mechanism for genetics in 1953. Physical models were vital in the first 3D structural determination of a protein in 1958, using X-ray crystallography (Kendrew *et al.*, 1958). Their myoglobin structure was modelled by brass Tinkertoy-like (or a Meccano-like) representations for the atoms and bonds, supported by thousands of vertical rods. At a scale of 5 cm/Å, the model occupied roughly a cube of nearly 2 m per side.

Interactive molecular graphics was pioneered by Leventhal in the 1960s (Levinthal, 1966). Several research groups in the late 1970s and early 1980s began efforts to replace the physical models used by crystallographers with computer graphics. In 1978, the first protein structure was solved with computer graphics using the GRIP system developed by Fred Brooks' computer science group at UNC (Tsernoglou *et al.*, 1977). An IBM mainframe performed the electron density map calculations while a DEC minicomputer drove the custom graphics terminal. All drawing consisted of white lines on a black background. The system displayed the electron density as a mesh of vectors and the molecular model as a stick figure that could be interactively manipulated to fit the density and 'regularized' (Hermans and McQueen, 1974) to maintain proper geometry. Most importantly, the adjusted coordinates could be easily saved to disk for another round of the iterative process of map calculation and model adjustment. The first widely used software of this type was FRODO, developed by the crystallographer Alwyn Jones (Jones, 1981) after this proof of concept.

The majority of current display representations and interactive techniques for computer-assisted molecular design were developed in the 1980s (Olson and Goodsell, 1992). The colour Evans and Sutherland PS300, a calligraphic vector-drawing device driven by a DEC VAX minicomputer, became the standard platform for macromolecular crystallographers and molecular modellers. A major drawback of these specialized display devices was a price over $50,000. Many of the techniques developed in universities were incorporated by newly-formed companies into commercial software for general molecular modelling.

UNIX workstations began to make inroads toward the end of the 1980s, dominating in the 1990s. Machines created by Silicon Graphics (SGI) became the preferred tool of the crystallographic and modelling communities. These workstations integrated a powerful general-purpose computer and a custom chip optimized for graphical display on a raster CRT monitor. While less expensive than a VAX/PS300, these workstations remained a costly investment. The emerging standards based on the X window system led to a system for 'push-button crystallography' by the crystallographer Duncan McRee (McRee, 1992).

At the same time, the power and capabilities of standard PCs increased dramatically. The Richardsons (Richardson and Richardson, 1992) at Duke University developed molecular graphics programs to run on the Macintosh. In a dramatic development, the World Wide Web emerged. Roger Sayle created RasMol as part of his PhD dissertation at Oxford, and made the source code freely available on the web. This program allowed raster display of molecules on UNIX or PCs. The RasMol code was converted by Molecular Design Limited into Chime (CHemical mIME), a widely used plug-in for molecular display on web browsers. VRML (Virtual Reality Modeling Language) was proposed as the standard for 3D presentation and interaction on the web.

### 13.6.2 Current state of molecular graphics

Several high-quality commercial molecular modelling software systems include crystallographic modules. Tripos offers solutions integrating chemistry and informatics targeting drug discovery (www.tripos.com). Accelrys provides software for pharmaceutical, chemical, and materials research (www.accelrys.com). These are certainly worth looking into if the budget will allow. The latter developed 'Quanta,' an older SGI product favoured by many of our local crystallographers.

Most crystallographic labs have made the transition from SGI workstations to commodity PCs running Linux. Most software shared among crystallographers has been converted. The conversion was relatively painless thanks to *de facto* standards, such as UNIX, the X windows system, and the OpenGL graphics API. A fully loaded Linux PC with a good graphics card and hardware stereo provides an excellent crystallographic workstation for under $3000. While the Linux/PC is likely to be dominant over the decade, a serious rival may be Apple PCs running OSX. Many programs are also available under Windows. Outstanding molecular graphics software for these machines is freely available to academic groups over the web.

## 13.7 Accessing software for electron density fitting and structure validation

This chapter will critique four software packages that are freely available to academics. No review of commercial software will be given. Each of the four packages will run under most flavours of UNIX, including Linux and Mac OSX, and most under Windows as well. Some require a license. Each is easily accessible over the web:

**O** (Kleywegt and Jones, 1997), the successor to FRODO, fits molecular models to maps. It is available through Jones' lab in Upsalla (xray.bmc.uu.se/~alwyn). A simple licensing procedure gives access to the binary version for Linux, and delivers a special access code specific for the PC on which it is run. The instructions make installation trivial. Version 9.07 is discussed here.

**XtalView** (McRee, 1999) is a complete system for solving a macromolecular crystal structure, with modules from isomorphous replacement through building the molecular model. Developed by Duncan McRee, it is now available through the San Diego Supercomputer Center (www.sdsc.edu/CCMS/Packages/XTALVIEW). Only the binary executable was acquired, but the source code is available with a separate license. An installation script is edited to set local directories and the environment. The model building facilities of version 4.0 of the 'xfit' module are discussed here.

**Coot** (Emsley and Cowtan, 2004), the crystallographic object-oriented toolkit, is the CCP4 module for model building, model completion, and validation. It is available from the home page of Paul Emsley (www.ysbl.york.ac.uk/~emsley/coot/). Anyone may download and use the package under the GNU GPL license. A new windows PC version was easily installed. The map fitting tools of Coot version 0.031 were tested.

**MolProbity** (Lovell *et al.*, 2003) is an interactive macromolecular structure validation and visualization tool from the Richardsons' laboratory (molprobity.biochem.duke.edu). The program may be run interactively using a Java-enabled web browser, or downloaded to your PC. The latest version was run over the web.

### 13.7.1 File formats

The PDB format, for better or worse, is the standard for macromolecular coordinate files. The handling of residues with alternate conformations and the 'chain id' versus 'segment id' distinction may cause small problems. Density map formats are much less standardized. Fortunately, the 'mapman' program from Uppsala can convert between many, but not all, types of maps. To convert to the 'fsfour' format (which can be read, but not written by MAPMAN) for XtalView, the 'cns2fsfour' available from San Diego is recommended. Coot uses the

CCP4 'MTZ' format for maps. A key advance in map representation beyond the conventional isosurface contours was extraction of 'ridge lines' or a 'skeleton' (Greer, 1974, 1985). This provides a pseudoatom structure with 'atoms' at local density peaks, connected through maximum density. Each program has its own format for this data structure.

### 13.7.2 Key software features

**O** is driven by a command line interface, and the ability to create user-defined macros. A general introduction, several short tutorials on specialized subjects, and a description of all possible commands are presented in html from the homepage. The most notable feature of O is the ability to build macromolecular models from scratch into a density map. The secondary structure templates make use of database fragments plus interactive real space refinement. It also allows analysis and easy modification of the structure, such as peptide oxygen flips. Another interesting aspect of O is inclusion of interaction via video game controllers. They note 'O has a bewildering number of tasks that may be used to make a Ca trace'. The p2map directory distributed with the program contains a step-by-step tutorial for fitting experimental density. Many useful O-accessories are available from the Uppsala Software Factory (xray.bmc.uu.se/usf). In particular, the OOPS program (Kleywegt and Jones, 1996) identifies potential errors and steps through them for user correction.

**Xfit** v4.0 is a large and complex program driven by 18 X-based GUI windows. A 30-plus-page manual is included. Notable features include model fitting to electron density, the simultaneous real-space fitting while maintaining accurate geometry, a built-in FFT to calculate omit maps on the fly, the ability to print the screen as PostScript, and a scripting language. XtalView offers a complete package for crystallography. Some locally use it only to locate heavy atoms and for phasing. The 'xfit' package recommends running without map files, rather generating them on demand from an ascii h,k,l,Fo,fom,phi file. They report that the shake option with sigma-A maps is about as good as it gets at removing model phase bias. One can identify errors, such as in a Ramachandran plot, generating a clickable list to position the user at the offending residue for analysis.

**Coot** is driven by standard pulldown menus from a menubar. A short tutorial and longer user's manual are available in PDF format. All the standard features of the other map fitting programs are available. Commands may be scripted in the Python language. The most attractive feature is the intuitiveness of the interface. The validation features are well integrated. For example, a density fit bar graph may be displayed in a separate window. Clicking on the graph automatically centres on the questionable residue.

**MolProbity** runs on a java-enabled browser from the well-documented homepage. The notable feature is web-based identification and viewing of potential problems in the model. Hydrogen atoms are added to assist in the contact analysis. Poor phi/psi angles are flagged only. Poor rotatamers are flagged but cannot be fixed interactively on line, but tools are provided to do this for both O and XtalView. Potential Asn, Gln, and His flips are visualized, and user-selected residues will be flipped. Bad contacts are visualized with colour-coded dots, plus a prioritized list. It does an excellent job of finding Thr residues that should be flipped.

### 13.7.3 Experimental assessment of software within the laboratory

Two common crystallographic tasks were examined: creation of a new structure and completing a refinement. The data are as follows: (1) A complete medium resolution structure after several rounds of refinement has a new round of maps calculated and is analysed. A range of residues in a turn and several leucine rotamers are deemed questionable. These were examined and corrected. (2) The sequence representing a novel fold with no known homolog was crystallized and solved (Li *et al.*, 2002) from an ISAS (iterative single wavelength anomalous signal) map (Wang, 1985) based only on native sulphur atoms with data to 1.8 Å. The majority of the main chain was automatically traced with arp/warp. For this test, the sequence (*C. elegans* Sequencing Consortium, 1998), the predicted secondary structure (Rost, 1996), the refined sulphur sites (Furey and Swaminathan, 1997), and an automated attempt at solution created by the TEXTAL (Ioerger and Sacchettini, 2003) server (http://textal.tamu.edu) was examined and corrected.

The results of the experimentation with the data and tasks above are based strictly on 'feel'. I found Coot to be by far the best map fitting program, generally being able to perform the task at hand without reading the manual. In my opinion, the interface to O presents a serious problem. Others in the lab. claim that once familiarity is gained that it is just right. MolProbity has the best aesthetics and gives a good final check. A disclaimer: the author last actively fit models over a decade ago, and was quite happy with a variant of FRODO.

Illustrations of two examples are included. Figure 13.2 captures the Java display inside a web browser running a MolProbity session. Figure 13.3 shows before and after screen shots captured while performing map fitting with Coot. These figures cannot do the respective software systems justice. The figures are in black and white – the programs use colour to convey vital information. The figures are static – the programs are interactive. It is hoped this chapter will encourage the gentle reader to run the software – it's easy, it's fun!

## 13.8 Future of crystallographic software

The graphics history section highlights the rapid change in computer hardware. Current platforms will certainly also become obsolete. Grid computing over the internet will probably become prevalent.

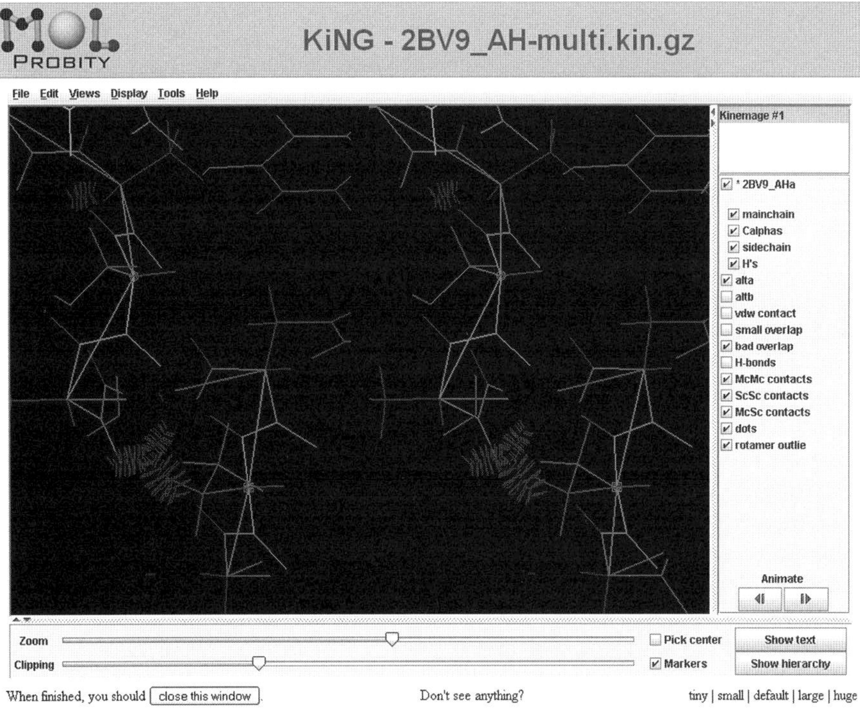


**Figure 13.2** Screen shot of MolProbity. The image was captured from a Java-enabled web browser, and converted to greyscale. Of course, colour and interactivity are vital to this application. The image shows a stick figure atomic model, including all hydrogens, in side-by-side stereo. Steric clashes are shown by dots (red in the original). The highest concentration of dots in the lower centre is due to a water–side chain clash. Just above the water is a serine residue that has been flagged (orange in the original) as an unlikely rotamer. One of its hydrogens connected to the CB atom clash with a neighbouring carbonyl. A separate window presents a colour-coded table of the properties of all residues.

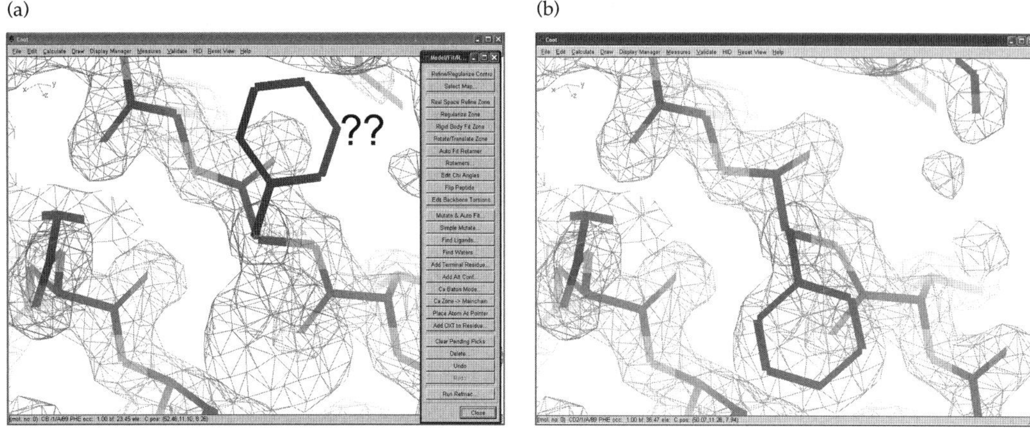


**Figure 13.3** Screen shot of density fitting with Coot. The images were captures from the screen of the Windows PC version and converted to greyscale. Of course, colour and interactivity are vital to this application. The images show the electron density map as a mesh of thin, grey lines. The atomic model is rendered as thick lines. The original model is shown in (a), with the '??' label added by an external program to draw the readers attention to the questionable residue that does not fit into the density. From the 'Calculate' option shown on the menu bar, the first item, 'Model/Fit/Refine', was selected. This invokes the submenu of choices shown overlaid on the right edge of (a). The corrected model is shown in (b), with the phenylalanine ring now completely inside the density. This is the near instantaneous result of clicking 'Auto Fit Rotamer', the seventh item on the submenu.

The future in high-powered graphics is no longer scientific workstations, but computer games. However, their internals are now jealously guarded secrets.

Most crystallographic programs arose in academic labs from the vision of their author and the help of a small number of collaborators. The authors maintain various levels of access to their source code. The 'open source' model is an alternative. According to the Open Source Initiative (www.opensource.org): 'The basic idea behind open source is very simple: When programmers can read, redistribute, and modify the source code for a piece of software, the software evolves. People improve it, people adapt it, people fix bugs. And this can happen at a speed that, if one is used to the slow pace of conventional software development, seems astonishing...'

Our structural genomics bioinformatics group (sgce.cbse.uab.edu) used only well-known, open-source programs: the Linux operating system; the Apache web server; a PostgreSQL database; and the Python programming language. PyMOL (sourceforge.net/projects/pymol) is an open source molecular graphics project. The CCP4 (Collaborative Computational Project, Number 4, 1994) is an excellent example of providing full source. The PHENIX project (Adams *et al.*, 2002) promises the same. This software is provided free or at a nominal cost.

While the authors of the software reviewed here continue to make incremental improvement, one wonders who will carry on their work. A perhaps unfortunate trend is that more students in structural biology come from a biology background, using software as a black box, with little interest in programming. Many computer scientists believe they should develop all software, and consider the scientist/programmer to be a dilettante. This may explain why many are willing to pay the high fees for commercial software. Of course I could not disagree more, and hope that talented crystallographers/developers will learn a little software engineering and continue the tradition of scientific software development.

## Acknowledgements

This chapter would not be possible without all the scientists who freely provide their software. Thanks to the crystallographers at the UAB/CBSE, Debasish Chattopadhyay, Champion Deivanayagam, Songlin Li, Krishna Murthy, Lisa Nagy, S. V. L.

Narayana, Norbert Schormann, Craig Smith, Henry Symersky, Mike Teale, and Mark Walter, for helpful discussion.

## References

Adams, P. D., Grosse-Kunstleve, R. W., Hung, L.-W., Ioerger, T. R., McCoy, A. J., Moriarty, N. W., Read, R. J., Sacchettini, J. C. and Terwilliger, T. C. (2002). PHENIX: Building new software for automated crystallographic structure determination. *Acta Crystallogr. D* **58**, 1948–1954.

Allen, F. H. and Motherwell, W. D.S. (2002). Applications of the Cambridge Structural Database in organic chemistry and crystal chemistry. *Acta Crystallogr. B* **58**, 407–422.

Allen, F. H., Bellard, S., Brice, M. D., Cartwright, B. A., Doubleday, A., Higgs, H., Hummelink, T., Hummelink-Peters, B. G., Kennard, O., Motherwell, W. D. S., Rodgers, J. R. and Watson, D. G. (1979). The Cambridge Crystallographic Data Centre: computer-based search, retrieval, analysis and display of information. *Acta Crystallogr. B* **35**, 2331–2339.

Berman, H. M., Westbrook, J., Feng, Z., Gilliland, G., Bhat, T. N., Weissig, H., Shindyalov, I. N. and Bourne, P. E. (2000). The Protein Data Bank. *Nucleic Acids Research* **28**, 235–242.

Bhat, T. N. and Cohen, G. H. (1984). OMITMAP: An electron density map suitable for the examination of errors in a macromolecular model. *J. Appl. Cryst.* **17**, 244–248.

Bowie J. U., Lüthy, R. and Eisenberg, D. (1991). A method to identify protein sequences that fold into a known three-dimensional structure. *Science* **253**, 164–170.

Brändén, C.-I. and Jones, T. A. (1990). Between objectivity and subjectivity. *Nature* **343**, 687–689.

Brünger, A. T. (1992). The Free R value: a novel statistical quantity for assessing the accuracy of crystal structures. *Nature* **355**, 472–474.

Brünger, A. T., Adams, P. D., Clore, G. M., Delano, W. L., Gros, P., Grosse-Kunstleve, R. W., Jiang, J.-S., Kuszewski J., Nilges N., Pannu, N. S., Read R. J., Rice L. M., Simonson T. and Warren G. L. (1998). Crystallography and NMR System (CNS): A new software system for macromolecular structure determination. *Acta Crystallogr. D* **54**, 905–921.

*C. elegans* Sequencing Consortium (1998). Genome sequence of the nematode *C. elegans*: a platform for investigating biology. *Science* **282**, 2012–2018.

Carson, M. (1997). Ribbons. *Method Enzymol.* **277**, 493–505.

Carson, M. and Bugg, C. E. (1988). Protein crystal graphics. In: *Proceedings of the Annual Meeting of the American Crystallographic Association*. American Crystallographic Association, Series 2, Vol. 16, abstract G2, p. 36.

Carson, M., Buckner, T. W., Yang, Z., Narayana, S. V. L. and Bugg C. E. (1994). Error detection in crystallographic models. *Acta Crystallogr. D* **50**, 900–909.

Chandonia, J. M. and Brenner, S. E. (2006). The impact of structural genomics: expectations and outcomes. *Science* **311**, 347–351.

Collaborative Computational Project, Number 4. (1994). The CCP4 suite: programs for protein crystallography. *Acta Crystallogr. D* **50**, 760–763.

Diamond, R. (1971). A real-space refinement procedure for proteins. *Acta Crystallogr. A* **27**, 436–452.

Emsley, P. and Cowtan, K. (2004). Coot: model-building tools for molecular graphics. *Acta Crystallogr. D* **60**, 2126–2132.

Engh, R. A. and Huber, R. (1991). Accurate bond and angle parameters for X-ray structure refinement. *Acta Crystallogr. A* **47**, 329–400.

Feigenbaum, E. A., Engelmore, R. S. and Johnson, C. K. (1977). A correlation between crystallographic computing and artificial intelligence research. *Acta Crystallogr. A* **33**, 13–18.

Furey, W. and Swaminathan, S. (1997). PHASES-95: A program package for the processing and analysis of diffraction data from macromolecules. *Method Enzymol.* **277**, 590–620.

Glasgow, J., Fortier, S. and Allen, F. (1993). Molecular scene analysis: crystal structure determination through imagery. In: Hunter, L., ed. *Artificial Intelligence and Molecular Biology*. MIT Press, Cambridge, MA.

Greer, J. (1974). Three-dimensional pattern recognition: an approach to automated interpretation of electron density maps of proteins. *J. Mol. Biol.* **82**, 279–301.

Greer, J. (1985). Computer skeletonization and automatic electron density map analysis. *Method Enzymol.* **115**, 206–224.

Hermans, J. and McQueen, J. E. (1974). Computer manipulation of (macro)molecules with the method of local change. *Acta Crystallogr. A* **30**, 730–739.

Hodel, A., Kim, S.-H. and Brünger, A. T. (1992). Model bias in macromolecular crystal structures. *Acta Crystallogr. A* **48**, 851–859.

Hooft, R. W. W., Vriend, G., Sander, C. and Abola, E. E. (1996). Errors in protein structures. *Nature* **381**, 272–272.

Ioerger, T. R. and Sacchettini, J. C. (2003). TEXTAL system: artificial intelligence techniques for automated protein model building. *Method Enzymol.* **374**, 244–270.

Jones, T. A. and Thirup, S. (1986). Using known substructures in protein model building and crystallography. *EMBO J.* **5**, 819–822.

Jones, T. A. (1981). FRODO: A graphics fitting program for macromolecules. In: *Computational Crystallography*, Sayre, D., ed., pp 303–317. Clarendon Press, Oxford.

Jones, T. A., Zou, J.-Y., Cowan, S. W. and Kjeldgaard, M. (1991). Improved methods for building protein models in electron density maps and the location of errors in these models. *Acta Crystallogr. A* **47**, 110–119.

Kendrew, J. C., Bodo, G., Dintzis, H. M., Parrish, R. G., Wyckoff, H. and Phillips, D. C. (1958). A three dimensional model of the myoglobin molecule obtained by x-ray analysis. *Nature* **181**, 662–666.

Kleywegt, G. J. (2000). Validation of protein crystal structures. *Acta Crystallogr. D* **56**, 249–265.

Kleywegt, G. J. and Jones T. A. (1996). Efficient rebuilding of protein structures. *Acta Crystallogr. D* **52**, 829–832.

Kleywegt, G. J. and Jones, T. A. (1997). Electron density map interpretation. *Method Enzymol.* **277**, 525–545.

Laskowski, R. A., MacArthur, M. W., Moss, D. S. and Thornton, J. M. (1993). PROCHECK: A program to check the stereochemical quality of protein structures. *J. Appl. Crystallogr.* **26**, 283–291.

Levinthal, C. (1966). Molecular model-building by computer. *Scientific American* **214**, 42–52.

Li, S, Finley, J., Liu, J., Qiu, S. H., Chen, H., Luan, C. H., Carson, M., Tsao, J., Johnson, D., Lin, G., Zhao, J., Thomas, W., Nagy, L. A., Sha, B., DeLucas, L. J., Wang, B. C. and Luo M. (2002). Crystal structure of the cytoskeleton-associated protein glycine-rich (CAP-Gly) domain. *J. Biol. Chem.* **277**, 48596–48601.

Lovell, S. C., Davis, I. W., Arendall III, B. W., de Bakker, P. I. W., Word, J. M., Prisant, M. G., Richardson, J. S. and Richardson, D. C. (2003). Structure validation by C-alpha geometry: phi, psi, and C-beta deviation. *Proteins* **50**, 437–450.

Lüthy, R., Bowie, J. U. and Eisenberg, D. (1992). Assessment of protein models with three-dimensional profiles. *Nature* **356**, 83–85.

McRee, D. E. (1992). A visual protein crystallographic software system for X11/XView. *J. Mol. Graphics* **10**, 44–46.

McRee, D. E. (1993). *Practical Protein Crystallography.* Academic Press, San Diego.

McRee, D. E. (1999). XtalView/Xfit – a versatile program for manipulating atomic coordinates and electron density. *J. Struct. Biol.* **125**, 156–165.

Murshudov, G. N., Vagin, A. A. and Dodson, E. J. (1997). Refinement of macromolecular structures by the maximum-likelihood method. *Acta Crystallogr. D* **53**, 240–255.

Olson, A. J. and Goodsell, D. S. (1992). Macromolecular graphics. *Curr. Opin. Struct. Biol.* **2**, 193–201.

Perrakis, A., Morris, R. and Lamzin, V. S. (1999). Automated protein model building combined with iterative structure refinement. *Nature Struct. Biol.* **6**, 458–463.

Ponder, J. W. and Richards, F. M. (1987). Tertiary templates for proteins. Use of packing criteria in the enumeration of allowed sequences for different structural classes. *J. Mol. Biol.* **193**, 775–791.

Ramachandran, G. N. and Sasisekharan, V. (1968). Conformation of polypeptides and proteins. *Adv. Protein Chem.* **23**, 283–438.

Rice, L. M. and Brünger, A. T. (1994). Torsion angle dynamics: reduced variable conformational sampling enhances crystallographics structure refinement. *Proteins* **19**, 277–290.

Richardson, D. C. and Richardson, J. S. (1992). The kinemage: a tool for scientific communication. *Protein Sci.* **1**, 3–9.

Richardson, J. S. and Richardson, D. C. (1985). Interpretation of electron density maps. *Method Enzymol.* **115**, 189–206.

Rost, B. (1996). PHD: predicting one-dimensional protein structure by profile based neural networks. *Method Enzymol.* **266**, 525–539.

Sayle, R. and Milner-White, E. J. (1995). RasMol: biomolecular graphics for all. *TIBS* **20**, 374.

Tsernoglou, D., Petsko, G. A., McQueen Jr., J. E. and Hermans J. (1977). Molecular graphics: application to the structure determination of a snake venom neurotoxin. *Science* **197**, 1378–1380.

Vaguine, A. A., Richelle, J. and Wodak, S. J. (1999). SFCHECK: a unified set of procedures for evaluating the quality of macromolecular structure-factor data and their agreement with the atomic model. *Acta Crystallogr. D* **55**, 191–205.

Vriend, G. (1990). WHAT IF: A molecular modeling and drug design program. *J. Mol. Graph.* **8**, 52–56.

Wang, B. C. (1985). Resolution of phase ambiguity in macromolecular crystallography. *Method Enzymol.* **115**, 90–112.

CHAPTER 14

# RNA crystallogenesis

**Benoît Masquida, Boris François, Andreas Werner, and Eric Westhof**

## 14.1 Introduction

During the previous few years, the number of RNA crystal structures has increased in an exponential manner. This is mainly due to the fact that RNA is increasingly viewed as a predominant part of biological processes such as translation, ribozyme catalysis, and gene regulation (RNAi (Kamath *et al.*, 2003), riboswitches (Barrick *et al.*, 2004), and mRNA–protein interactions (Lescure *et al.*, 2002)), for which the gap in structural knowledge is still deep despite the determination of the crystal structure of the ribosome (Clemons *et al.*, 2001; Ban *et al.*, 2000; Yusupov *et al.*, 2001).

Simultaneously, structural studies of the ubiquitous roles of RNA at all levels of cellular processes are starting to be supported by technological developments which enable high-throughput crystallography (HTC). Recently, robotics has entered the field of crystallization. HTC is built on the availability of robots capable of setting up crystallization trials automatically, employing both a robot and a specific dish to set up sitting drops. Specific plastic dishes, inspired by the 96-well ELISA plates, have been designed and robot-driving software has been adapted to the various formats of plates. These robots could be made quickly available to crystallographers because they have been developed for high-throughput screening (HTS) applications, such as PCR designed on 96-wells ELISA plates, a very suitable format for crystallization trays.

Despite the high cost of these robots, they are highly valuable because they significantly shorten the time to set up experiments as well as multiply the number of possible tests by a 100-fold, just by going to the nanolitre scale in terms of liquid sample handling. They also allow samples to be tested that are too scarce for the usual microlitre-scale techniques. Furthermore, they reduce handling time, which can then be spent on more valuable tasks such as macromolecule purification or structure solving.

In this review will be presented guidelines to purify and set up RNA oligonucleotides crystallization experiments using a robot. An overview of crystallization robots available on the market will also be given with their advantages and drawbacks.

## 14.2 Design of short RNA constructs

RNA structures can be seen as assemblies built from a construction set consisting of building blocks of various shapes and complexities, which obey conservation rules at the level of sequence and structure. In order to understand RNA architecture, it is therefore necessary to elucidate the structure of these building blocks. To achieve this goal, the motifs are analysed in their wild type contexts and, after alignment of the sequences, designed so as to ensure that they will conserve their original structure. It is worth noting that the best situation is when the secondary structure is well supported by biochemical data and that the edges of the motif (5′ and 3′ ends) are well located. Attention should be given to the design process in order to increase the probability for the structure to adopt the wild type conformation. The stability of the constructs can be evaluated using UV-melting techniques under various ionic strengths (Werner, 2003).

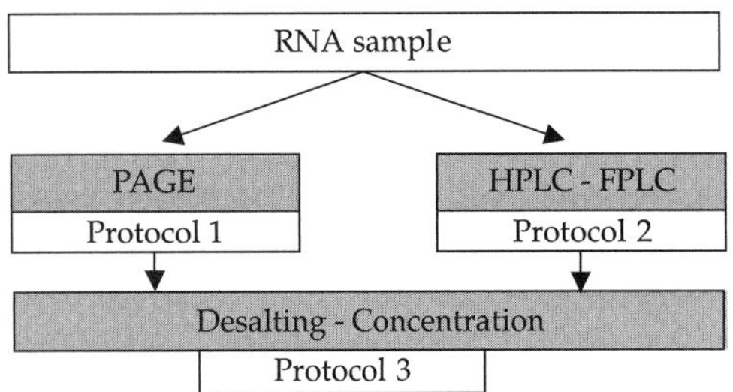


**Figure 14.1** Protocols used to purify RNA according to their nucleotide length. Note that HPLC–FPLC can be used after PAGE to cleanse an RNA sample whatever the length of the RNA.

The design of short RNA constructs is greatly helped by *in silico* folding programs (Zuker, 2003; Zuker and Jacobson, 1998; Isambert and Siggia, 2000; Ding and Lawrence, 2003). They allow for the testing of modifications to the secondary structure upon modifications to the various base pairs. This step is crucial because, sometimes, apparently insignificant events, such as the reversal of a G = C pair, can have major implications. A second point to think about is that the RNA motif may not reveal any propensity to pack and consequently to grow crystals. To address this problem it is advisable to design constructs exhibiting various edges, producing different situations regarding length and sequence, that would eventually be more favourable for crystal growth.

## 14.3 RNA purification

RNA oligonucleotides can be purified using polyacrylamide gel electrophoresis (PAGE) or liquid chromatography (HPLC: High Performance Liquid Chromatography or FPLC: Fast Protein Liquid Chromatography). Coupling these methods can improve the results in some cases. While PAGE is applicable to any RNA length, HPLC is dedicated to RNA up to about 35 nucleotides long, but the latter method is always useful to cleanse RNA preparations purified on gels (Fig. 14.1). Routine techniques mentioned in this chapter are described in more detail in Sambrook and Russell (2001). Since structured RNA molecules adopt a precise architecture, it is worth noting that purification under native conditions is also achievable (Kieft and Batey, 2004).

### 14.3.1 Gel electrophoresis

RNAs of any size (up to 500 nucleotides) can be purified efficiently using polyacrylamide gel electrophoresis under denaturing, semidenaturing, or native conditions. Various urea concentrations can be tried at analytical scale, before going to preparative scale, in order to identify the most appropriate protocol. A sequencing electrophoresis apparatus with an aluminium back to homogenize the glass plates' temperature, allowing the use of $30 \times 40\,\text{cm}^2$ gel plates, is recommended. The running temperature is usually set between 50 and 60°C. Of course, the gel thickness has to be significantly increased when going to preparative scale (use at least 1.5 mm thick spacers) to well separate the RNAs of different length in the sample. The volume of gel to be prepared is thus around 250 ml (Protocol 14.1).

### 14.3.2 HPLC purifiction

When the RNA oligonucleotide is shorter than around 35 nucleotides, it can be purified using FPLC or HPLC techniques. The best results are obtained using salt gradients on anion exchange columns bearing quaternary amines such as mono-Q matrices. HPLC has the advantage that the column can be heated in an oven to temperatures up to 90°C, thus promoting the unfolding of the RNA and increasing the retention time on the column to give a better separation. The addition of chaotropic agents, such as urea or formamide, enhances the effect of heating the sample. However, formamide should be used with caution in the presence of RNA. Heat leads to formamide decomposition into carbon monoxide and ammonia, and the latter can very quickly hydrolyse the RNA preparation. In such denaturing conditions, the RNA mix is fractionated according to the size of the species present, close to what can be achieved using gel electrophoresis. A typical method is described in Protocol 14.2. Other HPLC purification procedures for RNA have been described elsewhere (Anderson *et al.*, 1996).

Care should be taken to avoid several pitfalls. If the RNA has been produced by *in vitro* transcription, proteins should be removed by phenol/chloroform extraction. Otherwise the column bed may become coated with proteins and the column will lose its

### Protocol 14.1 Polyacrylamide gel electrophoresis (PAGE)

**Equipment and reagents for a 250 ml gel**

Electrophoresis apparatus
Comb and spacers (at least 1.5 mm thick)
Acrylamide 20% / urea 8 M stock solution
8 M urea solution
Ammonium persulphate
Siliconized glass plates ($\approx 30 \times 40\,cm^2$)
UV lamp and silica plate
10× TBE (Tris–borate–EDTA buffer)
TEMED (stored at 4°C)
Bromophenol blue–xylene cyanol
Required acrylamide percentage see (Sambrook, 2001)

**Method**

**1.** Mix the required amount of acrylamide/urea stock solution with TBE buffer and complete with millipore water to give a volume of 250 ml. Then add 10% (w/v) of ammonium persulphate as a catalyst. The polymerization reaction is started by adding 100 µl TEMED.
**2.** Polyacrylamide preparative gels usually run with a 0.5× or 1× TBE buffer and require a power value around 25 Watts in denaturing conditions. The gel temperature during migration must be around 50°C to ensure the denaturation of small RNA fragments.
**3.** Clean acrylamide residue from the loading well with TBE buffer.
**4.** Prepare the RNA solution by mixing with one volume of 8 M urea and load it onto the gel.
**5.** Prepare a small well containing a mix of bromophenol blue and xylene cyanol in order to follow the migration.
**6.** After migration, remove the glass plates and wrap the gel in plastic film.
**7.** Use a UV lamp and a silica plate as a screen to detect the RNA bands by UV shadowing.

RNA recovery:
**8.** Delineate the RNA bands on the plastic wrap and then cut them out from the gel using a sterile scalpel blade.

Elution of the RNA from the gel:
**9.** Crush the RNA bands in a mill (A11 basic analysis mill, IKA), and poured it into a 50 ml polypropylene conical tube with about 30 ml Millipore water. Place the tube in a rock and roll stirrer at 4°C overnight.
**10.** Filter the eluted RNA solution through a 0.22 µm sterile filtration unit (Nalgene) to remove the acrylamide particles.

### Protocol 14.2 HPLC–FPLC

**Equipment and reagents**

| | Buffer A | Buffer B |
|---|---|---|
| MES | 20 mM pH 6.2 | 20 mM pH 6.2 |
| Urea | 4 M | 4 M |
| $NaClO_4$ | 1 mM | 400 mM |

**Method**

**1.** Warm the anion exchange column up to 90°C.
**2.** Equilibrate the column in 85% buffer A / 15% buffer B.
**3.** Prepare the RNA solution by mixing with one volume of 8 M urea and heat it to 90°C for 2 min.
**4.** Inject the RNA sample and run the perchlorate gradient over 70 min from 15% to 70% of buffer B. Follow the RNA elution by UV absorbance at 260 nm. The flow rate is about 1 ml/h.
**5.** Increase the perchlorate gradient to 90% of buffer B in 10 min.
**6.** Wash the column with 90% of buffer B for 5 min.
**7.** Re-equilibrate the column in 85% of buffer A / 15% of buffer B to prepare for the next run.

loading capacity over time. The sample should be assayed for precipitation by mixing with the highest salt buffer that is going to be used for separation to avoid clogging the HPLC. Since the sample is going to be heated, divalent ion contamination should be avoided. Hence, the pKa value of the buffer should be in the acidic range so as to minimize spontaneous hydrolysis of phosphodiester bonds in the case of a contamination with divalent cations. To achieve this, we recommend the use of peek-coated (poly-ether-ether-ketone) pumps as well as peek tubing.

**Protocol 14.3 Concentration and desalting**

Plug the column inlet to the luer of a 10 ml syringe and fix it to a bench stand.
Equilibrate the column using 10 ml of methanol.
Pass through 10 ml of millipore water.
Load the sample.
Wash the sample with 10 ml of millipore water.
Elute the sample with 5 ml of water/acetonitril (1:4) in 1 ml fractions.

### 14.3.3 Concentration and desalting

Whatever the technique employed to purify the RNA, it is necessary to desalt and concentrate it prior to use in crystallization trials. A very efficient way of achieving this is to use reverse-phase Sep-Pak columns that can be used on the bench (Waters Sep-Pak C18 Classic short-body). These are operated by gravity or using a syringe (Protocol 14.3).

Three facts should be kept in mind when using Sep-Pak cartridges. The pH of the sample should not exceed 7 to guarantee efficient binding to the column bed. The loading step should not exceed 10 min to minimize loss of material due to driving by the mobile phase. If loading is expected to take longer, the sample should be fractionated on more than one column. The column should never run dry, to prevent the loss of the sample. Hence, the syringe luer should be removed with caution in intermediate steps. The next solution should be added when there is still a small volume (100 μl) of the previous phase in the syringe.

The RNA containing water/acetonitrile solution is then evaporated to dryness in a SpeedVac. The pellet can be resuspended in the solution of choice for further studies.

## 14.4 Setting up crystallization screens for RNA

The main technique employed to set up crystal screens is the vapour diffusion method, either in the hanging drop or sitting drop set up. This method is based on slowly concentrating the droplet solution against a reservoir solution of infinite volume (ml scale) compared to the volume of the droplet (μl scale, see Fig. 14.2). Other techniques based on diffusion or counter-diffusion in agarose gels (Biertumpfel *et al.*, 2002) can also be useful. The choice between the various types of commercially available plastic ware will be driven by the amount of RNA sample and the number of crystallization conditions to be tested. Nowadays, more and more laboratories have the opportunity to use crystallization robots that permit the drop volume to be decreased to hundreds of nanolitres and hence, with the same amount of material, to set up thousands of trays in very short time scales.

After purification of a sufficient amount of RNA, conditions for crystallization have to be found. An economical screening method should use the least possible amount of RNA. Therefore, it is recommended to start by screening a large number of combinations, and then switch to other methods to optimize crystal shape and size. For a broad, general screen, specific crystallization sparse matrices for proteins or nucleic acids have been published (Doudna *et al.*, 1993; Berger *et al.*, 1996; Scott *et al.*, 1995; Cate and Doudna, 1997) and some are commercially available (Hampton Research: http://www.hamptonresearch.com/, Decode Genetics: http://www.decode.com/, Nextal: http://www.nextalbiotech.com/). Their designs are based on extensive mining of previously published crystal-yielding conditions. Although the general considerations for crystal screens of proteins equally apply to RNA, some particularities have been identified. If no crystals have been obtained during the first trials, it is often more promising to vary the sequence instead of sampling a larger variety of combinations. In the crystallization process, sequence and shape of the molecules will drive the nucleation and subsequent crystal growth through a network of packing interactions mediated among symmetry-related molecules. Considering RNA, these factors have even more drastic effect than for proteins, since the former are usually less globular in shape than the latter. Thus, various RNA constructs with different

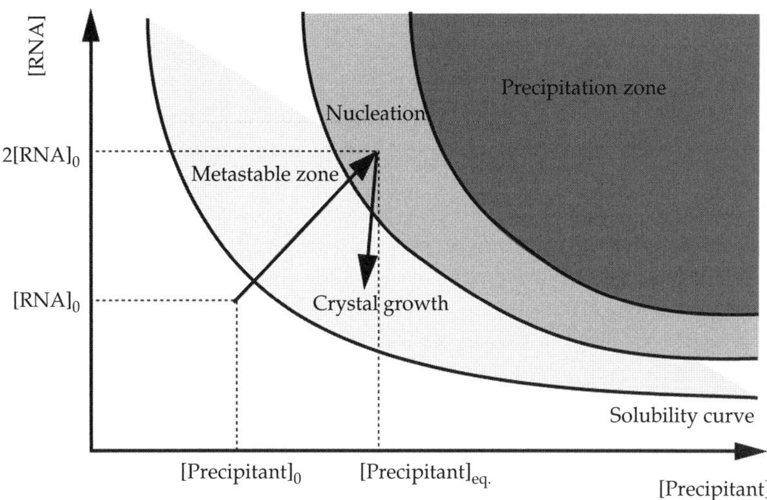


**Figure 14.2** Solubility diagram. During equilibration, the concentration of both precipitant and macromolecule increase until precipitation occurs. The formation of crystal nuclei reduces the amount of solvated macromolecule and allows the system to remain in the metastable zone where crystals can grow.

sequences and helix length, in other words with different shapes, should be tried when no crystals appear for a given construct (Scott *et al.*, 1995; Anderson *et al.*, 1996). Chemical synthesis makes this process relatively straightforward for small RNAs (<30 nt). Higher crystallization temperatures (37°C) also seem to favour formation of RNA crystals. Finally, the choice of crystallization method (vapour diffusion or batch crystallization) may also influence the success.

Here are discussed initial crystallization conditions for screening by robots using microplates. Three types of crystallization robots are presented. They were chosen for their distinct features, which give a picture of the technologies currently available. Crystallization conditions refinement methods based on statistical scoring have been discussed elsewhere (Francois *et al.*, 2004).

### 14.4.1 Robotics applied to crystallization

The requirements for handling liquids in high-throughput crystallography are identical to those for HTS: reproducibility, small volume handling, accuracy, and speed – to avoid evaporation – are required. Robots also offer parallelization through multineedle dispensing systems. Usually eight pipettes are used to simultaneously dispense liquids in one column of a 96-well plate, containing 12 rows.

In practice, five companies build robots: Cartesian dispensing system; Tecan (Genesis Station); Douglas instruments (Impax and Oryx robots); TTP Labtech (Mosquito); and Fluidigm (Topaz system). All, but the latter, have been designed for HTS applications for the pharmaceutical industry and subsequently adapted to crystallography due to the growing need for powerful means of condition screening in the field of drug discovery and genomics. They all consist of a pipetting robot linked to a computer to animate it. In general, these robots use different proprietary technologies, which allow scientists to use a wide panel of crystallization techniques from microbatch to hanging drops. In addition, companies marketing crystallization screens have started to offer 96-wells plates prefilled with screens (http://www.nextalbiotech.com/). Robots are best operated under controlled atmosphere in order to control the hygrometry and avoid evaporation. The humidity level should be kept at 85–90% with a temperature of 20°C. At the end of the screening plate preparation, it should be sealed with a plastic foil (viewseal from Greiner, for example). It is recommended to use a manual press, such as the Corning storage mat applicator, to avoid any drying of the nanodrops. Due to the high cost of these machines, it is advisable to dedicate specific persons to the robot operation and maintenance. Here, only the Tecan Genesis station, the Mosquito from TTP Labtech, and the Topaz system from Fluidigm will be described.

It is worth noting that HTC yields a considerable number of droplets that need to be visualized in the time course of crystal growth, that is from once a day to once a week. Robot manufacturers usually also offer visualization systems to automate this process. These tools are not discussed in this review.

#### *14.4.1.1 The Genesis Station from Tecan*

The Genesis Station from Tecan (Fig. 14.3) consists of a fixed deck supporting crystallization plates and deepwell-blocks overhung by an arm capable of moving in the horizontal plane. This arm is equipped with teflon pipettes which move up and down to aspirate and dispense liquids. In the conformation shown in the figure, nine slots can be used either for crystallization plates of the Greiner or Corning types and deepwell-blocks. In addition, the versatility of the Genesis Station allows for pipetting from 50 ml containers into deepwell blocks or crystallization plates. When using 96 wells plates, it is recommended to have eight pipettes so as to pipette one column each time. A second arm can move the plates when needed.

The Genesis Station proceeds through successive steps for setting up crystallization experiments. First, it transfers reservoir solutions, for example from a deepwell block to a screening plate. Second, the sample to crystallize, aliquoted in eight

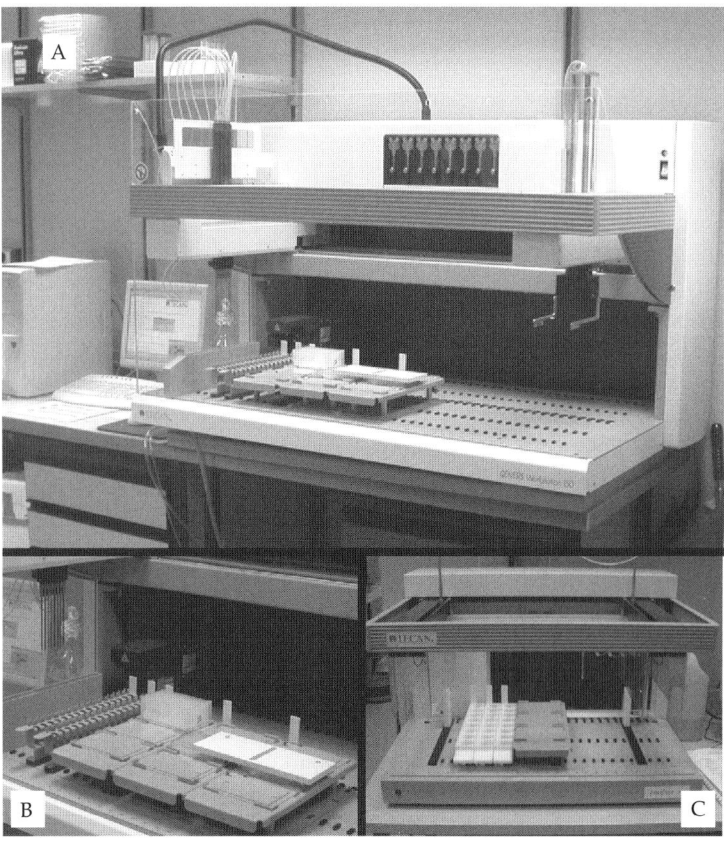


**Figure 14.3** The Genesis workstation from Tecan. (a) A general view of the workstation. The two mobile arms move above the deck supporting the microplates. (b) A close-up view of the deck supporting the microplates. (c) The Miniprep 75 workstation is very handy when one needs to prepare crystal screening solutions from mother solutions.

test-tubes, is pipetted directly followed by the reservoir solution, which is then separated from the RNA solution by a tiny air gap. At last, the pipette mixes the two solutions by dispensing them in one of the three wells on the drop shelf. To avoid evaporation, a home-made plastic mask, leaving access to one column of the crystallization plate at a time, is translated from column to column by the second arm each time reservoir solutions from a column are pipetted from a deepwell block. Moreover, the tubes containing the RNA sample are opened at the beginning of the second step.

The computer interface of the robot is complex to program but shows almost no limitation. The pipette tips are cleaned along the process by perfusing water through the upper part of the pipette. Setting up a microplate is thus fairly time-consuming and special care should be given to evaporation issues. The sample dead volume is usually about 30% and the optimal volume that can be handled reproducibly is 500 nl. A pressure detector in the pipettes notifies the user when a tube is empty and idles the program. The pipetting speed can be adapted to the viscosity of the solution (three speeds). When many users share the same robot, it can be useful to use a pipetting station such as the miniprep75 from Tecan in order to aliquote screens packed in 50 or 15 ml tubes into deepwell blocks.

*14.4.1.2 The Mosquito from TTP Labtech*

What is fixed in the Tecan Genesis is mobile in the Mosquito (Fig. 14.4) and *vice versa*. The pipette head only moves up and down, while the deck supporting the plates moves in the horizontal plane. On the robot, decks supporting two or more plates can be mounted, although the configuration ordered cannot be changed afterwards. The crystallography deck comprises two slots for 96-wells plates and, in between, a slot for a strip of eight tubes dedicated to the sample to crystallize.

The Mosquito uses the positive pipetting principle. Instead of pumping in using air pressure, a needle is used as a piston. Hence, there is no air in the pipette, so no need to circumvent fluid dilatation by changing the pumping speed. It is very practical when considering pipetting viscous liquid such as MPD or PEGs. The piston is longer than the pipette so it can be snapped by the pipetting head which then pumps up and down the liquid while the deck moves underneath in order to change the plates. The volumes handled by the Mosquito are between 0.05 and 1.2 μl. The pipettes are tethered perpendicularly to a tape which is wound round a reel. Pipettes are used in batches of eight and the reel unwinds when the tensioner arm is pushed due to the increasing tension of the tape.

The needles are strong enough to efficiently puncture the aluminium foil of prefilled screening boxes, which is the best way of using a Mosquito since it can only set up the drops and cannot fill the reservoir solutions. However, the screening plates can be transferred efficiently from a deepwell block using a manual 12-channel multipipette. The pipettes are disposable and bend easily in case of a crash caused by the misspecification or misplacement of a plate on the deck. Pipette crashes have thus no more consequences than the loss of few nanoliters of sample. The Mosquito can make sitting and hanging drops and operates very quickly. Programming is intuitive and easy.

*14.4.1.3 The Topaz system from Fluidigm*

The design of the Topaz from Fluidigm differs completely from the robots described in the previous paragraphs. Standard crystallization plates are replaced by elastomer chips integrating microchannels in which flow the reservoir and sample solution. These microchannels can be pinched off by nanoflex valves, a revolutionary concept (http://www.fluidigm.com/nanoflex.htm#). This valve is actually an elastomeric-made membrane which deflects when pressure is applied (Fig. 14.5).

Three sets of nanoflex valves are placed so as to control the connection between two sets of microchannels. One set of microchannels is dedicated to the reservoir solution and the other to the sample to crystallize. The first set of valves separates the two sets of microchannels (interface line), the second and the third (containment lines) hermetically seal the reservoir and sample solutions once they have been injected in the chip. A crystallization experiment proceeds as follows. First, pressure is applied to the interface line. Then the reservoir and sample solutions are injected. The two containment lines can then be pressurized while the interface line is relaxed to mix the solutions. The crystallization

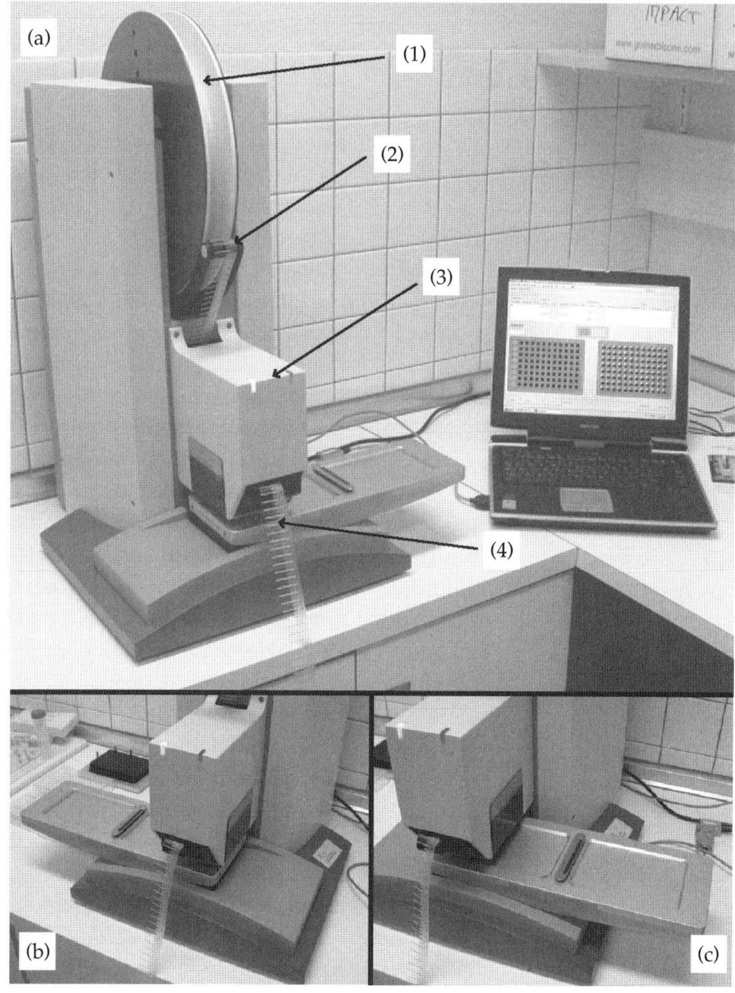


**Figure 14.4** The Mosquito from TTP Labtech. (a) A general view of the Mosquito. The pipette reel (1) is unwound when the tensioner arm (2) detects the pipette head (3) pulling the pipette tape. The pipette tape (4) is discarded at the outlet of the pipette head. (b) A view of the Mosquito showing the 'park left' position of the deck. (c) Position 'park right'.

thus occurs through free interface diffusion (FID), which could be regarded as a quasibatch method once thorough mixing has occurred. The chip is designed to set up 48 distinct conditions, each tested in three sample/ reservoir ratio.

### 14.4.2 Refining initial conditions

Once suitable starting conditions have been found, the strategy consists of a rational variation of conditions. The crystallization process, as visualized in the phase diagram (Fig. 14.2), is influenced by numerous variables $x_1, x_2, \ldots$ (called *factors*), namely: RNA sequence, crystallization temperature, buffer and pH, kind and percentage of precipitant and salts. Each of these factors can be adjusted to different *levels* (for example, factor [LiCl] to 150 mM and factor [MPD] to 25%). In order to quantitate each observation (clear drop, precipitate, spherulites, microcrystals, or crystals), an arbitrary score (*response*) is assigned to it and represented on a multidimensional *response surface* $f(x_1, x_2, \ldots)$. The aim of

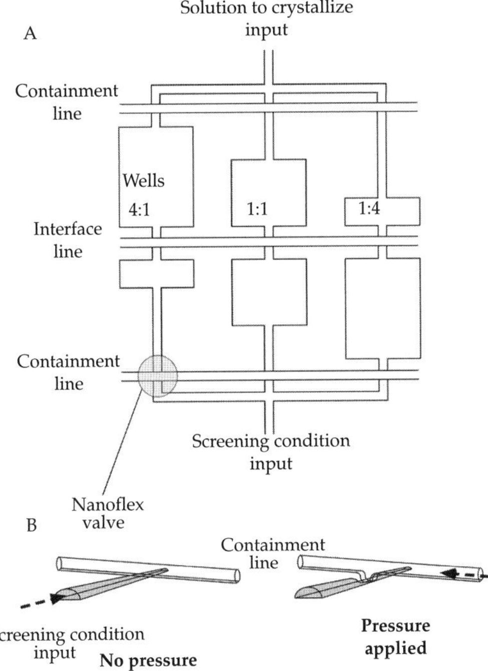


**Figure 14.5** The nanoflex valve used by the Topaz system from Fluidigm. A Fluidigm chip contains 48 sets of wells as represented in (a). Three subwells are designed so as to test various ratios of solution to crystallize versus screening solution. The process of feeding the wells is controlled by the nanoflex valve technology (b). Pressure is applied to the upper channel to inflate the valve, resulting in closing the input lower channel.

the crystallization screen is to explore this surface (Fig. 14.6) where the expected summit would yield the optimal result (best crystals). However, in practice, only a limited number of all possible factor combinations can be tested. The simplest approach would be varying a single factor at a time, while keeping all others constant. While it is possible to reach the optimum by sheer luck, the response surface shows that the score will more likely converge to a plateau or local maximum. Furthermore, the results of each testing series cannot be generalized, and interactions between different factors are neglected.

These problems are avoided by using an experimental design, where multiple factors are varied simultaneously between different crystallization trials. Each experiment $n$ represents a combination $Cn = (x_1, x_2, \ldots, x_k)$ of $k$ factors. For each combination, multiple factors are changed simultaneously according to a predefined plan. A detailed exploration of this powerful technique is beyond the scope of this text and instead can be found in books on engineering statistics, or see Petersen (1991) and Carter (1991) for its application to crystallization. In general, at least two levels are defined for each factor at a chosen, equal distance around a central point, staking out a well-balanced experimental space in which the response of the system (the quality of crystals) is noted. Rather than just trying out some extreme values, it is important to chose reasonable values within a well-known range to avoid veering towards some jagged region, or close to an asymptote. The different levels of each factor are coded in coefficients. In our two-factor example ($k = 2$), both 50 mM/250 mM [LiCl] and 10%/40% MPD would be listed in an experimental design matrix and coded as +1/−1. Following scoring, the response surface spread out over all combinations indicates the direction where the best score is to be expected (Fig. 14.6). If $k$ is not too large, one can also take into account possible interactions between factors. If factor x influences factor y, then the response surface $f(x_1, x_2)$ is not only a polynomial of $c_1 \cdot x_1$ and $c_2 \cdot y_2$, respectively, but the additional interaction term $c_3 \cdot x_1 x_2$ also has to be considered. To help with the design process, several computer programs have been made available (Potter, 1994). A number of different designs have been coined with the common goal of reducing the number of experiments without compromising the well-balanced exploration of the experimental space.

Initial screens can be distinguished between methods that are used to determine what factors are most important, and follow-up screens that allow optimization and improvement of crystal quality (Table 14.1). In experimental design, this is known as the *Box–Wilson strategy* (Box *et al.*, 1978). The first group of screens is generally based on a so-called factorial plan which determines the polynomial coefficients of a function with $k$ variables (factors) fitted to the response surface. It can be shown that the number of necessary experiments $n$ increases with $2^k$ if all interactions are taken into account. Instead of running an unrealistic, large number of initial experiments, the full factorial matrix can

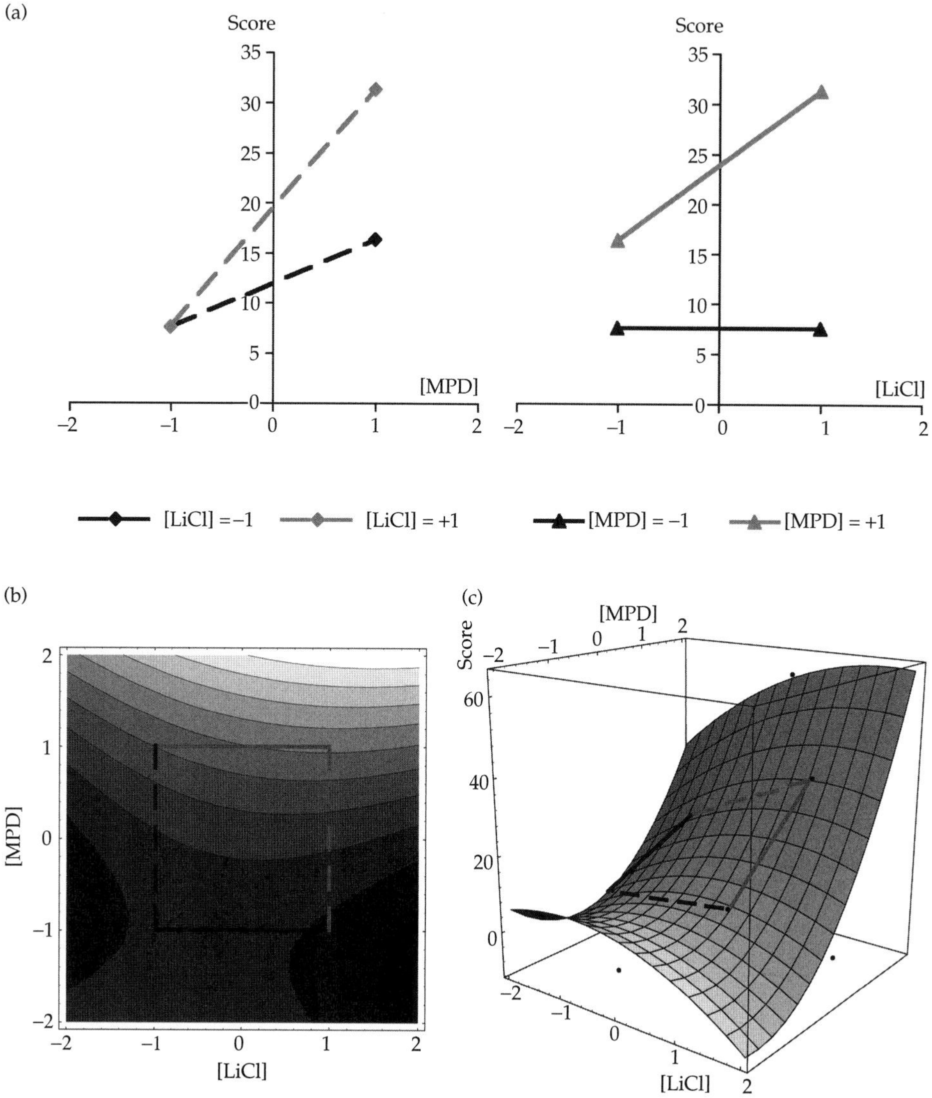

**Figure 14.6** (a) Scoring of the influence of two factors on crystal growth (see text, score is in arbitrary units). Response surface fitted to the result, top view (b) and three-dimensional view (c).

be advantageously replaced by a fractional factorial matrix or a *Plackett–Burman design* (Plackett and Burman, 1946). Here, interactions between factors are partially or completely neglected. For example, if a multiplicative effect of salt concentration and MPD can be ruled out, the interaction between these factors can be neglected, thereby reducing the number of necessary experiments.

Based on the obtained response surface, a second round of optimization follows, using the *steepest ascent method* where the direction of the steepest slope indicates the position of the optimum. Alternatively, a quadratic model can be fitted around a region known to contain the optimum somewhere in the middle. This so-called *central composite design* contains an imbedded factorial design with centre

**Table 14.1** Application of various experimental designs in crystallization

| Field of application | Experimental design | Factors and levels | Number of experiments | Comments |
|---|---|---|---|---|
| Initial screen: Which factors are most important ? | Full 2-level factorial design | $k$ factors at 2 levels | $2^k$ | Accounts for interactions between factors, but too many experiments necessary if $k > 5$ factors |
| | Incomplete 2-level factorial design | $k$ factors at 2 levels | $2^{k-p}$ | p factors are confounded, effect of interactions cannot be evaluated, but less experiments necessary |
| | Plackett–Burman design | $k$ factors at 2 levels | $k+1$ | Greatly reduced number of experiments; interaction bias neglected |
| Optimization | Steepest ascent | $k$ factors at 2 levels | | Follow-up on 2-level design; reduction of step size when approaching maximum |
| | Central composite Box–Behnken Hardin–Sloane | $k$ factors at 5 levels | | Follow-up on steepest ascent design; quadratic model, approach to optimum |
| | Randomized block designs | 3–5 factors, 3–4 levels | | One single factor of primary interest; interaction bias neglected |
| | Simplex matrix | $k$ factors at 3 levels | initially $k+1$ | Iterative triangulation towards optimum |

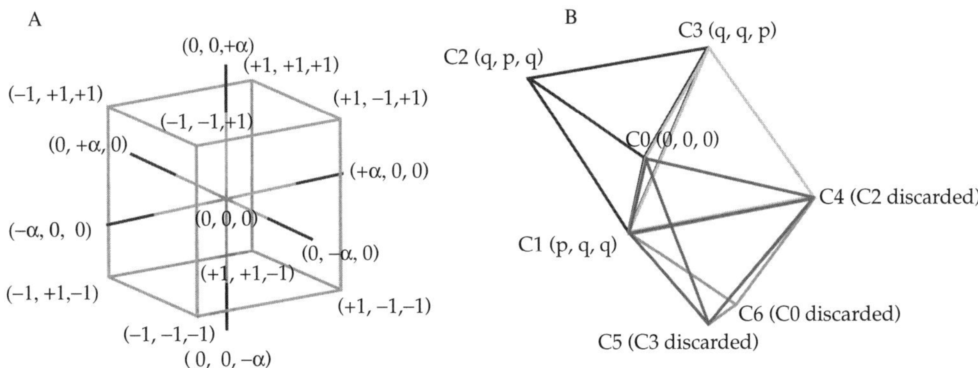


**Figure 14.7** Optimization designs. (a) Centred composite design with three factors. (b) Three factors optimized in four steps using a simplex design.

points, points at $-1/+1$, and an additional grou of outlying 'star points' $\alpha$ as upper and lower limits, which allows an estimation of the curvature (see Fig. 14.7 for an example with combinations of three factors). There are alternative designs, if the number of factors is small and optimization is the main goal. *Randomized block designs* (Latin squares and Graeco-Latin squares) are useful if there is one main factor to consider. The design helps to separate it from the influence of *nuisance factors* that may affect the measured result, but are not of primary interest.

Finally, the *simplex design* has also been adopted for crystallization purposes (Prater *et al.*, 1999). This is an iterative approach starting with one more combination than factors under investigation. In an example with three factors at three equally spaced levels, 0, p and q, the first set consists of combinations C0 (0, 0, 0), C1 (p, q, q), C2 (q, p, q), and C3 (q, q, p) (Fig. 14.7). Combination C2 giving the worst result, it is replaced in the following round by combination C4 with coordinates exactly opposite to C2, where the mirror plane is defined by C0, C1, and C3. Comparing these points with C4, the worst result is now C3 and it is therefore replaced by mirror point C5, and so forth. After several rounds of triangulation in the experimental space, the optimum is reached when no further improvements are observed. While multiple rounds of optimization are required, this extremely economical approach is especially useful when too little sample is available for extended factorial plans.

### 14.4.3 Renaturing the RNA

Prior to setting up crystallization experiments, the concentrated RNA has to be properly folded in the native state. This is performed by a heating step in a heating block for 1 min at 70 to 85°C (depending on the melting temperature), in the presence of monovalent salts only. Then, the solution is left in the switched-off heating-block to cool down slowly until room temperature is reached. In order to avoid self-cleavage of the RNA, the pH is usually chosen slightly acidic and divalent cations are added only around 35°C.

### 14.4.4 Forming complexes with organic ligands: the example of aminoglycosides

RNA molecules bind various organic ligands. Different RNA fragments based on the *E. coli* A site located in the penultimate helix of the 16S ribosomal RNA (Vicens and Westhof, 2001, 2002, 2003; Walter *et al.*, 1999) have been tested in the presence of their natural ligands, antibiotics of the aminoglycoside family. The RNA construct was designed as a self-complementary oligonucleotide so as to incorporate two A sites in a head-to-head manner (Fig. 14.8). This choice eliminates two drawbacks. Firstly, since the internal loop is asymmetric, one would otherwise need to synthesize, purify, and mix 1:1 two different RNA strands in order to obtain a single site. Secondly, one could also use a single site capped by a stable hairpin of the GNRA family, for example. However, in such cases, it is frequently observed that the crystallized structure reveals a full duplex with several non-Watson–Crick pairs (Kacer *et al.*, 2003). In order to monitor the effect of sequence variations

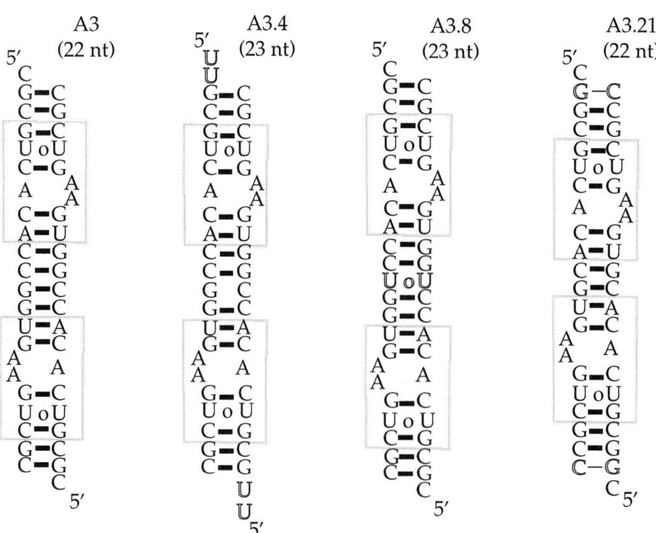


**Figure 14.8** Four self-complementary RNA fragments containing a tandem array of two *E. coli* 16S ribosomal A site modules.

for the crystallization of these complexes, various modifications have been performed, such as addition of a 5′ UU overhang, insertion of a U–U pair, or moving the two A sites closer to one another (Fig. 14.8).

Routinely, the purified oligoribonucleotides are solubilized in a solution containing 2 mM RNA, 25 mM NaCl, 5 mM $MgSO_4$, and 100 mM sodium cacodylate buffer pH 6.4. This solution is first heated to 85°C for 2 min. and then slowly cooled until a temperature of 37°C is reached. One volume of a 4 mM aminoglycoside solution is added to the RNA solution and incubated for 2 hours at 21°C. The two solutions should be at the same temperature. Since the RNA fragments contain two antibiotic binding sites the aminoglycoside concentration is twice the RNA concentration. A general rule is that the organic ligand concentration should be 100 times higher than its dissociation constant ($K_d$) to ensure binding site saturation, which is usually easily achieved in the mM range.

In the example of the aminoglycoside/A site complexes, different crystallization solutions were prepared to test various glycerol/MPD ratios: 5, 2, 1, 0.75, 0.67, and 0.5 (Table 14.2). All trials are performed at the optimal temperature of 37°C using the vapour diffusion method in the hanging drop set-up: 1 μl RNA–antibiotic complex solution was added to 1 μl crystallization solution and equilibrated over a 40% MPD reservoir.

### 14.4.5 Evaluate screening results

Since initial crystallization attempts will not automatically yield crystals or their quality may be poor for X-ray diffraction experiments, evaluation of screening results is required prior to proceeding to crystallization optimization. This is performed by using a binocular microscope hooked up to a digital camera to record observations. A numerical scoring value describing the content of the droplet (Fig. 14.9) is reported on a paper scoring sheet.

Two weeks are enough for droplets of about 200 nl to 3 μl to equilibrate under any conditions (Mikol *et al.*, 1990). During this period, droplets should be inspected daily to follow up the appearance of crystals. Crystals may still form after 2 weeks, but this is less likely in the case of oligonucleotides. Crystals can then be cryoprotected and frozen or capillary-mounted to be tested. Fluidigm markets crystallization chips dedicated to crystal growth optimization which can sustain and are transparent to X-rays in order to discriminate between salt and macromolecule crystals. Extracting the crystal from the chip is performed only for crystals deserving data collection.

### 14.4.6 The optimization process

Here are provided non-exhaustive guidelines to interpret the droplet content of crystallization screenings (Fig. 14.9) and possible ways to optimize positive hits. See also Ducruix and Giegé (1992) for more details.

*Clear drops:* indicates that the RNA supersaturation state has not been reached, the RNA concentration is outside the nucleation zone (Fig. 14.2). These experiments must be repeated with higher sample and/or salt concentrations. The temperature could also be lowered.

**Table 14.2** Crystallization conditions testing various glycerol/ MPD ratios

| Reagent | Stock | Crystallization condition (reservoir) | | | | | |
|---|---|---|---|---|---|---|---|
| | | 1 | 2 | 3 | 4 | 5 | 6 |
| MPD | 60% | 1 | 2 | 2 | 2 | 1.5 | 2 |
| Glycerol | 100% | 5 | 4 | 2 | 1.5 | 1 | 1 |
| Na cacodylate pH 6.4 | 0.85 M | 0.05 | 0.05 | 0.05 | 0.05 | 0.05 | 0.05 |
| KCl | 3 M | 0.15 | 0.15 | 0.15 | 0.15 | 0.15 | 0.15 |
| Glycerol/ MPD ration | | 5 | 2 | 1 | 0.75 | 0.67 | 0.5 |

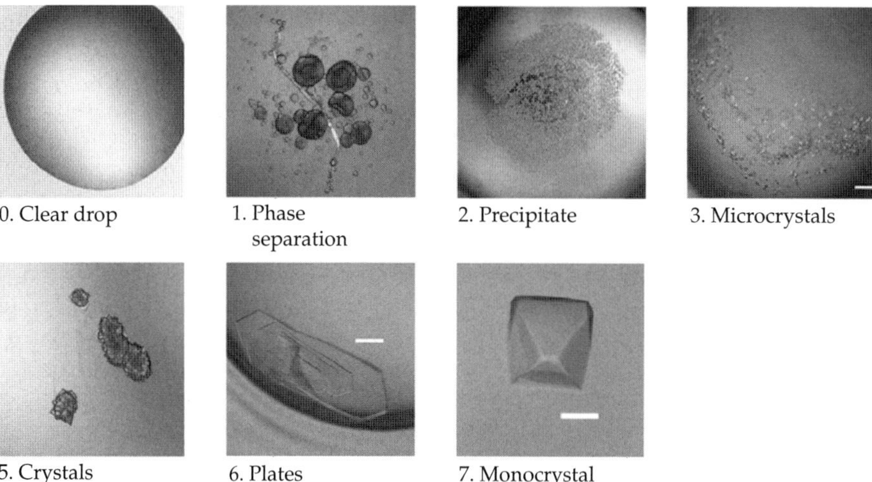


**Figure 14.9** Numerical scoring terms.

*Phase separation*: indicates a need to increase the monovalent salt concentration and/or to test a smaller precipitant concentration (MPD, PEG) to make the RNA sample more soluble.

*Light precipitates*: indicates that the relative supersaturation between sample and reagent is too high. Prepare new tests with a decreased RNA and/or precipitant concentration or dilute the droplet by vapour diffusion by adding water into the reservoir.

*Strong precipitates*: indicates that the sample has been partially denatured. The sample must be tested at a lower concentration or less salt should be used. Note that a fresh test should be prepared in this case.

*Small precipitates*: must be carefully inspected using polarized light because it may contain a microcrystalline shower. A microscope with a magnification factor greater than 100-fold can be useful in this case.

*Cluster of homogenous crystals*: try to slow down the nucleation; test the conditions at lower temperature, cover the reservoir solution with oil to slow down the water diffusion.

Different parameters can be tested to optimize the growth of single monocrystals: salt type, additives, temperature. Finally, if no crystal can be obtained with a given construct, new RNA sequences have to be designed so as to provide new, potential scaffoldings to help crystal packing. When crystals of apparently good quality have been obtained, they are flash-cooled (Yao *et al.*, 2004) prior to be exposed to the X-ray beam to potentially circumvent high mosaicity.

## 14.5 Conclusions

Protocols to purify and concentrate large amounts of RNA under controlled buffer and salt conditions for crystallization experiments have been described. The crucial points to preserve the RNA sample are: the use of slightly acidic pH and the avoidance of divalent cations. Usually RNases, feared by most RNA scientists, are introduced into the solution by an upstream experiment such as plasmid and protein preparations (T7-RNA polymerase for example). It is, thus, strictly recommended to assess the RNase activity of a solution before using it on the whole RNA sample by incubating an aliquot in the presence of the RNA for few hours and check the extent of the digestion by PAGE. Also, strategies to crystallize RNA oligonucleotides in the presence or absence of ligands have been presented. A peculiarity of crystallization experiments is the absence of a negative control. To circumvent this, we advise attempting to crystallize simultaneously several related RNA sequences in the same well. This is easy to achieve when using a robot and crystallization plates containing a shelf for three drops in each well. Only one or few RNAs, if any, will crystallize, leading to the conclusion that RNA crystals instead of salt crystals have been obtained. In cases where no crystal is observed, it is recommended to design a new set of oligonucleotides bearing slight sequence changes in order to enhance interactions between symmetry-related molecules that lead to regular crystal packing interactions.

## References

Anderson, A. C., Earp, B. E. and Frederick, C. A. (1996). Sequence variations as a strategy for crystallizing RNA motifs. *J. Mol. Biol.* **259**, 696–703.

Anderson, A. C., Scaringe, S. A., Earp, B. E. and Frederick, C. A. (1996). HPLC purification of RNA for crystallography and NMR. *RNA* **2**, 110–117.

Ban, N., Nissen, P., Hansen, J., Moore, P. B. and Steitz, T. A. (2000). The complete atomic structure of the large ribosomal subunit at 2.4 A resolution. *Science* **289**, 905–920.

Barrick, J. E., Corbino, K. A., Winkler, W. C., Nahvi, A., Mandal, M., Collins, J., Lee, M., Roth, A., Sudarsan, N., Jona, I., Wickiser, J. K. and Breaker, R. R. (2004). New RNA motifs suggest an expanded scope for riboswitches in bacterial genetic control. *Proc. Natl. Acad. Sci. USA* **101**, 6421–6426.

Berger, I., Kang, C. H., Sinha, N., Wolters, M. and Rich, A. (1996). A highly efficient 24-condition matrix for the crystallization of nucleic acids fragments. *Acta Crystallogr. D* **52**, 465–468.

Biertumpfel, C., Basquin, J., Suck, D. and Sauter, C. (2002). Crystallization of biological macromolecules using agarose gel. *Acta Crystallogr. D* **58**, 1657–1659.

Box, G. E. P., Hunter, W. G. and Hunter, J. S. (1978). *Statistics for Experimenters. An Introduction to Design, Data Analysis, and Model Building*. Wiley, New York.

Carter, C. W. J. (1999). Experimental design, quantitative analysis, and the cartography of crystal growth. In: *Crystallization of Nucleic Acids and Proteins. A practical approach*, Ducruix, A. and R. G., eds., pp. 75–120. IRL Press, Oxford.

Cate, J. H. and Doudna, J. A. (1997). A sparse matrix approach to crystallizing ribozymes and RNA motifs. *Method Mol. Biol.* **74**, 379–386.

Clemons, W. M., Jr., Brodersen, D. E., McCutcheon, J. P., May, J. L., Carter, A. P., Morgan-Warren, R. J., Wimberly, B. T. and Ramakrishnan, V. (2001). Crystal structure of the 30 S ribosomal subunit from *Thermus thermophilus*: purification, crystallization and structure determination. *J. Mol. Biol.* **310**, 827–843.

Ding, Y. and Lawrence, C. E. (2003). A statistical sampling algorithm for RNA secondary structure prediction. *Nucleic Acids Res.* **31**, 7280–7301.

Doudna, J. A., Grosshans, C., Gooding, A. and Kundrot, C. E. (1993). Crystallization of ribozymes and small RNA motifs by a sparse matrix approach. *Proc. Natl. Acad. Soc. USA.* **90**, 7829–7833.

Ducruix, A. and Giegé, R. (1992). *Crystallization of Nucleic Acids and Proteins: a Practical Approach*. IRL Press, Oxford.

Francois, B., Szychowski, J., Adhikari, S. S., Pachamuthu, K., Swayze, E. E., Griffey, R. H., Migawa, M. T., Westhof, E. and Hanessian, S. (2004). Antibacterial aminoglycosides with a modified mode of binding to the ribosomal-RNA decoding site. *Angew Chem. Int. Ed. Engl.* **43**, 6735–6738.

Isambert, H. and Siggia, E. D. (2000). Modeling RNA folding paths with pseudoknots: application to hepatitis delta virus ribozyme. *Proc. Natl. Acad. Sci. USA* **97**, 6515–6520.

Kacer, V., Scaringe, S. A., Scarsdale, J. N. and Rife, J. P. (2003). Crystal structures of r(GGUCACAGCCC)2. *Acta Crystallogr. D* **59**, 423–432.

Kamath, R. S., Fraser, A. G., Dong, Y., Poulin, G., Durbin, R., Gotta, M., Kanapin, A., Le Bot, N., Moreno, S., Sohrmann, M., Welchman, D. P., Zipperlen, P. and Ahringer, J. (2003). Systematic functional analysis of the *Caenorhabditis elegans* genome using RNAi. *Nature* **421**, 231–237.

Kieft, J. S. and Batey, R. T. (2004). A general method for rapid and nondenaturing purification of RNAs. *RNA* **10**, 988–995.

Lescure, A., Fagegaltier, D., Carbon, P. and Krol, A. (2002). Protein factors mediating selenoprotein synthesis. *Curr. Protein Pept. Sci.* **3**, 143–151.

Mikol, V., Rodeau, J.-L. and Giegé, R. (1990). Experimental determination of water equilibration rates in the hanging drop method of protein crystallization. *Anal. Biochem.* **186**, 332–339.

Petersen, H. (1991). *Grundlagen der Statistik und der statistischen Versuchsplanung*. Ecomed, Landsberg/Lech.

Plackett, R. L. and Burman, J. P. (1946). The design of optimal multifactorial experiments. *Biometrika* **33**, 305–325.

Potter, C. D. (1994). Experiment design software: better data, less work. *Scientist* **8**, 18.

Prater, B. D., Tuller, S. C. and Wilson, L. J. (1999). Simplex optimization of protein crystallisation conditions. *J. Crystal Growth* **196**, 674–684.

Sambrook, J. and Russell, D. W. (2001). *Molecular Cloning: A Laboratory Manual*, 3rd edn. Cold Spring Harbor Laboratory Press, Cold Spring Harbor, NY.

Scott, W. G., Finch, J. T., Grenfell, R., Fogg, J., Smith, T., Gait, M. J. and Klug, A. (1995). Rapid crystallization of chemically synthesized hammerhead RNAs using a double screening procedure. *J. Mol. Biol.* **250**, 327–332.

Vicens, Q. and Westhof, E. (2001). Crystal structure of paromomycin docked into the eubacterial ribosomal decoding A site. *Structure (Camb)* **9**, 647–658.

Vicens, Q. and Westhof, E. (2002). Crystal structure of a complex between the aminoglycoside tobramycin and an oligonucleotide containing the ribosomal decoding a site. *Chem. Biol.* **9**, 747–755.

Vicens, Q. and Westhof, E. (2003). Crystal structure of geneticin bound to a bacterial 16S ribosomal RNA A site oligonucleotide. *J. Mol. Biol.* **326**, 1175–1188.

Walter, F., Vicens, Q. and Westhof, E. (1999). Aminoglycoside-RNA interactions. *Curr. Opin. Chem. Biol.* **3**, 694–704.

Werner, A. (2003. UV melting, native gels and RNA conformation. In: *Handbook of RNA Biochemistry*, Hartmann, R.K., Bindereif, A., Schön, A., and Westhof, E., eds, pp. 415–427. Wiley-VCH, Weinheim.

Yao, M., Yasutake, Y. and Tanaka, I. (2004). Flash-cooling of macromolecular crystals in a capillary to overcome increased mosaicity. *Acta Crystallogr. D* **60**, 39–45.

Yusupov, M. M., Yusupova, G. Z., Baucom, A., Lieberman, K., Earnest, T. N., Cate, J. H. and Noller, H. F. (2001). Crystal structure of the ribosome at 5.5 A resolution. *Science* **292**, 883–896.

Zuker, M. and Jacobson, A. B. (1998). Using reliability information to annotate RNA secondary structures. *RNA* **4**, 669–679.

Zuker, M. (2003). Mfold web server for nucleic acid folding and hybridization prediction. *Nucleic Acids Res.* **31**, 3406–3415.

CHAPTER 15

# Crystallography in the study of protein–DNA interaction

**Maninder K. Sohi and Ivan Laponogov**

## 15.1 Introduction

In the living cell, protein–DNA interactions take place during genetic processes such as chromatin organization, recombination, replication, transcription, and DNA repair. In order to fully understand the mechanism of these processes, study of protein–DNA interactions at atomic level is required. Despite the wealth of knowledge and attempts of several investigators to classify protein–DNA complexes on the basis of structural analysis (Luscombe *et al.*, 2000) of well over 200 protein–DNA complexes, the prediction and modelling based on three-dimensional structure or amino acid sequence of a protein and nucleotide sequence of DNA is challenging. X-ray crystallography is the most powerful technique used for structural studies and the development of technology for the production and purification of large quantities of proteins and DNA is vital to its success. Advances in oligonucleotide synthesis have not only provided ease of synthesis of large quantities of pure DNA but also of any required DNA sequence. This has resulted in the crystallization and structure determination of a vast number of DNA oligonucleotides and protein–DNA complexes (Fig. 15.1). Since 1994 there has been a dramatic increase in the number of the protein–DNA complex structures solved annually. The first and most difficult step in the X-ray crystallographic study is the growth of well-diffracting crystals of the protein–DNA complex under investigation. The main aim of this chapter is to describe methods employed to purify proteins and DNA for cocrystallization, procedures for crystallization of protein–DNA, and characterization of cocrystals.

## 15.2 Methods

### 15.2.1 Protein purification

Protein to be crystallized should be as pure as possible and also homogeneous. Proteins are separated from each other and other macromolecules on the basis of their solubility, size, charge, and affinity for a specific ligand. General methods for protein purification have been described in Chapters 1 and 2 and therefore will not be discussed in this chapter. However, the techniques that are useful for obtaining high yields of relatively pure protein or have been developed specifically for the purification of DNA-binding proteins will be covered here. If a crude bacterial cell lysate containing a strongly basic target protein is passed through a cation-exchange column, most of the bacterial proteins and negatively charged molecules such as RNA and DNA will pass through the column whereas the target protein will bind. The column is washed thoroughly before eluting the target protein with a salt gradient. Sometimes binding of the target protein to the column is impaired due to strong interactions between DNA and the protein. In such cases DNA should be removed by precipitation with polyethyleneimine followed by centrifugation. Polyethyleneimine is a basic, linear polymer of ethyleneimine with a molecular weight of 30,000–90,000. The amount of protein that precipitates with polyethyleneimine depends on the pH and the ionic strength of the buffer used. At high pH most proteins carry a negative charge and will precipitate with polyethyleneimine. All acidic proteins precipitate at low ionic strength, whereas only highly negatively charged proteins precipitate at higher ionic strength

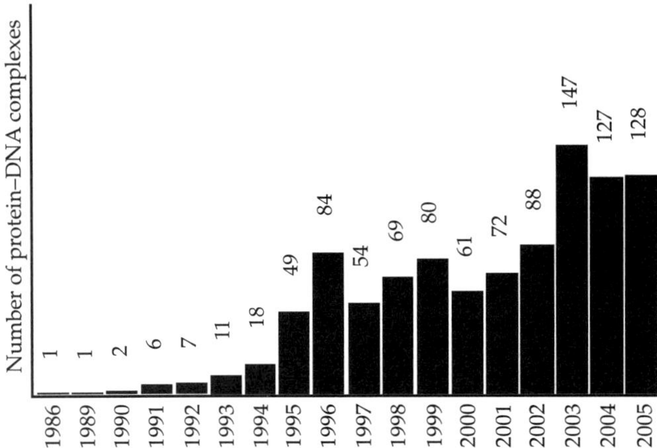


**Figure 15.1** Number of protein–DNA complexes solved and submitted to the Protein Data Bank annually since 1986.

on addition of polyethyleneimine. The DNA precipitates at both high and low ionic strength in the presence of polyethyleneimine. Burgess has described the principles and strategies of polyethyleneimine precipitation in detail (Burgess, 1991). In practice, either the target protein is kept in the supernatant by conducting polyethyleneimine precipitation at low to medium ionic strength to remove some proteins and nucleic acids or it is precipitated along with the nucleic acids and some proteins at low ionic strength and recovered from the precipitate at a higher ionic strength. *Drosophylla melanogaste*r Even-skipped homeodomain was purified after removing DNA by polyethyleneimine precipitation at high pH (7.8) and low salt concentration (0.1 M KCl) by Hirsch and Aggarwal (1995). Nelson and Sauer precipitated λ phage repressor with polyethyleneimine and redissolved in high salt buffer (Nelson and Sauer, 1986). Joyce and Grindley used this technique for purification of Klenow fragment of DNA polymerase 1 (Joyce and Grindley, 1983).

Protein purification by sequence-specific DNA affinity chromatography is based on the high affinity of DNA-binding proteins for specific sequences of DNA and yields highly purified protein. Some transcription factors and restriction enzymes have been purified using this technique (Gadgil *et al.*, 2001). The major advantage of using a single-stranded DNA affinity column is that it can be employed to discriminate between double-stranded-DNA-binding proteins from single-stranded-DNA-binding proteins. For example, *lac* repressor binds only to double-stranded DNA and will pass through a single-stranded DNA column whereas DNA polymerase and single-strand-binding proteins will bind due to their higher affinity for single-stranded DNA than for double-stranded DNA. However, some proteins, such as TFIIIA, bind both double and singe-stranded DNA (Joyce and Grindley, 1983). Single-strand DNA columns are also used for the purification of proteins that bind to DNA but do not exhibit affinity for any specific nucleotides. Other techniques, such as chromatography using phosphocellulose, hydroxipatite, heparin-sepharose, or heparin-agarose, MonoQ columns, are used in addition to DNA affinity chromatography to improve the purity of such proteins.

Genetically, adding an N-terminal or C-terminal affinity tag to the protein facilitates purification by affinity chromatography. However, these tags can affect the interaction of proteins with their target DNA. For example, a polyhistidine tag may stabilize protein–DNA complexes and interfere with the study of protein–DNA interactions. Büning and co-workers have demonstrated that p50 with a hexahistidine tail present at the N-terminus has a higher affinity than untagged p50 for DNA (Büning *et al.*, 1996). This problem can be overcome by introducing a proteolytic cleavage site between the protein sequence and the tag, which permits removal of the tag from purified protein. For example, crystals of *lac* repressor (residues 1–133) complexed with *lac* operator were grown from N-terminal

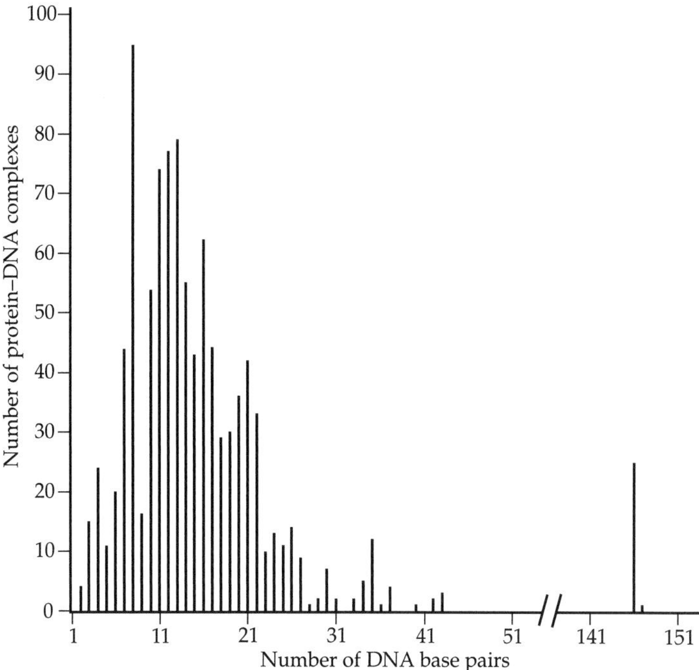


**Figure 15.2** A histogram of DNA length distribution used in cocrystallization with proteins. (This figure is based on the information available from the Protein Data Bank.)

histidine-tagged protein containing a tobacco etch virus protease site after purification of the protein on an immobilized $Ni^{2+}$ affinity column and removal of the tag with tobacco etch virus protease (Bell and Lewis, 2000). In most cases the histidine tag does not interfere with crystallization of the protein. Richmond's group expressed TF11Aβ (residues 303–376) as a glutathione-S-transferase (GST) fusion protein with tobacco etch virus NIa protease cleavage site between GST and N-terminal of the protein and purified by GSH-affinity chromatography; the GST tag was cleaved by protease treatment (Bleichenbacher *et al.*, 2003).

Purity and homogeneity of the purified protein is assessed by macromolecular exclusion chromatography, SDS-polyacrylamide gel electrophoresis, isoelectric focusing, and matrix-assisted laser desorption/ionization-time of flight mass spectrometry. The later technique, developed by Karas and Hillenkamp, ionizes and separates proteins on the basis of their mass-to-charge ratio (Karas and Hillenkamp, 1988).

### 15.2.2 DNA oligonucleotide synthesis and purification

In designing synthetic DNA oligonucleotides, the biological significance of the sequence, stability of the DNA moiety, and stability of the complex are the main consideration. A survey of the Protein Data Bank revealed that oligonucleotides containing up to 43 base pairs, with the exception of 32, 38, 39, and 41 base pairs, have been successfully cocrystallized (Fig. 15.2). There are no cocrystals with DNA oligonucleotide length above 43 base pairs, except for the nucleosome core particle containing 146–147 base pairs (Davey *et al.*, 2002; Luger *et al.*, 1997). There do not appear to be any precise rules that specify the terminal structure or optimum length of the oligonucleotides to be used in crystallizing protein–DNA complexes. DNA sequences with blunt ends as well as with overhangs have been crystallized as complexes (see Tables 15.1–15.4).

DNA to be crystallized should be as pure as possible and available in large quantities. Most

**Table 15.1** Crystallization conditions for DNA–protein complexes using polyethylene glycol (PEG) as a precipitant

| Protein | Synthetic oligonucleotide (5′ – 3′) | Res (Å) | References | Space group | P:DNA ratio | T°C | Buffer and additives |
|---|---|---|---|---|---|---|---|
| **PEG 400** | | | | | | | |
| Arc repressor | ATAGTAGAGTGCTTCTATCAT<br>TATCATCTCACGAAGATAGTA | 2.6 | Raumann *et al.*, 1994 | $P2_1$ | 1:2 | 20 | 30 mM BTP, pH 7.4, 100 mM $MgCl_2$, 1 mM $NaN_3$, 15% PEG 400 |
| C/EBP $\beta_{vc}$ | AATGTGGCGCAATCCT<br>TACACCGCGTTAGGAT | 2.1 | Tahirov *et al.*, 2001b | $C222_1$ | 1:1.2 | 25 | 50 mM MES, pH 6.0, 100 mM KCl, 10 mM $MgCl_2$, 10% PEG 400 |
| c-Fos-c-Jun | TTCTCCTATGACTCATCCAT<br>AGAGGATACTGAGTAGGTAA | 3.0 | Glover and Harrison, 1995 | $P2_12_12$ | 1:1.6 | 20 | 50 mM BTP, pH 6.7, 2.0 mM spermine, 300 mM NaCl, 50 mM $MgCl_2$, 5–10 mM DTT, 11–15% PEG 400. The drop contained in addition 100–200 mM $CH_3COONH_4$ |
| *Eco*RI | TCGCGAATTCGCG<br>GCGCTTAAGCGCT | 2.6 | Grable *et al.*, 1984; McClarin *et al.*, 1986a, 1986b | P321 | 2:9 | 4 | 40 mM BTP, pH 7.4, 15% Dioxane, 0.5M $CH_3COONH_4$ |
| GCN4 | TTCCTATGACTCATCCAGTT<br>AGGATACTGAGTAGGTCAAA | 2.9 | Ellenberger *et al.*, 1992 | $P2_12_12_1$ | 1.5:1 | 22 | 50 mM MES, pH 5.75, 1 mM spermine, 60 mM $MgCl_2$, 24% PEG 400 |
| GLI | ACGTGGACCACCCAAGACGAA<br>GCACCTGGTGGGTTCTGCTTT | 2.6 | Pavletich and Pabo, 1993 | $P2_12_12_1$ | 1:1.2 | 22 | 50 mM BTP-HCl, pH 7.0, $CoCl_2$ 2 molar equivalent $Co^{2+}$ per finger, 60–100 mM $MgCl_2$, 20–25% PEG 400 |
| HAP1-PC7 | fACCCTCGCTATTATCGCTATTA<br>TGGGAGCGATAATAGCGATAAT | 2.8 | Lukens *et al.*, 2000 | $P2_12_12_1$ | 3:4 | 20 | 50 mM MES, pH 5.6, 4 mM $MgSO_4$, 200 mM KCl, 5.5% PEG 400 |
| Human Topo I | AAAAAGACTCAGAAAAATTTTT<br>TTTTTCTGAGTCTTTTTAAAAA | 5.4,<br>2.8 | Redinbo *et al.*, 2000; Stewart *et al.*, 1998 | $P2_1$ | | 22 | 5 mM Tris, pH 6.0, 20 mM MES, pH 6.8, 145 mM $MgCl_2$, 30 mM DTT, 27.5% PEG 400 |
| Human Topo I | 22-base pair DNA duplex containing 10 5-iodo-deoxyuridine nucleotides | $2.1^{c}$<br>$2.5^{nc}$ | Redinbo *et al.*, 1998 | $P2_1$ | | 22 | Tris-HCl, pH 7.7, 24% PEG 400, 100 mM $MgCl_2$, 10 mM DTT |
| *lac* dimer-ONPF | GAATTGTGAGCGCTCACAATT<br>TTAACACTCGCGAGTGTTAAG | 2.6 | Bell and Lewis, 2000 | R32 | 1:1.5:10 | | 0.1 M HEPES, pH 7.5, 14% glycerol, 2 M $(NH4)_2SO_4$, 10% PEG 400 |
| Mat$\alpha$2 | ACATGTAATTCATTTACACGC<br>GTACATTAAGTAAATGTGCGT | 2.7 | Wolberger *et al.*, 1991a, 1991b | $P2_1$ | 4:1 | | 15 mM Tris-HCl, pH 7.5, 25 mM NaCl, 75 mM $CaCl_2$ 10% PEG 400 |
| p53 | ATAATTGGGCAAGTCTAGGAA<br>ATTAACCCGTTCAGATCCTTT | 2.2 | Cho *et al.*, 1994 | C2 | 1:1 | 4 | 50 mM BTP, pH 6.4, 10 mM DTT, 100 mM MES, 12–15% PEG 400 |

| | | | | | | | |
|---|---|---|---|---|---|---|---|
| RepE54 | CCTGTGACAAATTGCCCTCAGT<br>TGGACACTGTTTAACGGGAGTC | 2.0 | Komori *et al.*, 1999 | C2 | | | 100 mM Tris-HCl, pH 8.0, 12% PEG 400, 200 mM $MgCl_2$ |
| Sso7d | GCGT[5I]UCGC<br>CGCA AGCG | 1.6 | Gao *et al.*, 1998 | $P2_12_12_1$ | 1.3:1.0 | | 2 mM Tris (pH 6.5), 2.6% PEG 400, equilibrated against 15% PEG 400 |
| TBP/TFB | TATA-box-containing promoter | 2.4 | Littlefield *et al.*, 1999 | C121 | 2:2:3 | | 50 mM Tris, pH 8.5, 200 mM sodium citrate, 25 mM $SrCl_2$, 30% PEG 400 |
| Transcription factor Gambif1 | CTGGGAAAAACCCAG<br>ACCCTTTTTGGGTCG | 2.70 | Cramer *et al.*, 1999 | $P4_32$ | | | 2–5% PEG 400, 50 mM MES pH 5.6, 3 mM DTT |
| USF | CACCCGGTCACGTGGCCTACA<br>TGGGCCAGTGCACCGGATGTG | 2.9 | Ferré-D'Amaré *et al.*, 1994 | $P2_12_12_1$ | 2:1 | 4 | 100 mM $CH_3COONa$, pH 4.75, 100 mM KCl, 2.8 mM $MgCl_2$, 1.4 mM $(CH_3COO)_2Cd$, 5 mM HEPES, 15% PEG 400, 15% glycerol |
| Zif268 | AGCGTGGGCGT<br>CGCACCCGCAT | 2.1 | Pavletich and Pabo, 1991 | $C222_1$ | 1:1 | | 25 mM BTP, pH 8.0, 0.35–0.65 M NaCl, 0–10% PEG 400, In addition drop contained: 1.5 M equivalence $ZnCl_2$ |
| λ Repressor | TATATCACCAGTGGTAT<br>TATAGTGGTCACCATAA | 1.8 | Jordan and Pabo, 1988, Jordan *et al.*, 1985, Beamer and Pabo, 199 | $P2_1$ | 1:2 | 20 | 15 mM BTP, pH 7.0, 1 mM $NaN_3$, 20% PEG 400 |
| **PEG 600–PEG 1540** | | | | | | | |
| *Bgl*II | TTATAGATCTATAA<br>AATATCTAGATATT | 1.5 | Lukacs *et al.*, 2000 | $P2_12_12_1$ | 1.6 mg/ml: 0.6 M | 20 | 0.1 M MES, pH 5.2, 0.2 M $(NH4)_2SO_4$, 17–20% PEG 1540 |
| Hin recombinase | TGTTTTTGATAAGA<br>CAAAAACTATTCTA | 3.0 | Feng *et al.*, 1993 | $C222_1$ | 3:1 | 20 | 50 mM Tris-HCl, pH 7.0, 20–50 mM NaCl, 20 mM $MgCl_2$, 15–18% PEG 1500 |
| Max | GTGTAGGCCACGTGACCGGGTG<br>CACATCCGGTGCACTGGCCCAC | 2.9 | Ferré-D'Amaré *et al.*, 1993 | $P6_122$ | 2:1 | 4 | 100 mM sodium cacodylate, pH 5.5, 100 mM KCl, 2 mM $MgCl_2$, 5–10% glycerol, 4.5–7.5% PEG 1000 |
| paired homeodomain | AACGTCACGGTTGAC<br>GCAGTGCCAACTGTT | 2.5 | Xu *et al.*, 1995 | $P2_12_12_1$ | 1:1.3 | 20 | 10 mM BTP, pH 7.0, 5 mM DTT, 10 mM $MgCl_2$, 10% PEG 1000 |
| SAP-1-SRF | CACACCGGAAGTCCTAATTAGGCCAT<br>TGTGGCCTTCAGGATTAATCCGGTAG | 3.15 | Hassler and Richmond, 2001 | $P2_12_12$ | 1:1:1 | 4 | 50 mM bis-Tris, pH 6.3–6.8, 100 mM $NH_4NO_3$, 1 mM DTT, 16–20% PEG 1500 |
| Transcription factor Pu.1 | AAAAAGGGGAAGTGGG<br>TTTTCCCCTTCACCCT | 2.1 | Kodandapani *et al.*, 1996; Pio *et al.*, 1996 | $C1_1$ | 1:1 | | 100 mM sodium cacodylate pH 6.5, 3% to 10% PEG 600, 200 mM $(CH_3COO)_2Zn$ The hanging drop: 5 μl of 0.5mM complex + 5 μl of reservoir solution. |

*(Continued)*

**Table 15.1** (Continued)

| Protein | Synthetic oligonucleotide (5′ – 3′) | Res (Å) | References | Space group | P:DNA ratio | T°C | Buffer and additives |
|---|---|---|---|---|---|---|---|
| **PEG 3000** | | | | | | | |
| 434 Repressor/$O_R$1 | TATACAAGAAAGTTTGTACT<br>TATGTTCTTTCAAACATGAA | 2.5 | Aggarwal *et al.*, 1988 | $P2_12_12_1$ | 2:1 | 4 | 100 mM NaCl, 120 mM $MgCl_2$, 2 mM spermine, 12–14% PEG 3000 |
| 434 Repressor/$O_R$2 | ACAAACAAGATACATTGTAT<br>GTTTGTTCTATGTAACATAT | 2.5 | Shimon and Harrison, 1993 | $P2_12_12_1$ | 2:1 | 4 | 100 mM MES, pH 5.7, 120 mM NaCl, 120 mM $MgCl_2$, 2 mM cobaltic hexamine, 19% PEG 3000 |
| 434 Repressor/$O_R$3 | TATACAAGAAAAACTGTACT<br>TATGTTCTTTTTGACATGAA | 2.5 | Rodgers and Harrison, 1993 | $P2_12_12_1$ | 1:1 | 4 | 100 mM MES, pH 5.5, 80 mM NaCl, 100 mM $MgCl_2$, 2 mM spermine, 15–17% PEG 3000 |
| NFAT/Fos/Jun | TTGGAAAATTTGTTTCATAG<br>CCTTTTAAACAAAGTATCAA | 2.7 | Chen *et al.*, 1998 | $P2_1$ | 3:3:3:2 | RT | 10 mM HEPES, pH 7.5, 1mM DTT, 100 mM NaCl, 20% glycerol, 300 mM $CH_3COONH_4$, 10% PEG 3000 |
| NF-*k*B | AGATGGGGAATCCCCTAGA<br>AGATCCCCTAAGGGGTAGA | 2.6 | Müller *et al.*, 1995 | $P4_12_12$ | 1:1.1 | | 50 mM $CH_3COONa$, pH 4.7, 50 mM $MgCl_2$, 1 mM DTT, 2 mM spermine, 1–4% PEG 3000 |
| **PEG 3350** | | | | | | | |
| BPV E2 | CCGACCGACGTCGGTCG<br>GCTGGCTGCAGCCAGCC | 1.7 | Hegde *et al.*, 1992 | R32 | 1:1.2 | 20 | 40 mM Tris-HCl, pH 7.5, 75 mM NaCl, 50 mM KCl, 1.5 mM lanthanide chloride, 2 mM $CaCl_2$ 35–37% PEG 3350 |
| CAP Repressor 30mer | † GCGAAAAGTGTGACAT·AT<br>GCTTTTCACACTG | 3.0 | Schultz *et al.*, 1990, 1991 | $C222_1$ | 1:1.5 | | 50 mM MES, pH 5–6, 0.2 M NaCl, 0.1 M $CaCl_2$, 2 mM cAMP, 0.02% $NaN_3$, 2 mM DTT, 2 mM spermine, 0.3% *N*-octyl-β-D-glucopyranoside, 5–10% PEG 3350 |

| | | | | | | | |
|---|---|---|---|---|---|---|---|
| IFN β-κB | TGGGAAATTCCT<br>CCCTTTAAGGAA | 2.75 | Berkowitz *et al.*, 2002 | C2 | | 18 | 50 mM $CH_3COONa$, pH 5.5, 50 mM $CaCl_2$, 0.125% β-octyl glucopyranoside, 1.0 mM spermine, 10 mM DTT, 7–8% PEG 3350 |
| Ig/HIV-2-κB | TGGGACTTTCCT<br>CCCTGAAAGGAA | 3.0 | Berkowitz *et al.*, 2002 | C2 | | 18 | 50 mM $CH_3COONa$, pH 5.5, 50 mM $CaCl_2$, 0.125% β-octyl glucopyranoside, 1.0 mM spermine, 10 mM DTT, 7–8% PEG 3350 |
| POU | TGTATGCAAATAAGG<br>CATACGTTTATTCCA | 3.0 | Klemm *et al.*, 1994 | $C222_1$ | 2:1 | RT | 50 mM sodium citrate, pH 5.4, 100–150 mM $(NH_4)_2SO_4$, 30–40% PEG 3350 |
| RXR-RAR | CCAGGTCAAAGGTCAG<br>GTCCAGTTTCCAGTCC | 1.7 | Rastinejad *et al.*, 2000 | $P2_12_12_1$ | 2:1 | 8 | 25 mM Tris, pH 7.5, 5 mM $MgCl_2$, 0.4 M $NH_4Cl$, 18–23% PEG 3350 |
| Tdp1/Vanadate/<br>Topo I-Derived Peptide | AGAGTT | 2.3 | Davies *et al.*, 2003 | $P2_12_12_1$ | | | 100 mM HEPES, pH 7.8, 200 mM NaCl, 10 mM spermine, 22% PEG 3350 |
| γδ resolvase | † GCAGTGTCCGATAATTTATAAA·<br>GTCACAGGCTATT | 3.0 | Yang and Steitz, 1995 | $P2_12_12_1$ | 1:2 | RT | 100 mM MES, pH 6.0, 0.5 mM EDTA, 0.2 M $(NH4)_2SO_4$, 5% ethylene glycol, 30% PEG 3350 |
| **PEG 3500** | | | | | | | |
| 434 Cro Repressor/$O_R$1 | TATACAAGAAAGTTTGTACT<br>TATGTTCTTTCAAACATGAA | 2.5 | Mondragon and Harrison, 1991 | $P2_1$ | 2:1 | 4 | 100 mM MES, pH 6.2, 160 mM NaCl, 120 mM $MgCl_2$, 2 mM spermine, 12% PEG 3500 |
| *Hae*III | ACCAGCAGGC$^F$CACCAGTG<br>GGTCGTCC$^M$GGTGGTCACT | 2.8 | Reinisch *et al.*, 1994, 1995 | $P2_12_12_1$ | | | 100 mM MES, pH 6.5, 120 mM $CaCl_2$, 1 mM DTT, 13% glycerol, 9–13% PEG 3500 |
| MyoD | TCAACAGCTGTTGA<br>AGTTGTCGACAACT | 2.8 | Ma *et al.*, 1994 | $P2_12_12_1$ | 2:1 | 22 | 100 mM Tris-HCl, pH 8.5, 100 mM sodium citrate, 20 mM $BaCl_2$, 10–15% PEG 3500 |

(Continued)

**Table 15.1** (Continued)

| Protein | Synthetic oligonucleotide (5′ – 3′) | Res (Å) | References | Space group | P:DNA ratio | T°C | Buffer and additives |
|---|---|---|---|---|---|---|---|
| PPR1 | TCGGCAATTGCCGA<br>AGCCGTTAACGGCT | 3.0 | Marmorstein and Harrison, 1994 | I222 | 1:1.2 | | 20 mM sodium cacodylate, pH 6.8, 2% MPD, 87.5 mM NaCl, 62.5 mM $CaCl_2$, 50 μM $Zn(CH_3COO)_2$, 15% PEG 3500 |
| **PEG 4000** | | | | | | | |
| *Eco*RV | GGGATATCCC<br>CCCTATAGGG | 3.0 | Winkler *et al.*, 1991, 1993 | $C222_1$ | 1:1.7 | 20 | 10 mM sodium phosphate, pH 7.0, 80 mM NaCl, 2.5–3.5% PEG 4000 |
| *Hha*II | GGAGTCC<br>CCTCAGG | 4.0 | Chandrasegaran *et al.*, 1986 | C2 | 2:1 | 20 | 10 mM potassium phosphate, pH 7.2, 0.2M NaCl, 1 mM EDTA, 0.1 mM DTT, 10% PEG 4000 |
| NF-*k*B | TGGGAATTCCC<br>CCCTTAAGGGT | 2.3 | Ghosh *et al.*, 1995 | C2 | 1.1:1 | 18 | 50 mM HEPES, pH 7.5, 100 mM $NH_4Cl$, 10 mM DTT, 10 mM $CaCl_2$, 0.1 mM spermine, 0.1% β-octyl glucoside, 8% PEG 4000 |
| CBFα | AAACTCTGTGGTTGCG<br>TTGAGACACCAACGCT | 2.65 | Tahirov *et al.*, 2001a, 2001c | $P2_12_12_1$ | 1:1 | 25 | 50 mM HEPES, pH 7.0, 200 mM $CH_3COONH_4$, 150 mM $(CH_3COO)_2Mg$, 5% PEG 4000 |
| CBFα-C/EBPβ | GAAGATTTCCAAACTCTGTGGTTGCG<br>TTCTAAAGGTTTGAGACACCAACGCC | 3.0 | Tahirov *et al.*, 2001a, 2001c | $P2_12_12$ | 1:1:1 | 25 | 50 mM MES, pH 5.6, 5 mM $MgSO_4$, 3% PEG 4000, 2% dioxane |
| E47 | AACACCTGGCT<br>TGTGGACCGAT | 2.8 | Ellenberger *et al.*, 199 | C2 | 1:1.4 | | 20 mM MES, pH 6.5, 50 mM $CaCl_2$, 1 mM spermine, 4% PEG 4000 |
| *Eco*RV | CGAGCTCG<br>GCTCGAGC | 3.0 | Winkler *et al.*, 1991, 1993 | $P2_1$ | 1:5 | 20 | 20 mM MES, pH 6.0, 100 mM NaCl, 1.0–1.5% PEG 4000 |
| *Hinc*II | GCCGGTCGACCGG<br>GGCCAGCTGGCCG | 2.5 | Horton *et al.*, 199 | I222 (or $I2_12_12_1$) | 1:1.5 | 23 | 0.1 M sodium citrate, pH 5.5, 0.12 M NaCl, PEG 4000 (variable conc.) |

| | | | | | | | |
|---|---|---|---|---|---|---|---|
| MutS | GCGACGCTAGCGTGCGGCTCGTC<br>GCTGCGATCGC·CGCCGAGCAGG | 2.5 | Obmolova *et al.*, 200 | $P2_12_12_1$ | 1:1.2 | 20 | 100 mM sodium cacodylate, pH 6.2, 200 mM $(NH4)_2SO_4$, 1 mM DTT, 18% PEG 4000 |
| NFAT1-RHR (microbatch method) | TTGAGGAATTTCCA<br>CTCCTTAAAGGTAA | 3.1 | Jin *et al.*, 2003 | $P2_12_12_1$ | 2:1 | 19 | 50 mM Tris-HCl, pH 8.0, 100 mM NaCl, 10 mM $MgCl_2$, 2.5 mM spermine, 12–15% PEG 4000 |
| NFAT1-RHR | TTGAGGAATTTCCA<br>CTCCTTAAAGGTAA | 3.9 | Jin *et al.*, 2003 | $P3_121$ | 1:1 | 19 | 50 mM bis-Tris propane, pH 6.3, 100 mM NaCl, 10 mM $MgCl_2$, 2.5 mM spermine, 6% glycerol, 6% PEG 4000 |
| PcrA DNA helicase | CGAGCACTGC<br>TTTTTTTGCTCGTGACG | 2.9 | Velankar *et al.*, 1999 | $P2_1$ | 1:1 | | 100 mM Tris, pH 8.5, 50 mM $Li_2SO_4$, 20% PEG 400, 20–25% PEG 4000 |
| *Pvu*II | TGACCAGCTGGTC<br>CTGGTCGACCAGT | 3.0 | Balendiran *et al.*, 1994; Cheng *et al.*, 1994 | $P2_12_12_1$ | 2.7:1 | 16 | 50 mM $CH_3COONa$, pH 4.5, 0.3 mM EDTA, 8% PEG 4000 |
| TFIIIA | ACGGGCCTGGTTAGTACCTGGATGGGAGACC<br>GCCCGGACCAATCATGGACCTACCCTCTGGT | 3.1 | Nolte *et al.*, 1998 | P1 | | 18 | 165 mM NaCl, 35 mM $CH_3COONa$, 3.2 mM DTT, 9.2% glycerol, 1.8 mM $NaN_3$, 1.8 mM cadaverine-2 HCl, 5.5 mM Tris-HCl, pH 8.0, 22.5% PEG 4000 |
| **PEG 350 monomethylether (MME)** | | | | | | | |
| RB69 DNA polymerase | GCGGAACTACTGCTTACG<br>GCGCCTTGATGACGAAT | 2.7 | Shamoo and Steitz, 1999 | R3 | 1:1 | 12 | 100 mM sodium cacodylate, pH 6.5, 150 mM $CaCl_2$, 12% PEG350 MME |
| RB69 DNA polymerase | ACAGGTAAGCAGTCCGCG<br>dTTP+ ddCCATTCGTCAGGCG | 2.6 | Franklin *et al.*, 2001 | $P2_12_12_1$ | 1:1 | 16 | 100 mM sodium cacodylate, pH 6.25, 30% PEG350 MME |
| **PEG 2000 MME** | | | | | | | |
| TFIIIB | CTATAAAAAAATGTTTTTT<br>GATATTTTTTTACAAAAAA | 2.95 | Juo *et al.*, 2003 | $P2_12_12$ | | 4 | 50 mM MES, pH 6.5, 100 mM NaCl,or LiCl, 10 mM $MgCl_2$, 2 mM 2-mercaptoethanol, 25% glycerol, 10% PEG 2000 MME |

*(Continued)*

**Table 15.1** (Continued)

| Protein | Synthetic oligonucleotide (5′ – 3′) | Res (Å) | References | Space group | P:DNA ratio | T°C | Buffer and additives |
|---|---|---|---|---|---|---|---|
| **PEG 5000 MME and 0.2–0.4 M $(NH_4)_2SO_4$** | | | | | | | |
| hPol*ι* | GGGGGAAGGACC<br>CCCCCTTCCTGGGATCTT | 2.6 | Nair *et al.*, 2004 | $P6_5$ | 1:1.2 | | 0.1 M MES, pH 6.5, 5 mM ddTTP, 5 mM $MgCl_2$, 10–15% PEG 5000 MME, 0.2–0.4 M $(NH4)_2SO_4$ |
| **PEG 6000** | | | | | | | |
| DNase I | GCGATCGC<br>CGCTAGCG | 2.0 | Suck *et al.*, 1988; Lahm and Suck, 1991 | $C222_1$ | – | 4 | 15 mM EDTA, 8% PEG 6000 |
| Eco57I-sinefungin | TTCGATGTGCTGAAGTTTAGACCTG<br>GCTACACGACTTCAAATCTGGACTT | 4.2 | Tamulaitiene *et al.*, 2004 | $P2_12_12$ | 1:(1–2): 2 | | 0.1 M MES, pH 6.5, 2 M NaCl, 0–4% ethylene glycol, 8–10% PEG 6000 |
| GCN4-bZIP | ATF/CREB recognition sequence | 2.2 | Keller *et al.*, 1995 | $P4_12_12$ | 1:1 | 4 | 30 mM sodium citrate, pH 4.6, 50 mM $CH_3COONa$, 5 mM $MgCl_2$, 1 mM $NaN_3$, 10% PEG 6000 |
| human TFIIA/TBP | AdMLP TATA-box | 2.1 | Bleichenbacher *et al.*, 2003 | $P2_12_12_1$ | 1:1:1 | 22 | 50 mM bis-Tris, pH 6.5, 100 mM $LiNO_3$, 10 mM $CaCl_2$, 1 mM DTT, 12% PEG 6000 |
| **PEG 8000** | | | | | | | |
| *Bam*HI | TATGGATCCATA<br>ATACCTAGGTAT | 2.2 | Strzelecka *et al.*, 1994; Newman *et al.*, 1995 | $P2_12_12_1$ | 1:2 | | 3.9 mM Tris, pH 7.6, 6.1 mM potassium phosphate, pH 6.9, 150 mM KCl, 3% glycerol, 3.9 mM NaCl, 0.4 mM EDTA, 12% PEG 8000 |
| C/EBPβ$_{ve}$ | AATGTGGCGCAATCCT<br>TACACCGCGTTAGGAT | 1.7 | Tahirov *et al.*, 2001b | $C222_1$ | 1:1.2 | 25 | 50 mM sodium cacodylate, pH 6.5, 200 mM KCl, 100 mM $(CH_3COO)_2Mg$, 10% PEG 8000 |
| CBFα – β-C/EBPβ | GAAGATTTCCAAACTCTGTGGTTGCG<br>TTCTAAAGGTTTGAGACACCAACGCC | 3.0 | Tahirov *et al.*, 2001a,2001c | $C222_1$ | 1:1:1:1 | 25 | 50 mM MES, pH 5.6, 200 mM KCl, 10 mM $MgCl_2$, 20 mM DTT, 4.5% PEG 8000, 1% glycerol, 1% MPD |

| | | | | | | | |
|---|---|---|---|---|---|---|---|
| Even-skipped homeodomain | TTCAGCACCGTTCAGCACCG<br>AGTCGTGGCAAGTCGTGGCA | 3.5 | Hirsch and Aggarwal, 1995 | $P6_3$ | 1.8:1 | 20 | 0.4 M $CH_3COOK$, pH 6.2, 3 mM DTT, 5–9% PEG 8000 |
| hTFIIBc-hTBPc | GGGGGGCGCCTATAAAAGGGGGGCG<br>CCCCCCGCGGATATTTTCCCCCCGC | 2.65 | Tsai and Sigler, 2000 | $P2_1$ | 1:1.5 | 4 | 50 mM Tris-HCl, pH 8.0, 50 mM $MgCl_2$, 100 mM sodium citrate, pH 5.6, 10% glycerol |
| TFIIB/TBP | AGTTGCTTTAAAAAGTAAGTTGCTT<br>TCAACGAAATTTTTCATTCAACGAA | 2.1 | Kosa *et al.*, 1997 | C2 | 1:1:1 | 18 | 200 mM potassium phosphate, pH 7.4, 8% PEG 8000 |
| Vsr endonuclease | AGC<u>T</u>AGGTACGT<br>TCG<u>G</u>TCCATGCA | 2.3 | Tsutakawa *et al.*, 1999 | $P6_5$ | duplex | 20 | 25 mM MES, pH 5.6, 6.5 mM Tris, pH 7.5, 100 mM KCl, 15 mM $MgCl_2$, 31 mM NaCl, 2.5% PEG 8000 |
| yTBP | GTATATAAAACGGGTGG<br>CATATATTATGC | 2.5 | Kim, Y. *et al.*, 1993 | $P4_3$ | 1:2 | 22 | 40 mM BTP, pH 7.5, 500 mM NaCl, 2.5% glycerol, 2% ethylene glycol, 30% PEG 8000 |

† -half site, M-Methylated base, F-Fluorinated base, f-fluorescein label, ddC-dideoxycytosine, $^{5I}$U - 5-iodo-deoxyuridine, RT-room temperature.

**Table 15.2** Crystallization conditions for DNA–protein complexes using 2-methylpentan-2,4-diol (MPD) as a precipitant

| Protein | Synthetic oligonucleotide (5′ – 3′) | Res (Å) | References | Space group | P:DNA ratio | T°C | Buffer and additives |
|---|---|---|---|---|---|---|---|
| *Bam*H I | BA27-49 (Nonspecific DNA) | 1.9 | Viadiu *et al.*, 2000 | $P2_12_12_1$ | | | 10 mM $CH_3COONa$, pH 4.8, 5 mM $CaCl_2$, 16–20% MPD |
| CENP-B | 21mer oligonucleotides containing the CENP-B box sequence | 2.5 | Tanaka *et al.*, 2001 | $P3_112$ | 1:1 | 20 | 50 mM sodium cacodylate, pH 6.5, 100 mM $CH_3COONa$, 15% MPD |
| Cre recombinase | Holiday Junction 1 (HJ-1) | 2.7 | Gopaul *et al.*, 1998 | $C222_1$ | 5:2 | 18 | 50 mM $CH_3COONa$, pH 5.0, 80 mM $MgCl_2$, 26% MPD |
| Cre recombinase | loxS6 | 2.5 | Gopaul *et al.*, 1998 | $C222_1$ | 6.5:8.1 | 4 | 50 mM $CH_3COONa$, pH 5.0, 20 mM $CaCl_2$, 200 mM NaCl, 24% MPD |
| Cre recombinase | loxA | 2.4 | Guo *et al.*, 1997 | $C222_1$ | | 18 | 100 mM $CH_3COONa$, pH 5.0, 20 mM $CaCl_2$, 24% MPD |
| *Eco*RI | † GAATTC<br>CTTAAG | 2.8 | Young *et al.*, 85 | $P42_12$ | | 4 | 15 mM potassium phosphate, pH 7.4, 0.075 mM DTT, 0.75 mM EDTA, 0.23 M NaCl, 20% MPD |
| estrogen receptor | CCAGGTCACAGTGACCTG<br>GTCCAGTGTCACTGGACC | 2.4 | Schwabe *et al.*, 1993 | $P2_12_12_1$ | 2.25:1 | 20 | 20 mM MES, pH 6.0, 1.8 mM spermine, 2 μM $ZnCl_2$, 10% MPD, 2–8 mM $CaCl_2$, 30–80 mM NaCl |
| GAL4 | CCGGAGGACAGTCCTCCGG<br>GGCCTCCTGTCAGGAGGCC | 2.7 | Marmorstein *et al.*, 1992 | $P4_32_12$ | 1:1.5 | | 20 mM sodium cacodylate, pH 6.8, 125 mM $CaCl_2$, 40 mM NaCl, 27% MPD |
| glucocorticoid receptor | CCAGAACATCGATGTTCTG<br>GTCTTGTAGCTACAAGACC | 2.9 | Luisi *et al.*, 1991 | $P2_12_12_1$ | 2:1 | 8 | 24 mM sodium cacodylate, pH 6.0, 0.2 M NaCl (25 mM NaCl in drop), 8% MPD<br>In addition the drop contained: 1.8 mM $MgSO_4$, 2 μM $ZnCl_2$, 1.8 mM spermine |

| | | | | | | | |
|---|---|---|---|---|---|---|---|
| *met* repressor | TTAGACGTCTAGACGTCTA<br>ATCTGCAGATCTGCAGATT | 2.8 | Somers and Phillips, 1992 | $P6_222$ | | | 10 mM sodium cacodylate, pH 7.0, 1 mM $NaN_3$,30–35% MPD. In addition drop contained:15–30 mM $CaCl_2$, 6 mM NaCl, 1 mg/ml SAM |
| *Sfi*I | ATGTGGCCAACAAGGCCTATT<br>TACACCGGTTGTTCCGGATAA | 3.0 | Viadiu *et al.*, 2003 | $P3_121$ or $P3_221$ | | | 100 mM $CH_3COONa$, pH 4.6–5.0, 30–32.5% MPD, 5–10 mM $CaCl_2$ |
| *trp* repressor | TGTACTAGTTAACTAGTAC<br>CATGATCAATTGATCATGT | 2.4 | Joachimiak *et al.*, 1987; Otwinowski *et al.*, 1988 | $P2_1$ | 1:2 | 20 | 10 mM sodium cacodylate, pH 6.8, 11 mM $CaCl_2$, 50 mM NaCl, 35% MPD, 2 mM L-tryptophan |
| yeast RAP1 | CCGCACACCCACACACCAG<br>GCGTGTGGGTGTGTGGTCC | 2.25 | König *et al.*, 1996 | $P3_1$ | | 20 | 20 mM MES, pH 6.0, 20 mM KCl, 2 mM spermine, 40% MPD |

† -halfsite.

**Table 15.3** Crystallization conditions for DNA–protein complexes using ammonium sulphate as a precipitant

| Protein | Synthetic oligonucleotide (5′ – 3′) | Res (Å) | References | Space group | P:DNA ratio | T°C | Buffer and additives |
|---|---|---|---|---|---|---|---|
| 434 Cro Repressor/$O_L$2 | ACAATATATATTGT<br>TGTTATATATAACA | 3.2 | Wolberger and Harrison, 1987;<br>Wolberger *et al.*, 1988 | C2 | 2:1 | 20 | 10 mM Tris-HCl, pH 8.0, 1 mM EDTA, 1 mM DTT, 0.7 M NaCl, 10 mM cobaltic hexamine, 1 mM $MgCl_2$, 36% (w/v) $(NH4)_2SO_4$ |
| 434 Repressor/$O_L$2 | ACAATATATATTGT<br>TGTTATATATAACA | 3.2 | Anderson *et al.*, 1984, 1987 | I422 | 2:1 | 4 | 5 mM sodium phosphate, pH 4.7, 1mM EDTA, 10 mM NaCl, 1.3 M $(NH4)_2SO_4$ |
| ADAR1 (Zα domain) | TCGCGCG<br>GCGCGCT | 2.4 | Schwartz *et al.*, 1999a, 1999b | $P42_12$ | 1:1 | RT | 1.8 M $(NH4)_2SO_4$, 10% glycerol |
| DNA Polymerase I | pdT4 | 2.8 | Freemont *et al.*, 1988 | | | 18 | 0.2 M sodium citrate, pH 5.6, 1 mM EDTA, 38% $(NH4)_2SO_4$ |
| *Fok* I | TCGGATGATAACGCTAGTCA<br>GCCTACTATTGCGATCAGTA | 2.8 | Wah *et al.*, 1997 | $P2_1$ | | | 2.2 M $(NH4)_2SO_4$ |
| *Hha*I methyl-transferase | TGATAGC$^F$ GCTATC<br>CTATCG CGATAGT | 2.8 | Klimasauskas *et al.*, 1994 | $R3_2$ | 1:1 | 16 | 50 mM sodium citrate, pH 5.6, 1.2–1.4 M $(NH4)_2SO_4$, 25 mM NaCl, 0.25 mM EDTA |
| HIV RT | ATGGCGCCCGAACAGGGAC<br>ACCGCGGGCTTGTCCCTG | 3.5 | Jacobo-Molina *et al.*, 1991,<br>Arnold *et al.*, 1992 | $P3_112$/<br>$P3_212$ | 1:2 | 4 | 100 mM sodium cacodylate, pH 5.6, 30% saturated $(NH4)_2SO_4$ |
| HIV-1 RT p66/p51-Fab | ACAGTCCCTGTTCGGG*CGCC<br>GTGTCAGGGACAAGCCC GCGGTACGTA | 3.1 | Sarafianos *et al.*, 2002 | $P3_212$ | | 4 | 100 mM sodium cacodylate, pH 5.6, 31–34% saturated $(NH4)_2SO_4$ |

Fluorinated base, *-dG residue with a disulphide linker at the exocyclic $N^2$ amino group, RT-room temperature.

**Table 15.4** Crystallization conditions for DNA–protein complexes using other precipitants (NaCl, Na Formate, $CH_3COONH_4$, $NH_4OH$, sodium citrate, $CaCl_2$, $CH_3COONa$, potassium phosphate . . . .)

| Protein | Synthetic oligonucleotide (5′ – 3′) | Res (Å) | References | Space group | P:DNA ratio | T°C | Buffer and additives |
|---|---|---|---|---|---|---|---|
| CAP–tαCTD | † AGATCACATTTTAGGAAAAAAG TCTAGTGTAAAATCCTTTTTTC | 3.1 | Benoff *et al.*, 2002 | $P6_222$ | 1:2 | 20 | 100 mM $CH_3COONa$, pH 4.5, 500 mM NaCl |
| engrailed homeodomain | TTTTGCCATGTAATTACCTAA AAACGGTACATTAATGGATTA | 2.8 | Liu *et al.*, 1990; Kissinger *et al.*, 1990 | C2 | | | 30 mM Tris-HCl, pH 6.7. In the drop in addition: $NH_4OH$ to pH 8.0–9.0 |
| HNF-3/*fork head* | † GACTAAGTCAACC CTGATTCAGTTGG | 2.5 | Clark *et al.*, 1993 | P31 | 1:1 | 4 | 100 mM $CH_3COOK$, pH 5.5, 50 mM KCl, 2 mM $MgCl_2$, 20 mM DTT. In addition in the drop: 550 mM $CH_3COONH_4$ |
| TBP2 | GCTATAAAAGGGCA CGATATTTTCCCGT | 2.2 | Kim, J. *et al.*, 1993 | $P2_1$ | 1:1 | 4 | 25 mM MES, pH 6.2, 100 mM KCl, 4 mM $MgCl_2$, 14% glycerol, 15 mM DTT In addition in the drop: 250 mM $CH_3COONH_4$ |
| TFIIB-TBP | GGCTATAAAAGGGCTG CCGATATTTTCCCGAC | 2.7 | Nikolov *et al.*, 1995 | $P2_12_12_1$ | | 4 | 40 mM Tris-HCl, pH 8.5, 40 mM KCl, 5 mM $MgCl_2$, 5 mM $CaCl_2$, 10 mM DTT, 10 $\mu$M $(CH_3COO)_2Zn$, 10% glycerol, 2% ethylene glycol, 300 mM $CH_3COONH_4$ |
| TrpR | TAGCGTACTAGTACGCT TCGCATGATCATGCGAT | 2.4 | Carey *et al.*, 1993 | C2 | 2.2:1 | 23 | 0.1 M $CH_3COONa$, pH 4.8, 2.0 M sodium formate |
| $\lambda$–cro repressor | TATCACCGCGGGTGATA ATAGTGGCGCCCACTAT | 3.9 | Brennan *et al.*, 1990, 1986 | $P3_2$ | 1:2 | | 20 mM sodium cacodylate, pH 6.9, 3.5–4.0 M NaCl |
| a1/MAT $\alpha$2 homeodomain | TACATGTAATTTATTACATCA GTACATTAAATAATGTAGTAT | 2.7 | Li *et al.*, 1995 | $P6_1/P6_5$ | 1:1:1.4 | 20 | 150–200 mM HEPES, pH 5.0, 20 mM $CaCl_2$, 5–10 mM cobaltic hexamine |
| DNase I | GGTATACC CCATATGG | 2.3 | Weston *et al.*, 1992 | $P2_12_12$ | 1:4 | 20 | pH 4.0–5.0, 0.3 M NaCl, 20 mM EDTA, 50–60% saturated sodium citrate |

*(Continued)*

**Table 15.4** (Continued)

| Protein | Synthetic oligonucleotide (5′ – 3′) | Res (Å) | References | Space group | P:DNA ratio | T°C | Buffer and additives |
|---|---|---|---|---|---|---|---|
| Even-skipped homeodomain | TCAATTAAATTCAATTAAAT<br>GTTAATTTAAGTTAATTTAA | 2.0 | Hirsch and Aggarwal, 1995 | $P2_1$ | 1.8:1 | 20 | pH 6.1, KCl, 7–15% 4 M $CH_3COONa$ |
| *lac* Repressor | TTGTGAGCGCTCACAA<br>AACACTCGCGAGTGTT | 6.5 | Pace *et al.*, 1990 | C2 | 1:2 | | 0.1 M *N*-(2-acetamido)-2-iminoiacetic acid, pH 6.5, 1.4 M $CH_3COONa$ |
| MAT**a**1/MATα2 | TACATGTAAAAATTTACATCA<br>GTACATTTTTAAATGTAGTAT | 2.4 | Li *et al.*, 1998 | $P6_1$ | | | 100 mM HEPES, pH 7.0, 20 mM $CaCl_2$, 5 mM $Co(NH_3)_6Cl_3$ |
| RecG | GCAGTGCTCGCATGGAGCTG<br>ACGGTTACTAGTACCTCGAC | 3.25 | Singleton *et al.*, 2001 | C2 | 1:1.2 | | 500 mM potassium phosphate, pH 5.0, 5 mM $MgCl_2$, 1 mM ADP |
| Recombinant *X.laevis* histones | 147bp palindromic DNA fragment from human α -satellite DNA | 1.9 | Davey *et al.*, 2002; Luger *et al.*, 1997 | $P2_12_12_1$ | | | 10–20 mM potassium cacodylate, pH 6.0, 40–46 mM $MnCl_2$, 30–40 mM KCl |
| TRAMTRAK | CTAATAAGGATAACGTCCG<br>ATTATTCCTATTGCAGGCT | 2.8 | Fairall *et al.*, 122 | $P2_12_12_1$ | 1:1 | 20 | 20 mM MES, pH 6.0, 5–20 mM NaCl, 1.75–3.5 mM spermine |

crystallographers have their DNA oligomers synthesized commercially or use automatic in-house DNA synthesizers. Large-scale synthesis (1 and 10 μmol) is performed using the solid-phase phosphoramidite method (Caruthers, 1985). This procedure starts with the 3′-hydroxyl nucleoside attached to solid support such as controlled pore glass through a long spacer arm. The dimethoxytritol (DMT) group is removed with trichloroacetic acid to make the 5′-hydroxyl accessible for the coupling reaction. The next nucleotide is added as a phosphoramidite derivative, along with tetrazole, to the reaction chamber. The tetrazole makes the phosphoramidite susceptible to nucleophilic attack by protonating its nitrogen. The coupling reaction is complete within 30 seconds. The chains that fail to undergo the coupling reaction are capped by acetylation, whereas the DMT group of the successful coupling step protects the 5′-OH end from being capped. The internucleotide linkage is oxidized with iodine and water to the phosphotriester. The next step is the removal of DMT with trichloroacetic acid. The cycle is repeated until the complete oligonucleotide sequence is synthesized. The oligonucleotide is removed from the column matrix by incubating the matrix with ammonium hydroxide at room temperature for 90 min. The protecting groups can be removed by an overnight incubation or by boiling the DNA in ammonium hydroxide.

Chemically synthesized DNA preparations contain impurities, which may hamper crystallization experiments. Therefore, it is important to remove impurities before mixing the DNA with the purified protein sample. Electrophoresis on polyacrylamide gel, reverse-phase HPLC (high-pressure liquid chromatography), or anion-exchange chromatography is employed for this purpose. If the oligonucleotide is to be purified by gel electrophoresis or ion exchange chromatography the DMT group should be removed at this stage. The advantages of denaturing polyacrylamide gels are speed and high resolution. The percentage of polyacrylamide in a gel is adjusted according to the size of the oligonucleotide to be resolved. The oligonucleotide bands are visualized by exposing the gel to UV light. The bands appear as black shadows against a green background. The nucleotide is extracted by cutting out the bands with a clean blade, crushing into fine particles, and shaking in a suitable buffer for several hours or overnight at room temperature.

Reverse-phase HPLC relies on hydrophobic interactions between the column matrix and oligonucleotides (Protocol 15.1). This technique is useful for the separation of tritylated oligonucleotide from untritylated contaminants and also untritylated oligonucleotides of interest from other sequences. The oligonucleotide with the DMT group attached is loaded on to a reverse-phase HPLC column in the presence of 100 mM triethylamine acetate pH 7.0. Elution is accomplished by decreasing the polarity by a gradient of acetonitrile. The desired level of purity (>99%) can be achieved by two cycles of purification. The final detritylation step is performed on the reverse phase HPLC column by passing 0.5% trifluoroacetic acid before eluting the oligonucleotide from the column. Jordan *et al.* (1985) purified each strand of λ operator site $O_LI$ with DMT group attached on a reverse-phase HPLC, removed the DMT group and then purified the detritylated sequence by gel electrophoresis, followed by a second cycle of reverse-phase HPLC for obtaining diffraction-quality crystals of the λ repressor-operator complex.

The complementary strands are mixed in 1:1 ratio and annealed by heating to 80°C followed by slow cooling (Protocol 15.2). The annealed double-stranded oligomer can be further purified by ion-exchange chromatography (Protocol 15.3) to remove single strands. Protocols 15.1 and 15.3 are modified from the methods described by Aggarwal (1990) and Joachimiak *et al.* (1987).

### 15.2.3 Protein–DNA complex formation

The protein and annealed DNA are mixed together in an appropriate ratio and screened for crystallization conditions. A large number of DNA–protein complexes have been crystallized using this method. The DNA:protein ratio ranges from stoichiometric up to a two-fold excess of DNA (see Tables 15.1–15.4). The CAP–αCTD–DNA complex was crystallized from solutions containing 0.1 to 0.2 mM CAP, 0.2 to 0.4 mM αCTD, 0.1 to 0.3 mM DNA, and 0.8 mM cAMP (Benoff *et al.*, 2002). The crystals of RecG and the substrate DNA were grown after mixing these

## Protocol 15.1 DNA purification by reverse-phase HPLC

### Materials

DNA oligonucleotide to be purified
HPLC machine
Reverse phase Perkin Elmer Reverse Phase Cartridge/Varian PureDNA/Hamilton PRP-3 (Polystyrene-divinylbenzene)/Vydac C4 column for DNA purification
Tubes for collecting fractions
100 mM triethylammonium acetate, pH 7.0, in water
10 mM triethylammonium bicarbonate, pH 7.0, in water
5% Acetonitrile in 100 mM triethylamine acetate pH 7.0
70% Acetonitrile in 100 mM triethylamine acetate pH 7.0
0.5% Trifluoroacetic acid
1000 MW cut-off dialysis tubing

### Method

**1.** Dissolve synthetic DNA oligonucleotide in 100 mM triethylamine acetate, pH 7.0.
**2.** Remove any undissolved material by centrifugation at 14000–15000 rpm for 5 min in a microcentrifuge.
**3.** Pass the DNA solution through a 0.2 μm pore filter.
**4.** Equilibrate the column with 100 mM triethylamine acetate, pH 7.0.
**5.** Inject an appropriate volume of DNA solution (2–200 μl) onto the column.
**6.** Elute the DNA with a 5–70% gradient of acetonitrile in 100 mM triethylamine acetate, pH 7.0 and collect fractions.
**7.** Pool the peak fractions together and dialyse against 10 mM triethylamine bicarbonate (pH 7.0) using a 1000 MW cut-off dialysis bag.
**8.** Lyophilize the DNA oligonucleotide.
**9.** Dissolve the DNA in an appropriate volume of 100 mM triethylamine acetate, pH 7.0.
**10.** Remove any undissolved material by centrifugation at 14000–15000 rpm for 5 min in a microcentrifuge.
**11.** Filter the DNA solution as in Step 3.
**12.** Equilibrate the column with 100 mM triethylamine acetate pH 7.0.
**13.** Inject an appropriate volume of DNA solution (2–200 μl) onto the column.
**14.** Wash the column with 0.5% trifluoroacetic acid for at least 10 min.
**15.** Wash the column with 100 mM triethylamine acetate, pH 7.0.
**16.** Elute the DNA with a 5–70% gradient of acetonitrile in 100 mM triethylamine acetate pH 7.0.
**17.** Pool the peak fractions together and dialyse against 10 mM triethylamine bicarbonate (pH 7.0) using a 1000 MW cut-off dialysis tubing.
**18.** Lyophilize the DNA oligonucleotide.

## Protocol 15.2 Annealing the DNA strands

### Materials

DNA oligonucleotides to be annealed
Appropriate buffer, pH 7.0
Spectrophotometer
Water bath at 80–90°C or PCR machine
Microcentrifuge tubes
Parafilm
Refrigerator

### Method

**1.** Dissolve DNA strands separately in an appropriate buffer at neutral pH.
**2.** Calculate the extinction coefficients of oligomers by adding coefficients of the constituent nucleotides.
**3.** Determine $OD_{260}$ of each solution.
**4.** Calculate molar concentration of each solution.
**5.** Mix the two DNA solutions in 1:1 molar ratio in a microcentrifuge tube.
**6.** Wrap the tube containing DNA in parafilm.
**7.** Incubate at 80–90°C for 10 min in a water bath or PCR machine.
**8.** Allow the water to cool slowly to room temperature followed by slow cooling to 5°C.

components in 1:1.2 ratio (Singleton *et al.*, 2001). The TBP/TFB/DNA complex was prepared simply by adding TBP and TFB to DNA in 2:2:3 ratio (Littlefield *et al.*, 1999). The crystals of λ repressor with 20-mer operator containing one DNA duplex per protein dimer were grown from solutions containing two DNA duplexes per one protein dimer (Jordan *et al.*, 1985).

**Protocol 15.3 Purification of DNA oligonucleotides by anion-exchange chromatography**

**Materials**

DNA oligonucleotide to be purified
Pharmacia MonoQ column
HPLC machine
Solution A: 0.1 M NaCl, 10 mM NaOH
Solution B: 1.0 M NaCl, 10 mM NaOH
1 M Hepes or Tris-HCl, pH 7.5
Microcentrifuge
Tubes for collecting fractions
Desalting column

**Method**

1. Dissolve DNA in an appropriate volume (0.5–1.0 ml) of solution A.
2. Remove any undissolved material by centrifugation at 14,000–15,000 rpm for 5 min in a microcentrifuge.
3. Equilibrate a column connected to HPLC machine with buffer A.
4. Inject the DNA sample onto the column.
5. Wash the column with solution A.
6. Elute the DNA with a gradient of A–B and collect fractions directly into tubes containing appropriate volume of 1 M Hepes or Tris-HCl, pH 7.5.
7. Pool appropriate fractions together.
8. Pass through a desalting column.

A second approach is to prepare a stable complex and separate it from the unbound components by size exclusion chromatography prior to conducting crystallization trials. For example, CENP-B and DNA were mixed in 1:1 ratio under denaturing conditions, the denaturant was removed by dialysis and complex was purified using a Hiload Superdex 75 column (Tanaka *et al.*, 2001). The TFIIA, TBP, and DNA were mixed in 1:1:1 ratio under non-denaturing conditions and the TFIIA/TBP/DNA complex was purified on a Superdex 200 column (Bleichenbacher *et al.*, 2003). Human HIV-1 RT (p66 and p51 subunits) –DNA complex was prepared by mixing RT and template-primer in 1:1.5 ratio in the presence of AZTTP and dATP, cross-linking RT-DNA with $MgC1_2$,and separating the cross-linked complex on an immobilized $Ni^{2+}$affinity column followed by a heparin column (Sarafianos *et al.*, 2002).

### 15.2.4 Crystallization of protein–DNA complexes

Factors that affect the growth of protein crystals also affect the growth of protein–DNA cocrystals. These factors include the nature of the buffering system, pH, ionic strength, temperature, concentration and nature of precipitant, and concentration of the complex (McPherson, 1990). Examples of crystallization conditions used for some of the protein–DNA complexes have been listed in Tables 15.1–15.4 as it is beyond the scope of this chapter to list crystallization conditions for all the protein–DNA complexes crystallized to date.

Tris (Tris(hydroxymethyl)methylamine)-HCl, BTP (bis-Tris-propane-HCl), MES (2(*N*-morpholino) ethanesulphonic acid), HEPES (*N*-(2-hydroxiethyl) piperazine-*N'*-(sulphonic acid), sodium phosphate, sodium citrate, and sodium cacodylate are the common buffering systems. The precipitants include PEG (polyethylene glycol) with molecular weight ranging from 100 to 8000, PEG MME (monomethyl ether) 350, PEG MME 2000, PEG MME 5000, MPD (2-methylpentan-2,4-diol), sodium formate, ammonium acetate, ammonium hydroxide, sodium citrate, sodium acetate, calcium chloride, and potassium phosphate. PEG (Table 15.1), MPD (Table 15.2), and ammonium sulphate (Table 15.3) are most widely used and each has produced crystals of complexes containing DNA with overhanging as well as blunt ends. Salts other than ammonium sulphate have been used successfully but less frequently (Table 15.4). The majority of the cocrystals were grown in the presence of a divalent ($Mg^{2+}$, $Ca^{2+}$, $Zn^{2+}$, $Ba^{2+}$, $Cd^{2+}$) or a polycationic (spermine) additive. The largest number of cocrystals has grown from PEG. Although some cocrystals have grown at pH as low as 3.5 and as high as 9.0, the most of the complexes appear to have preference for pH between 5.5 and 8.5 (Fig. 15.3).

The technique of choice for screening crystallization conditions is the vapour diffusion 'hanging drop' described in Chapter 3. In this method drops containing different concentrations of protein–DNA mixtures, buffer, additive, and precipitant are suspended from a siliconized coverslips placed over sealed reservoirs containing different precipitant solutions. A very large number of parameters such a pH, temperature, precipitant concentration, and additives can be explored using this method as the volume of each drop may be as little as 1–2 μl (see Protocol 15.4).

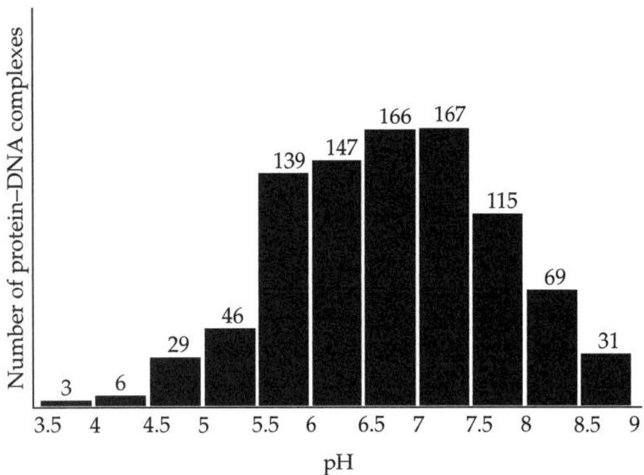


**Figure 15.3** A histogram of pH distribution used for cocrystallization of DNA and proteins. (This figure is based on the information available from the Protein Data Bank.)

### 15.2.5 Analysis and characterization of cocrystals

Visual examination of crystals using a light microscope does not indicate whether the crystals consist of only the protein or the protein–DNA complex. Therefore, the crystals are washed free of any uncrystallized DNA and protein several times with a solution containing the precipitant and any additives, etc. at the concentration and pH used for growing crystals (mother liquor). Finally, the crystals are separated from the mother liquor by microcentrifugation, dissolved in a suitable buffer, and analysed biochemically. The protein content is determined by SDS polyacrylamide gel electrophoresis, the protein concentration by BIO-RAD assay, and amino acid composition by mass-spectroscopy. The DNA can be detected by staining the gel with ethidium bromide or methylene blue (Jordan *et al.*, 1985), whereas

## Protocol 15.4 DNA–protein cocrystallization

### Materials

- Protein solution at 5–10 mg/ml in a suitable buffer; sodium azide may be added as a preservative
- Purified oligonucleotides (from 1 or 10-μmol scale synthesis)
- Precipitants: polyethylene glycol (PEG) 200–8000, 2-methyl-2,4-pentanediol (MPD), ammonium sulphate, calcium chloride, ammonium acetate, and other salts (see Tables 15.1–15.4 for more information)
- Additives such as spermine, spermidine, $Mg^{2+}$, $Ca^{2+}$, $Zn^{2+}$, $Ba^{2+}$, and $Cd^{2+}$
- 24-well tissue culture plates (Linbro, Flow Laboratories, Inc.), XRL plates (Molecular Dimensions Ltd) or VDX crystallization plates (Hampton Research Ltd)
- Siliconized, 22-mm circular microscope coverslips (Hampton Research Ltd)
- Vacuum grease, white petroleum jelly or Vaseline
- Microscope (maximum magnification 40×) with analyser and polarizer
- Incubators and a cold room

### Method

1. Prepare a series of precipitant solutions at various concentration and pH 3.5–9.0 in an appropriate buffer containing the additives to be tested.
2. Mix appropriate volumes of the protein and the DNA.
3. Grease rims of the crystallization plate.
4. Pipette 0.5 or 1 ml of precipitant solution into the well.
5. Place1–2 μl protein–DNA solution on a clean 22-mm siliconized coverslip.
6. Add 1–2 μl reservoir solution to the drop.
7. Invert the coverslip, place on the well, and seal.
8. Place one set of plates at 4°C and the other at 18 or 22°C.
9. Examine the drops for the presence of crystals using a microscope.

the protein is detected by staining with Coomasie Blue (Laemmli, 1970). Both qualitative and quantitative analysis of DNA can also be carried out by reverse-phase HPLC. For this purpose the washed crystals are dissolved in a suitable buffer and boiled to dissociate the DNA from the protein before loading onto the column. Absorption is monitored to obtain an elution profile which is compared with the elution profile of the standard oligomer in order to calculate the concentration of DNA (Brown and Freemont, 1996).

Diffraction properties of crystals are determined by X-ray analysis which is covered in Chapters 4 and 5. Imperfections within the crystal are indicated by high mosaicity exhibited by broadening of diffraction spots and diffuse scattering. Prolonged exposure of protein and protein–DNA crystals to X-rays causes loss of diffraction due to radiation damage.

### 15.2.6 Flash-freezing and cryocooling of cocrystals

It is often necessary to flash-freeze crystals in order to prevent radiation damage during preliminary characterization and data collection. Conducting X-ray diffraction experiments at cryogenic temperatures usually results in obtaining better quality data (see Protocol 15.5).

The crystal is passed through a drop of cryoprotectant to prevent formation of ice crystals and flash-cooled in a nitrogen stream at 100 K. Polyethylene glycol, glycerol, and MPD are the commonly used cryoprotectants. The exact concentration and the type of the cryoprotectant required to stabilize a given protein–DNA cocrystal depend on the crystallization conditions. For crystals grown from MPD or PEG, increments in concentrations of these compounds should be tested in a stepwise manner to avoid cracking of the crystals due to changes in osmotic pressure. Raising concentrations of PEGs of molecular weight greater than 4000 to levels that are effective as cryoprotectants is not always possible in the case of crystals grown from very low concentrations of these compounds. Increased concentrations of glycerol in the stabilizing solutions should be tested for these crystals. Alternatively, crystals can be grown from precipitants such as PEGs and salts in the presence of glycerol. A crystal flash-frozen under cryogenic conditions can be stored indefinitely in a liquid nitrogen dewar and transported to synchrotron stations. The methods

#### Protocol 15.5 Flash-freezing of cocrystals grown from PEGs or salts

**Materials**

X-ray generator
Goniometer head
Area detector
Cryocooling system
Siliconized, 22-mm circular microscope coverslips (Hampton Research Ltd)
Pins and loops (Hampton Research Ltd or Molecular Dimensions Ltd)
Microscope (maximum magnification 40×) with analyser and polarizer
Computer and molecular graphic system (See Chapter 5 for details)
Cryoprotectant solutions containing all the components of crystallization solution plus 20%, 30%, and 35% glycerol

**Method**

1. Mount a pin with a loop on the goniometer head attached to the gonoistat.
2. Centre the loop.
3. Choose a tray with crystals of approximately the same size and morphology.
4. Prepare a set of cryoprotectant solutions.
5. Place 10–50 μl cryoprotectant on a siliconized coverslip.
6. Stir the liquid around the crystal upwards with a loop.
7. Pick a crystal with the loop, pass through a solution of cryoprotectant and very quickly freeze the crystal in the nitrogen stream above the goniometer and remount the loop on the goniometer keeping the crystal in the nitrogen stream.
8. Centre the crystal.
9. Collect several frames of diffraction data.
10. Repeat Steps 1 to 7 using the rest of the cryoprotectants.
11. Compare frames for ice rings, mosaicity, and resolution.
12. Choose the cryoprotectant that gives no ice rings, the lowest mosaicity, and highest resolution.

for cooling and mounting protein crystals described by Pflugrath (2004) are also applicable to cooling and mounting of protein–DNA crystals.

The X-ray crystallographic techniques for characterizing crystals, data collection, and solving structures of proteins and protein–DNA complexes have been discussed elsewhere in this book.

## References


Aggarwal, A. K. (1990). Crystallization of DNA binding proteins with oligodeoxynucleotides. *Methods* **1**, 83–90.

Aggarwal, A. K., Rodgers, D. W., Drottar, M., Ptashne, M. and Harrison, S. C. (1988). Recognition of a DNA operator by the repressor of phage 434: a view at high resolution. *Science* **242**, 899–907.

Anderson, J. E., Ptashne, M. and Harrison, S. C. (1984). Cocrystals of the DNA-binding domain of phage 434 repressor and a synthetic phage 434 operator. *Proc. Natl. Acad. Sci. USA* **81**, 1307–1308.

Anderson, J. E., Ptashne, M. and Harrison, S. C. (1987). Structure of the repressor-operator complex of bacteriophage 434. *Nature* **326**, 846–852.

Arnold, E., Jacobo-Molina, A., Nanni, R. G., Williams, R. L., Lu, X., Ding, J., Clark, A. D., Jr., Zhang, A., Ferris, A. L., Clark, P., Hizi, H. and Hughes, S. H. (1992). Structure of HIV-1 reverse transcriptase/DNA complex at 7 Å resolution showing active site locations. *Nature* **357**, 85–89.

Balendiran, K., Bonventre, J., Knott, R., Jack, W., Benner, J., Schildkraut, I. and Anderson, J. E. (1994). Expression, purification, and crystallization of restriction endonuclease PvuII with DNA containing its recognition site. *Proteins* **19**, 77–79.

Beamer, L. J. and Pabo, C. O. (1992). Refined 1.8 Å crystal structure of the λ repressor-operator complex. *J. Mol. Biol.* **227**, 177–196.

Bell, C. E. and Lewis, M. (2000). A closer view of the conformation of the Lac repressor bound to operator. *Nat. Struct. Biol.* **7**, 209–214.

Benoff, B., Yang, H., Lawson, C. L., Parkinson, G., Liu, J., Blatter, E., Ebright, Y. W., Berman, H. M. and Ebright, R. H. (2002). Structural basis of transcription activation: the CAP-αCTD-DNA complex. *Science* **297**, 1562–1566.

Berkowitz, B., Huang, D.-B., Chen-Park, F. E., Sigler, P. B. and Ghosh, G. (2002). The X-ray crystal structure of the NF-κB p50.p65 heterodimer bound to the interferon β – κB site. *J. Biol. Chem.* **277,** 24694–24700.

Bleichenbacher, M., Tan, S. and Richmond, T. J. (2003). Novel interactions between the components of human and yeast TFIIA/TBP/DNA complexes. *J. Mol. Biol.* **332,** 783–793.

Brennan, R. G., Roderick, S. L., Takeda, Y. and Matthews, B. W. (1990). Protein-DNA conformational changes in the crystal structure of a λ Cro-operator complex. *Proc. Natl. Acad. Sci. USA* **87**, 8165–8169.

Brennan, R. G., Takeda, Y., Kim, J., Anderson, W. F. and Matthews, B. W. (1986). Crystallization of a complex of Cro repressor with a 17 base-pair operator. *J. Mol. Biol.* **188**, 115–118.

Brown, D. G. and Freemont, P. S. (1996). Crystallography in the study of protein-DNA interactions. *Method Mol. Biol.* **56**, 293–318.

Büning, H., Gärtner, U., von Schack, D., Baeuerle, P.A and Zorbas, H. (1996). The histidine tail of recombinant DNA binding proteins may influence the quality of interaction with DNA. *Anal. Biochem.* **234**, 227–230.

Burgess, R. R. (1991). The use of polyethyleneimine in the purification of DNA binding proteins. *Method Enzymol.* **208**, 3–10.

Carey, J., Combatti, N., Lewis, D. A.E. and Lawson, C. L. (1993). Cocrystals of *Escherichia coli* Trp repressor bound to an alternative operator DNA sequence. *J. Mol. Biol.* **234**, 496–498.

Caruthers, M. H. (1985). Gene synthesis machines: DNA chemistry and its uses. *Science* **230**, 281–285.

Chandrasegaran, S., Smith, H. O., Amzel, M. L. and Ysern, X. (1986). Preliminary X-ray diffraction analysis of HhaII endonuclease-DNA cocrystals. *Proteins* **1**, 263–266.

Chen, L., Glover, J. N.M., Hogan, P. G., Rao, A. and Harrison, S. C. (1998). Structure of the DNA-binding domains from NFAT, Fos and Jun bound specifically to DNA. *Nature* **392**, 42–48.

Cheng, X. D., Balendiran, K., Schildkraut, I. and Anderson, J. E. (1994). Structure of PvuII endonuclease with cognate DNA. *EMBO J.* **13**, 3927–3935.

Cho, Y. J., Gorina, S., Jeffrey, P. D. and Pavletich, N. P. (1994). Crystal structure of a p53 tumor suppressor-DNA complex: understanding tumorigenic mutations. *Science* **265**, 346–355.

Clark, K. L., Halay, E. D., Lai, E. and Burley, S. K. (1993). Co-crystal structure of the HNF-3/fork head DNA-recognition motif resembles histone H5. *Nature* **364**, 412–420.

Cramer, P., Varrot, A., Barillas-Mury, C., Kafatos, F. C. and Müller, C. W. (1999). Structure of the specificity domain of the Dorsal homologue Gambif1 bound to DNA. *Structure (London)* **7**, 841–852.

Davey, C. A., Sargent, D. F., Luger, K., Maeder, A. W. and Richmond, T. J. (2002). Solvent mediated interactions in the structure of the nucleosome core particle at 1.9 Å resolution. *J. Mol. Biol.* **319**, 1097–1113.

Davies, D. R., Interthal, H., Champoux, J. J. and Hol, W. G.J. (2003). Crystal structure of a transition state mimic for Tdp1 assembled from vanadate, DNA, and a topoisomerase I-derived peptide. *Chem. Biol.* **10,** 139–147.

Ellenberger, T., Fass, D., Arnaud, M. and Harrison, S. C. (1994). Crystal structure of transcription factor E47: E-box recognition by a basic region helix-loop-helix dimer. *Genes Devel.* **8**, 970–980.

Ellenberger, T. E., Brandl, C. J., Struhl, K. and Harrison, S. C. (1992). The GCN4 basic region leucine zipper binds DNA as a dimer of uninterrupted αhelices: crystal structure of the protein-DNA complex. *Cell* **71**, 1223–1237.

Fairall, L., Schwabe, J. W. R., Chapman, L., Finch, J. T. and Rhodes, D. (1993). The crystal structure of a two zinc-finger peptide reveals an extension to the rules for zinc-finger/DNA recognition. *Nature* **366,** 483–487.

Feng, J.-A., Simon, M., Mack, D. P., Dervan, P. B., Johnson, R. C. and Dickerson, R. E. (1993). Crystallization and preliminary X-ray analysis of the DNA binding domain of the Hin recombinase with its DNA binding site. *J. Mol. Biol.* **232**, 982–986.

Ferré-D'Amaré, A. R., Pognonec, P., Roeder, R. G. and Burley, S. K. (1994). Structure and function of the b/HLH/Z domain of USF. *EMBO J.* **13,** 180–189.

Ferré-d'Amaré, A. R., Prendergast, G. C., Ziff, E. B. and Burley, S. K. (1993). Recognition by Max of its cognate DNA through a dimeric b/HLH/Z domain. *Nature* **363**, 38–45.

Franklin, M. C., Wang, J. and Steitz, T. A. (2001). Structure of the replicating complex of a pol α family DNA polymerase. *Cell* **105,** 657–667.

Freemont, P. S., Friedman, J. M., Beese, L., Sanderson, M. R. and Steitz, T. A. (1988). Cocrystal structure of an editing complex of Klenow fragment with DNA. *Proc. Natl. Acad. Sci. USA* **85**, 8924–8928.

Gadgil, H., Oak, S. A. and Jarrett, H. W. (2001). Affinity purification of DNA-binding proteins. *J. Biochem. Biophys. Meth.* **49**, 607–624.

Gao, Y. G., Su, S. Y., Robinson, H., Padmanabhan, S., Lim, L., McCrary, B. S., Edmondson, S. P., Shriver, J. W. and Wang, A. H.-J. (1998). The crystal structure of the hyperthermophile chromosomal protein Sso7d bound to DNA. *Nat. Struct. Biol.* **5**, 782–786.

Ghosh, G., Van Duyne, G., Ghosh, S. and Sigler, P. B. (1995). Structure of NF-κB p50 homodimer bound to a κB site. *Nature* **373,** 303–310.

Glover, J. N. M. and Harrison, S. C. (1995). Crystal structure of the heterodimeric bZIP transcription factor c-Fos-c-Jun bound to DNA. *Nature* **373**, 257–261.

Gopaul, D. N., Guo, F. and Van Duyne, G. D. (1998). Structure of the Holliday junction intermediate in Cre-loxP site-specific recombination. *EMBO J.* **17**, 4175–4187.

Grable, J., Frederick, C. A., Samudzi, C., Jen-Jacobsen, L., Lesser, D., Greene, P., Boyer, H. W., Itakura, K. and Rosenberg, J. M. (1984). Two-fold symmetry of crystalline DNA-EcoRI endonuclease recognition complexes. *J. Biomol. Struct. Dynam.* **1**, 1149–1160.

Guo, F., Gopaul, D. N. and Van Duyne, G. D. (1997). Structure of Cre recombinase complexed with DNA in a site-specific recombination synapse. *Nature* **389**, 40–46.

Hassler, M. and Richmond, T. J. (2001). The B-box dominates SAP-1-SRF interactions in the structure of the ternary complex. *EMBO J.* **20**, 3018–3028.

Hegde, R. S., Grossman, S. R., Laimins, L. A. and Sigler, P. B. (1992). Crystal structure at 1.7 Å of the bovine papillomavirus-1 E2 DNA-binding domain bound to its DNA target. *Nature* **359**, 505–512.

Hirsch, J. A. and Aggarwal, A. K. (1995). Purification, crystallization, and preliminary X-ray diffraction analysis of even-skipped homeodomain complexed to DNA. *Proteins* **21**, 268–271.

Horton, N. C., Dorner, L. F., Schildkraut, I. and Perona, J. J. (1999). Crystallization and preliminary diffraction analysis of the HincII restriction endonuclease-DNA complex. *Acta Crystallogr. D* **55**, 1943–1945.

Jacobo-Molina, A., Clark, A. D., Jr., Williams, R. L., Nanni, R. G., Clark, P., Ferris, A. L., Hughes, S. H. and Arnold, E. (1991). Crystals of a ternary complex of human immunodeficiency virus type 1 reverse transcriptase with a monoclonal antibody Fab fragment and double-stranded DNA diffract X-rays to 3.5- Å resolution. *Proc. Natl. Acad. Sci. USA* **88**, 10895–10899.

Jin, L., Sliz, P., Chen, L., Macián, F., Rao, A., Hogan, P. G. and Harrison, S. C. (2003). An asymmetric NFAT1 dimer on a pseudo-palindromic κB-like DNA site. *Nat. Struct. Biol.* **10**, 807–811.

Joachimiak, A., Marmorstein, R. Q., Schevitz, R. W., Mandecki, W., Fox, J. L. and Sigler, P. B. (1987). Crystals of the trp repressor-operator complex suitable for X-ray diffraction analysis. *J. Biol. Chem.* **262**, 4917–4921.

Jordan, S. R. and Pabo. C. O. (1988). Structure of the lambda complex at 2.5 Å resolution: details of the repressor-operator interactions. *Science* **242**, 893–899.

Jordan, S. R., Whitcombe, T. V., Berg, J. M. and Pabo, C. O. (1985). Systematic variation in DNA length yields highly ordered repressor-operator cocrystals. *Science* **230**, 1383–1385.

Joyce, C. M. and Grindley, N. D. F (1983). Construction of a plasmid that overproduces the large proteolytic fragment (Klenow fragment) of DNA polymerase I of *Escherichia coli*. *Proc. Natl. Acad. Sci. USA* **80**, 1830–1834.

Juo, Z. S., Kassavetis, G. A., Wang, J., Geiduschek, E. P. and Sigler, P. B. (2003). Crystal structure of a transcription factor IIIB core interface ternary complex. *Nature* **422**, 534–539.

Karas, M. and Hillenkamp, F. (1988). Laser desorption ionization of proteins with molecular masses exceeding 10,000 daltons. *Anal. Chem.* **60**, 2299–2301.

Keller, W., König, P. and Richmond, T. J. (1995). Crystal structure of a bZIP/DNA complex at 2.2 Å: determinants of DNA specific recognition. *J. Mol. Biol.* **254**, 657–667.

Kim, Y., Geiger, J. H., Hahn, S. and Sigler, P. B. (1993). Crystal structure of a yeast TBP/TATA-box complex. *Nature* **365**, 512–520.

Kim, Y., Grable, J. C., Love, R., Greene, P. J. and Rosenberg, J. M. (1990). Refinement of EcoRI endonuclease crystal structure: a revised protein chain tracing. *Science* **249**, 1307–1309.

Kim. J., Nikolov, D. B. and Burley, S. K. (1993). Co-crystal structure of TBP recognizing the minor groove of a TATA element. *Nature* **365**, 520–527.

Kissinger, C. R., Liu, B., Martin-Blanco, E., Kornberg, T. B. and Pabo, C. O. (1990). Crystal structure of an engrailed homeodomain-DNA complex at 2.8 Å resolution: a framework for understanding homeodomain-DNA interactions. *Cell* **63**, 579–590.

Klemm, J. D., Rould, M. A., Aurora, R., Herr, W. and Pabo, C. O. (1994). Crystal structure of the Oct-1 POU domain bound to an octamer site: DNA recognition with tethered DNA-binding modules. *Cell* **77**, 21–32.

Klimasauskas, S., Kumar, S., Roberts, R. J. and Cheng, X. (1994). HhaI methyltransferase flips its target base out of the DNA helix. *Cell* **76**, 357–369.

Kodandapani, R., Pio, F., Ni, C. Z., Piccialli, G., Klemsz, M., McKercher, S., Maki, R. A. and Ely, K. R. (1996). A new pattern for helix-turn-helix recognition revealed by the PU.1 ETS-domain-DNA complex. *Nature* **380**, 456–460.

Komori, H., Matsunaga, F., Higuchi,Y., Ishiai, M., Wada, C. and Miki, K. (1999). Crystal structure of a prokaryotic replication initiator protein bound to DNA at 2.6 Å resolution. *EMBO J.* **18**, 4597–4607.

König, P., Giraldo, R., Chapman, L. and Rhodes, D. (1996). The crystal structure of the DNA-binding domain of yeast RAP1 in complex with telomeric DNA. *Cell* **85**, 125–136.

Kosa, P. F., Ghosh, G., DeDecker, B. S. and Sigler, P. B. (1997). The 2.1-Å crystal structure of an archaeal preinitiation complex: TATA-box-binding protein/transcription factor (II)B core/TATA-box. *Proc. Nat. Acad. Sci. USA* **94**, 6042–6047.

Laemmli, U. (1970). Cleavage of structural proteins during the assembly of the head of bacteriophage T4. *Nature* **227**, 680–685.

Lahm, A. and Suck, D. (1991). DNase I-induced DNA conformation. 2 Å structure of a DNase I-octamer complex. *J. Mol. Biol.* **221**, 645–667.

Li, T., Jin, Y., Vershon, A. K. and Wolberger, C. (1998). Crystal structure of the MATa1/MATα2 homeodomain heterodimer in complex with DNA containing an A-tract. *Nucleic Acids Res.* **26**, 5707–5718.

Li, T., Stark, M., Johnson, A. D. and Wolberger, C. (1995). Crystallization and preliminary X-ray diffraction studies of an a1/α2/DNA ternary complex. *Proteins* **21**, 161–164.

Littlefield, O., Korkhin, Y. and Sigler, P. B. (1999). The structural basis for the oriented assembly of a TBP/TFB/promoter complex. *Proc. Nat. Acad. Sci. USA* **96**, 13668–13673.

Liu, B., Kissinger, C. R., Pabo, C. O., Martin-Blanco, E. and Kornberg, T. B. (1990). Crystallization and preliminary X-ray diffraction studies of the engrailed homeodomain and of an engrailed homeodomain/DNA complex. *Biochem. Biophys. Res. Commun.* **171**, 257–259.

Luger, K., Mäder, A. W., Richmond, R. K., Sargent, D. F. and Richmond, T. J. (1997). Crystal structure of the nucleosome core particle at 2.8 Å resolution. *Nature* **389**, 251–260.

Luisi, B. F., Xu, W. X., Otwinowski, Z., Freedman, L. P., Yamamoto, K. R. and Sigler, P. B. (1991). Crystallographic analysis of the interaction of the glucocorticoid receptor with DNA. *Nature* **352**, 497–505.

Lukacs, C. M., Kucera, R., Schildkraut, I. and Aggarwal, A. K. (2000). Understanding the immutability of restriction enzymes: crystal structure of BglII and its DNA substrate at 1.5 Å resolution. *Nat. Struct. Biol.* **7**, 134–140.

Lukens, A., King, D. and Marmorstein, R. (2000). Structure of HAP1-PC7 bound to DNA: implications for DNA recognition and allosteric effects of DNA-binding on transcriptional activation. *Nucleic Acids Res.* **28**, 3853–3863.

Luscombe, N. M., Austin, S. E., Berman, H. M. and Thornton, J. M. (2000). An overview of the structures of protein-DNA complexes. *Genome Biol.* **1**, reviews 001.1–001.37.

Ma, P. C. M., Rould, M. A., Weintraub, H. and Pabo, C. O. (1994). Crystal structure of MyoD bHLH domain-DNA complex: perspectives on DNA recognition and implications for transcriptional activation. *Cell* **77**, 451–459.

Marmorstein, R. and Harrison, S. C. (1994). Crystal structure of a PPR1-DNA complex: DNA recognition by proteins containing a Zn2Cys6 binuclear cluster. *Genes Devel.* **8**, 2504–2512.

Marmorstein, R., Carey, M., Ptashne, M. and Harrison, S. C. (1992). DNA recognition by GAL4: structure of a protein-DNA complex. *Nature* **356,** 408–414.

McClarin, J. A., Frederick, C. A., Wang, B.-C., Greene, P., Boyer, H. W., Grable, J. and Rosenberg, J. M. (1986). Structure of the DNA-EcoRI endonuclease recognition complex at 3 Å resolution. *Science* **234,** 1526–1541.

McPherson, A. (1990). Current approaches to macromolecular crystallization. *Eur. J. Biochem.* **189**, 1–23.

Mondragon, A. and Harrison, S. C. (1991). The phage 434 Cro/OR1 complex at 2.5 Å resolution. *J. Mol. Biol.* **219**, 321–334.

Müller, C. W., Rey, F. A., Sodeoka, M., Verdine, G. L. and Harrison, S. C. (1995). Structure of the NF-κB p50 homodimer bound to DNA. *Nature* **373**, 311–317.

Nair, D. T., Johnson, R. E., Prakash, S., Prakash, L. and Aggarwal, A. K. (2004). Replication by human DNA polymerase-iota occurs by Hoogsteen base-pairing. *Nature* **430**, 377–380.

Nelson, H. C. and Sauer, R. T. (1986). Interaction of mutant lambda repressors with operator and non-operator DNA. *J. Mol. Biol.* **192**, 27–38.

Newman, M., Strzelecka, T., Dorner, L. F., Schildkraut, I. and Aggarwal, A. K. (1995). Structure of BamHI endonuclease bound to DNA: partial folding and unfolding on DNA binding. *Science* **269**, 656–663.

Nikolov, D. B., Chen, H., Halay, E. D., Usheva, A. A., Hisatake, K., Lee, D. K., Roeder, R. G. and Burley, S. K. (1995). Crystal structure of a TFIIB-TBP-TATA-element ternary complex. *Nature* **377**, 119–128.

Nolte, R. T., Conlin, R. M., Harrison, S. C. and Brown, R. S. (1998). Differing roles for zinc fingers in DNA recognition: structure of a six-finger transcription factor IIIA complex. *Proc. Nat. Acad. Sci. USA* **95**, 2938–2943.

Obmolova, G., Ban, C., Hsieh, P. and Yang, W. (2000). Crystal structures of mismatch repair protein MutS and its complex with a substrate DNA. *Nature* **407**, 703–710.

Otwinowski, Z., Schevitz, R. W., Zhang, R.-G., Lawson, C. L., Joachimiak, A., Marmorstein, R. Q., Luisi, B. F. and Sigler, P. B. (1988). Crystal structure of Trp repressor/operator complex at atomic resolution. *Nature* **335**, 321–329.

Pace, H. C., Lu, P. and Lewis, M. (1990). lac repressor: crystallization of intact tetramer and its complexes with inducer and operator DNA. *Proc. Natl. Acad. Sci. USA* **87**, 1870–1873.

Pavletich, N. P. and Pabo, C. O. (1991). Zinc finger-DNA recognition: crystal structure of a Zif268-DNA complex at 2.1 Å. *Science* **252,** 809–817.

Pavletich, N. P. and Pabo, C. O. (1993). Crystal structure of a five-finger GLI-DNA complex: new perspectives on zinc fingers. *Science* **261,** 1701–1707.

Pflugrath, J. W. (2004). Macromolecular cryocrystallography – methods for cooling and mounting protein crystals at cryogenic temperatures. *Methods* **34**, 415–425.

Pio, F., Kodandapani, R., Ni, C. Z., Shepard, W., Klemsz, M., McKercher, S. R., Maki, R. A. and Ely, K. R. (1996). New insights on DNA recognition by ETS proteins from the crystal structure of the PU.1 ETS domain-DNA complex. *J. Biol. Chem.* **271**, 23329–23337.

Rastinejad, F., Wagner, T., Zhao, Q. and Khorasanizadeh, S. (2000). Structure of the RXR-RAR DNA-binding complex on the retinoic acid response element DR1. *EMBO J.* **19**, 1045–1054.

Raumann, B. E., Rould, M. A., Pabo, C. O. and Sauer, R. T. (1994). DNA recognition by β-sheets in the Arc repressor-operator crystal structure. *Nature* **367**, 754–757.

Redinbo, M. R., Champoux, J. J. and Hol, W. G.J. (2000). Novel insights into catalytic mechanism from a crystal structure of human topoisomerase I in complex with DNA. *Biochemistry* **39**, 6832–6840.

Redinbo, M. R., Stewart, L., Kuhn, P., Champoux, J. J. and Hol, W. G.J. (1998). Crystal structures of human topoisomerase I in covalent and noncovalent complexes with DNA. *Science* **279**, 1504–1513.

Reinisch, K. M., Chen, L., Verdine, G. L. and Lipscomb, W. N. (1994). Crystallization and preliminary crystallographic analysis of a DNA (cytosine-5)-methyltransferase from Haemophilus aegyptius bound covalently to DNA. *J. Mol. Biol.* **238**, 626–629.

Reinisch, K. M., Chen, L., Verdine, G. L. and Lipscomb, W. N. (1995). The crystal structure of HaeIII methyltransferase convalently complexed to DNA: an extrahelical cytosine and rearranged base pairing. *Cell* **82,** 143–153.

Rodgers, D. W. and Harrison, S. C. (1993). The complex between phage 434 repressor DNA-binding domain and operator site OR3: structural differences between consensus and non-consensus half-sites. *Structure* **1**, 227–240.

Sarafianos, S. G., Clark, A. D., Jr., Das, K., Tuske, S., Birktoft, J. J., Ilankumaran, P., Ramesha, A. R., Sayer, J. M., Jerina, D. M., Boyer, P. L., Hughes, S. H. and Arnold, E. (2002). Structures of HIV-1 reverse transcriptase with pre- and post-translocation AZTMP-terminated DNA. *EMBO J.* **21**, 6614–6624.

Schultz, S. C., Shields, G. C. and Steitz, T. A. (1990). Crystallization of *Escherichia coli* catabolite gene activator protein with its DNA binding site. The use of modular DNA. *J. Mol. Biol.* **211**, 159–166.

Schultz, S. C., Shields, G. C. and Steitz, T. A. (1991). Crystal structure of a CAP-DNA complex: the DNA is bent by 90 degrees. *Science* **253,** 1001–1007.

Schwabe, J. W. R., Chapman, L., Finch, T. and Rhodes, D. (1993). The crystal structure of the estrogen receptor DNA-binding domain bound to DNA: how receptors discriminate between their response elements. *Cell* **75**, 567–578.

Schwartz, T., Rould, M. A., Lowenhaupt, K., Herbert, A. and Rich, A. (1999a). Crystal structure of the Zα domain of the human editing enzyme ADAR1 bound to left-handed Z-DNA. *Science* **284**, 1841–1845.

Schwartz, T., Shafer, K., Lowenhaupt, K., Hanlon, E., Herbert, A. and Rich, A. (1999b). Crystallization and preliminary studies of the DNA-binding domain Zα from ADAR1 complexed to left-handed DNA. *Acta Crystallogr. D* **55**, 1362–1364.

Shamoo, Y. and Steitz, T. A. (1999). Building a replisome from interacting pieces: sliding clamp complexed to a peptide from DNA polymerase and a polymerase editing complex. *Cell* **99**, 155–166.

Shimon, L. J. W. and Harrison, S. C. (1993). The phage 434 OR2/R1-69 complex at 2.5 Å resolution. *J. Mol. Biol.* **232**, 826–838.

Singleton, M. R., Scaife, S. and Wigley, D. B. (2001). Structural analysis of DNA replication fork reversal by RecG. *Cell* **107**, 79–89.

Somers, W. S. and Phillips, S. E.V. (1992). Crystal structure of the met repressor-operator complex at 2.8 Å resolution reveals DNA recognition by β-strands. *Nature* **359**, 387–393.

Stewart, L., Redinbo, M. R., Qiu, X., Hol, W. G.J. and Champoux, J. J. (1998). A model for the mechanism of human topoisomerase I. *Science* **279**, 1534–1541.

Strzelecka, T., Newman, M., Dorner, L. F., Knott, R., Schildkraut, I. and Aggarwal, A. K. (1994). Crystallization and preliminary X-ray analysis of restriction endonuclease BamHI-DNA complex. *J. Mol. Biol.* **239**, 430–432.

Suck, D., Lahm, A. and Oefner, C. (1988). Structure refined to 2 Å of a nicked DNA octanucleotide complex with DNase I. *Nature* **332**, 464–468.

Tahirov, T. H., Inoue-Bungo, T., Morii, H., Fujikawa, A., Sasaki, M., Kimura, K., Shiina, M., Sato, K., Kumasaka, T., Yamamoto, M., Ishii, S. and Ogata, K. (2001a). Structural analyses of DNA recognition by the AML1/Runx-1 Runt domain and its allosteric control by CBFβ. *Cell* **104**, 755–767.

Tahirov, T. H., Inoue-Bungo, T., Sasaki, M., Fujikawa, A., Kimura, K., Sato, K., Adachi, S., Kamiya, N. and Ogata, K. (2001b). Crystallization and preliminary X-ray analysis of the C/EBP β C-terminal region in complex with DNA. *Acta Crystallogr. D***57**, 854–856.

Tahirov, T. H., Inoue-Bungo, T., Sasaki, M., Shiina, M., Kimura, K., Sato, K., Kumasaka, T., Yamamoto, M., Kamiya, N. and Ogata, K. (2001c). Crystallization and preliminary X-ray analyses of quaternary, ternary and binary protein-DNA complexes with involvement of AML1/Runx-1/CBFα Runt domain, CBFβ and the C/EBPβ bZip region. *Acta Crystallogr. D***57**, 850–853.

Tamulaitiene, G., Grazulis, S., Janulaitis, A., Janowski, R., Bujacz, G. and Jaskolski, M. (2004). Crystallization and preliminary crystallographic studies of a bifunctional restriction endonuclease Eco57I. *Biochim. Biophy. Acta* **1698**, 251–254.

Tanaka, Y., Nureki, O., Kurumizaka, H., Fukai, S., Kawaguchi, S., Ikuta, M., Iwahara, J., Okazaki, T. and Yokoyama, S. (2001). Crystal structure of the CENP-B protein-DNA complex: the DNA-binding domains of CENP-B induce kinks in the CENP-B box DNA. *EMBO J.* **20**, 6612–6618.

Tsai, F. T.F. and Sigler, P. B. (2000). Structural basis of preinitiation complex assembly on human Pol II promoters. *EMBO J.* **19**, 25–36.

Tsutakawa, S. E., Jingami, H. and Morikawa, K. (1999). Recognition of a TG mismatch: the crystal structure of very short patch repair endonuclease in complex with a DNA duplex. *Cell* **99**, 615–623.

Velankar, S. S., Soultanas, P., Dillingham, M. S., Subramanya, H. S. and Wigley, D. B. (1999). Crystal structures of complexes of PcrA DNA helicase with a DNA substrate indicate an inchworm mechanism. *Cell* **97**, 75–84.

Viadiu, H., Kucera, R., Schildkraut, I. and Aggarwal, A. K. (2000). Crystallization of restriction endonuclease BamHI with nonspecific DNA. *J. Struc. Biol.* **130**, 81–85.

Viadiu, H., Vanamee, E. S., Jacobson, E. M., Schildkraut, I. and Aggarwal, A. K. (2003). Crystallization of restriction endonuclease SfiI in complex with DNA. *Acta Crystallogr. D* **59**, 1493–1495.

Wah, D. A., Hirsch, J. A., Dorner, L. F., Schildkraut, I. and Aggarwal, A. K. (1997). Structure of the multimodular endonuclease FokI bound to DNA. *Nature* **388**, 97–100.

Weston, S. A., Lahm, A. and Suck, D. (1992). X-ray structure of the DNase I-d(GGTATACC)2 complex at 2.3 Å resolution. *J. Mol. Biol.* **226**, 1237–1256.

Winkler, F. K., Banner, D. W., Oefner, C., Tsernoglou, D., Brown, R. S., Heathman, S. P., Bryan, R. K., Martin, P. D., Petratos, K. and Wilson, K. S. (1993). The crystal structure of EcoRV endonuclease and of its complexes with cognate and non-cognate DNA fragments. *EMBO J.* **12**, 1781–1795.

Winkler, F. K., d'Arcy, A., Blöcker, H., Frank, R. and van Boom, J. H. (1991). Crystallization of complexes of EcoRV endonuclease with cognate and non-cognate DNA fragments. *J. Mol. Biol.* **217**, 235–238.

Wolberger, C. and Harrison, S. C. (1987). Crystallization and X-ray diffraction studies of a 434 Cro-DNA complex. *J. Mol. Biol.* **196**, 951–954.

Wolberger, C., Dong, Y., Ptashne, M. and Harrison, S. C. (1988). Structure of a phage 434 Cro/DNA complex. *Nature* **335**, 789–795.

Wolberger, C., Pabo, C. O., Vershon, A. K. and Johnson, A. D. (1991a). Crystallization and preliminary X-ray diffraction studies of a MAT α2-DNA complex. *J. Mol. Biol.* **217**, 11–13.

Wolberger, C., Vershon, A. K., Liu, B., Johnson, A. D. and Pabo, C. O. (1991b). Crystal structure of a MAT α2 homeodomain-operator complex suggests a general model for homeodomain-DNA interactions. *Cell* **67**, 517–528.

Xu, W. G., Rould, M. A., Jun, S., Desplan, C. and Pabo, C. O. (1995). Crystal structure of a paired domain-DNA complex at 2.5 Å resolution reveals structural basis for Pax developmental mutations. *Cell* **80**, 639–650.

Yang, W. and Steitz, T. A. (1995). Crystal structure of the site-specific recombinase γδ resolvase complexed with a 34 bp cleavage site. *Cell* **82**, 193–207.

Young, T.-S., Modrich, P., Beth, A., Jay, E. and Kim, S.-H. (1981). Preliminary X-ray diffraction studies of EcoRI restriction endonuclease-DNA complex. *J. Mol. Biol.* **145**, 607–610.

CHAPTER 16

# Virus crystallography

**Elizabeth E. Fry, Nicola G. A. Abrescia, and David I. Stuart**

## 16.1 Introduction

The genetic efficiency of viruses is manifest in their structural simplicity, thus many copies of relatively few proteins (and sometimes lipids) are used to construct a protective layer housing the genetic material and mediating the interactions of the virus with host cells. There is a great deal of variation in the physical appearance of viruses but they can be broadly classified as spherical, rod, or bullet shaped. Whilst protein crystallography was in its infancy as a scientific discipline, crystalline tobacco mosaic virus (TMV) was obtained (Stanley, 1935) and the recording of diffraction patterns from single crystals of tomato bushy stunt virus (TBSV) (Bernal and Fankuchen, 1941) and tobacco necrosis virus (TNV) (Crowfoot and Schmidt, 1945) mark the start of the historical timeline for structural virology (Rossmann, 1998; http://medicine.wustl.edu/virology/). At that time, no protein structures had been solved and viruses had only been visualized by low-resolution electron microscopy (EM). It wasn't until the late 1950s that the first high-resolution protein structure was obtained (Cullis *et al.*, 1962; Kendrew *et al.*, 1960), demonstrating the potential of crystallography, but this was the culmination of many years work and the methodology could not be applied at that time to the far more complex virus diffraction patterns. Nevertheless the X-ray diffraction studies of TMV had, by then, led to a proposal for the disk-like arrangement of protein subunits in this rod-shaped virus (Franklin, 1955). The first high-resolution crystal structure of a virus, tomato bushy stunt virus (an isometric virus), was published in 1978 (Harrison *et al.*, 1978). This achievement was made feasible by a series of improvements: in X-ray beam focusing, data recording and measurement, the development of software to capitalize upon the virus symmetry (all icosahedral particles are composed of 60 identical asymmetric units arranged with 532 symmetry), and advances in computational hardware. The presence of icosahedral symmetry leads to a minimum of five-fold non-crystallographic redundancy, which occurs when a virus lies on a point of 23 crystallographic symmetry. This redundancy is of enormous importance in a crystallographic analysis, providing information which often facilitates the determination of the particle position and then provides powerful constraints for phase improvement and model refinement. Obversely, the high symmetry can confuse the process of space group determination and the application of icosahedral constraints will obscure features that do not obey the capsid symmetry. In the early days of structural virology it became obvious that many viruses were composed of more than 60 subunits, the maximum number that can be accommodated in symmetrical positions in an icosahedron. In 1962, it was proposed that the basic triangular icosahedral building block could be broken down into a number (T) of equilateral subtriangles, leading to an icosahedral shell composed of $60 \times T$ chemically identical subunits which might theoretically make similar chemical contacts to form the capsid (Caspar and Klug, 1962). This theory, of so-called quasiequivalence, provides a basis for describing the architecture of many virus particles but breaks down at the level of molecular interactions.

Twenty five years on from the first spherical virus structures, the complexity of the isometric virus particles tackled by crystallography has correspondingly increased from the relatively simple 300 Å diameter $T = 3$ unenveloped single-layer capsids to 700 Å diameter $T = 13$ bilayer particles (Grimes *et al.*, 1998) and even to particles containing lipid membranes (Abrescia *et al.*, 2004; Cockburn *et al.*, 2004). The number of structures available (exceeding 200 PDB entries) now requires systematic organization in databases (website http://viperdb.scripps.edu/) and the information derived from them is driving our understanding of virus evolution (Bamford *et al.*, 2005), as well as assembly, host–cell interaction, host-adaptation, antigenic variation, and providing the underpinning knowledge for novel therapeutic strategies. In this chapter we will discuss all aspects of the methodology used to solve isometric virus structures with separate case study examples for two more demanding examples, the blue tongue virus core (BTV) and bacteriophage PRD1, the first structure of an intact virus with an internal membrane.

## 16.2 Crystallization

Crystallization may be enhanced by the isometric nature of the particles (providing they form a genuinely homogeneous population), thus a genetically homogeneous (cloned), clean, and concentrated virus preparation provides a good starting point. Viruses may be grown in suspension or monolayer cell culture, the latter sometimes requiring a vast number of roller bottles. The key step in purification is usually ultracentrifugation using CsCl or sucrose gradients, taking care to avoid damage through pelleting. Alternatively, media are available for separating high molecular weight species using FPLC/HPLC techniques, for example Hepatitis B virus cores coexist as both $T = 3$ and $T = 4$ particles which require careful separation prior to crystallization (Zlotnick *et al.*, 1999). Delicate viruses and/or those with spikes or protrusions will also require particular care, for example the PRD1 bacteriophage containing a lipid membrane required very rapid purification and concentration (Bamford *et al.*, 2002), as described in the PRD1 case study below. Concentration may be achieved using 100-kDa microconcentrators, using an appropriate dilute buffer (avoiding phosphate gives fewer salt crystals!). Standard spectrophotometric concentration measurements should be corrected for the nucleic acid content; an approximate formula is:

$$\text{Virus concentration (mg/ml)} = (1.55A_{280} - 0.76A_{260}) \times \text{dilution} \qquad (1)$$

where $A_{280}$ and $A_{260}$ are the absorbance readings at 280 and 260 nm, respectively.

High concentrations can often be achieved and, by analogy with protein concentration, it is often appropriate to start crystallization trials with a concentration of ~10 mg/ml; however, rather than be guided by an arbitrary rule-of-thumb, it is more sensible to perform partial concentration and then perform a precrystallization screen (PCT) (Jancarik *et al.*, 2004) to rapidly find the correct ball-park concentration. Do not be put off if the solubility of the virus is limited, we have had success at as low as 2 mg/ml. It can be helpful to check the virus preparation for aggregation prior to crystallization trials – readily detected using electron microscopy or dynamic light scattering. This is one of the major hindrances to crystallization and if present may be overcome by the addition of detergents, salts, or other additives, such as divalent cations, that might interact with the capsid. Commercially available crystallization kits use a sparse matrix to cover a wide variety of conditions (precipitants, salts, detergents, and other additives) and are hence useful starting trials. The virus preparation may be combined with the precipitants in the form of sitting or hanging drops or via microdialysis (sitting drops permit larger drops and facilitate crystal manipulation (Harlos, 1992)). A further check on virus concentration should be made after the first screens have been set up; thus at the optimum concentration there should not be rapid precipitation in more than ~50% of the drops. With a high sample concentration it may be useful to dilute the kit conditions. Our standard optimization method now involves varying the ratio of the volume of virus solution to precipitant, and this usually obviates the need for fine-tuning protein or virus concentrations (Walter, 2005). It is possible that the virus might disintegrate in some way prior to formation of crystals (as is true of any macromolecular assembly, especially if crystallization takes place over a period of weeks rather than hours). This can result in the crystallization of a substructure, such as an assembly intermediate. The crystallization of the intact structure can most easily be verified by SDS-PAGE to demonstrate the correct complement of proteins or tests to show recovery of infectivity from dissolved crystals (mass spectrometry is also possible (Nettleship, 2005)). Difficulties in obtaining crystals may be alleviated by protein engineering if an infectious clone is available or by particle cross-linking (see PRD1 discussion) (Bamford *et al.*, 2002) or by measuring the thermostability of the virus under varying conditions (Geerlof *et al.*, 2006). There is no reliable way of predicting crystallization conditions for viruses or any protein although a crystallization database exists to help rationalize starting points for searches and focused crystallization screens have proved useful when applied at the level of protein families (Walter, 2005; Gilliland *et al.*, 2002). Sometimes seeding techniques may be used to improve the crystals. Recently smaller volume crystallization methods have been developed (see, for example, Walter *et al.*, 2003), which are very effective for protein crystallization (Brown *et al.*, 2003), and our experience suggests that they work equally well for viruses, and the more rapid equilibration of such methods may be useful for less stable samples.

Crystals are examined optically to check size and morphology. Viruses often produce highly-ordered crystals with low mosaic spread, possibly attributable to their isometric nature and perfection in assembly and have a tendency to crystallize in space groups which reflect the underlying particle symmetry, for instance we find that the space group I23 is 30-fold over-represented in virus crystals compared to those reported for all proteins and viruses (see Table 16.1). As expected some of the capsid symmetry axes often coincide with those of the crystal in such higher symmetry space groups. Whilst most proteins form crystals with optically anisotropic properties, virus crystals tend to be optically isotropic and do not, therefore, demonstrate birefringence.

**Table 16.1** Representative crystal space group analysis: viruses versus proteins

| Space group | Virus total (%) | Total (%) |
|---|---|---|
| $P2_1$ | 15.5 | 12.9 |
| C2 | 4.4 | 8.2 |
| $P2_12_12_1$ | 8.9 | 23.7 |
| R32 | 4.4 | 1.4 |
| $P6_122$ | 0.0 | 1.8 |
| I23 | 13.3 | 0.3 |
| $I2_13$ | 2.2 | 0.4 |
| F432 | 2.2 | 0.0 |

Based on an analysis of the protein data bank (Berman *et al.*, 2000) January 2005, the entries considered for the virus column were selected as containing 'virus' in the PDB header and manually checked to eliminate component proteins and duplicates. The total column represents all crystal structure entries then in the PDB, including viruses and proteins.

## 16.3 Mounting crystals

At present, the lower size limit for virus crystals suitable for room temperature data collection at a synchrotron beamline is, in our hands, of the order of $0.03 \times 0.03 \times 0.03\,mm^3$. This stems from the relatively long exposure times required to obtain useful diffraction data and the radiation sensitivity of the crystals. However, if the virus crystals can be successfully frozen then it should be feasible to use smaller crystals, especially at a very finely collimated or microfocus beamline. Cryogenic data collection from viruses is complicated since freezing tends to increase the apparent mosaic spread of the crystal, often rendering the, already closely spaced, diffraction lunes hopelessly overlapped. For certain space groups, for example I23, lattice centring causes adjacent lunes to interlace so that, given a detector with adequate spatial resolution, it may still be possible to measure the diffraction data; however if the individual spots are also spread out, the situation may be irretrievable. Furthermore, disease security precautions may require that the crystals are frozen in contained conditions rather than the open cryoloops currently routinely used for flash freezing. An alternative procedure would be to automate the process entirely so that the whole data collection environment is contained. It is proposed to implement the latter approach at a Category 3 contained beamline now under construction at the Diamond Synchrotron in the UK (E. Duke personal communication). Hand-pulled tapered quartz tubes with extremely thin walls that fit snugly around the crystal and a slow freezing protocol permitted the collection of a 2 Å data set from a single crystal of foot-and-mouth disease virus with acceptable increase in mosaic spread (0.05° unfrozen, 0.2° frozen) (E. Fry unpublished). In practice, although freezing extends the lifetime of an irradiated crystal by some two orders of magnitude, there are still rather few reports where virus crystals have been successfully frozen (Wien *et al.*, 1997; Dokland *et al.*, 1998; Wynne *et al.*, 1999; Naitow *et al.*, 2001; Verdaguer *et al.*, 2004). Techniques for obtaining cryogenic data from virus crystals need further refinement if we are to benefit from the enormous gain in economy of crystals required for a typical structure determination (thus for BTV over 1000 crystals were examined (Grimes *et al.*, 1998)) and in the time and resources used in data collection. We note that in this respect viruses are representative of any delicate crystal lattice and there is an urgent need for improved cryocooling technologies.

In contrast, it is often possible to achieve a modest amount of cooling of crystals mounted in capillaries using an air cooling system, such as the FTS air jet cooler (FTS Systems Inc., P.O. Box 158, Rt 209 Stone Ridge, NY 12484, USA), which allows investigation of the effect of temperatures close to 273 K on a crystal's lifetime and diffraction (Diprose *et al.*, 1999).

For room temperature data collection the once-standard procedures for protein crystallography can be used to mount virus crystals in quartz capillaries (which are far more robust than glass) although for animal or human pathogens this may need to be performed with the appropriate safety precautions, for example in a fume hood. The outside of the capillaries may need to be washed in an appropriate antiviral solution (for instance acid and detergent or bleach), hence they should be particularly well sealed. Crystals with unit cells in excess of 400 Å tend to be very fragile so that handling them can be very difficult (the effect of particle size on crystal stability is discussed in Section 16.10 below). This may be largely overcome by growing the crystals in capillaries and by stabilizing them, for example cross-linking (see PRD1 case study) or employing novel methods whereby the crystals can be irradiated within the drops in which they formed. A useful generic method of growing virus crystals in open capillary tubes has been described (Cockburn *et al.*, 2003).

## 16.4 Data recording

A major problem for crystals with extremely large unit cells is that of radiation damage, since the diffracted radiation will be divided across far more diffraction spots, each of which must therefore be observed longer to give adequate signal-to-noise, with the result that relatively few images can be collected before radiation damage becomes severe. In the absence of routine freezing methods, virus structure determinations therefore usually gather data from many crystals.

Data-collection from large unit cells can be performed 'in-house' using CuK$\alpha$ radiation, if suitable focusing mirrors or confocal multilayer optics are in place to improve the beam geometry. In practice, however, synchrotron radiation has overwhelming advantages for virus projects: the beam can be focused to <0.2 mm to improve the signal-to-noise ratio, the radiation is tuneable, permitting the use of wavelengths <1 Å with a reduced absorption effect, smaller angle of incidence on the detector, and possibly proportionately lower radiation damage (Fry *et al.*, 1993; Hendrickson, 2000) as well as allowing the choice of optimum wavelengths for anomalous dispersion phasing (although this has, to date, been little exploited for virus crystallography), very narrow energy bandwidths, and finally very high brilliance, meaning faster data collection. Third-generation synchrotron sources (such as: the European Synchrotron Radiation Facility (ESRF), Grenoble, France; Advanced Photon Source (APS), Argonne, USA; and SPring-8, Himeji, Japan) are specifically designed to take advantage of the properties of undulator insertion devices. In one well documented case (Wikoff *et al.*, 2000) the properties of third-generation synchrotron radiation led to a significant increase in the observed resolution of diffraction (from crystals of the phage HK97) by improving the signal-to-noise ratio. This was attributed to the high X-ray flux, beamline geometry, including the use of focusing optics rather than apertures to limit the beam-size, a helium path to reduce air scatter, and beamline shielding combining to produce low background scatter. In addition, more exposures were obtained per crystal, possibly because the convergent beam geometry reduced the power density deposited in the crystal while increasing the diffracting volume. To put these observations on a quantitative basis we consider below the effect of certain aspects of experimental design, starting from an observed PRD1 diffraction pattern as defining the current 'state of the art'.

### 16.4.1 The sources of signal and noise in virus diffraction data

For a 'good' crystal (with modest mosaic spread and with uniform cell dimensions throughout its volume, i.e. a typical room temperature virus crystal), the diffracted spots diverge from the effective X-ray source (which at a third-generation synchrotron is usually several tens of metres before the virus crystal), so that the size of the diffraction spot observed on the detector is almost independent of the (far smaller) crystal to detector distance. It is even possible to use optics, such as Kirkpatrick–Baez mirrors, to focus the reflection to a tiny size on the detector. In contrast, the background originates in large part from scatter from the disordered components of the virus crystal and hence falls off with the square of the crystal to detector distance. This means that experimental design can be tailored to optimize the signal-to-noise for the diffraction data, well beyond what is currently routinely achieved at third-generation synchrotrons. Taking a PRD1 image collected at the ESRF as a start point Box 16.1 demonstrates that it should be possible to achieve an aggregate gain of an order of magnitude in signal-to-noise for the crucial higher resolution reflections.

Using a synchrotron source with crystals of infectious viral particles is only possible where disease security requirements can be met using an agreed protocol (Fry *et al.*, 1993). Newer synchrotrons may have beamlines designed to comply with disease security specifications (a website for biologists engaged in research at synchrotrons is http://biosync.sdsc.edu/). Robot technology (Muchmore *et al.*, 2000; Snell *et al.*, 2004) allowing automated crystal mounting, orientation and retrieval generally works with flash cooled crystals but could be adapted and indeed could prove very useful for the handling of human pathogens (such a system has been implemented at the new UK synchrotron, Diamond). As demonstrated above, it is important to have a large detector with high quantum efficiency and large dynamic range, especially as larger structures and hence larger unit cells are attempted. CCD detectors are currently used almost exclusively, since large, fast-read-out area detectors can be built with no inherent size limit (Pokric *et al.*, 2002). This is achieved by 'tiling' commercial detectors, the only limitation being cost. At present the largest array commercially available is 4 × 4 from MARResearch (with detector diameter approximately 30 cm). Even an earlier generation of detector, such as the ADSC Q4, consisting of four CCD detectors arranged in a square mosaic provides an 18.8 cm square active area (useful pixel size 81.6 μm) and is quite sufficient to resolve diffraction spots to 3.7 Å resolution from a 925 Å unit cell (as in the PRD1 case). New detector technologies that would be particularly applicable to virus studies involve the use of large-area amorphous semiconductors and so called pixel detectors. The potential of such detectors is considerable, they promise to provide large, high-spatial resolution detectors sensitive to hard X-rays. For both technologies there is a marked improvement in point spread function over a CCD detector, illustrated for an amorphous selenium detector in Fig. 16.1.

Note also that the possibility of continuous read-out will be very important for large unit cells, allowing reflection overlap to be minimized and signal-to-noise to be further improved. At present, neither technology is quite mature.

Data collection protocols which incur unnecessary radiation exposure, for example crystal alignment, are usually dispensed with when collecting room temperature virus data. In fact random orientation is an extremely efficient way of collecting data if a partially complete data set is adequate, as it is for virus crystallography, where it is compensated for by the inevitable oversampling of the icosahedral virion transform (Table 16.2) (Rossmann and Erickson, 1983). Nevertheless when collecting only a few images from each of many crystals some pre-alignment of each crystal on an optical goniometer is worthwhile, since it speeds up crystal centring in the X-ray hutch and facilitates the application of systematic offsets from visually appealing views

## Box 16.1 Signal-to-noise in virus diffraction data

Let us assume that we are measuring diffraction data by simple integration of X-ray counts and let us assume no instrumental errors, then the error in any measurement will be given simply by the square root of the total number of photons contributing to that measurement (including background scatter). A further error is introduced when the background is subtracted from the signal; however we ignore this since it will be small and will broadly follow the rest of the error model. We perform a simple quantitative analysis, based on numbers from an actual diffraction image from the PRD1 virus collected at a third-generation synchrotron (the ESRF) using a CCD detector and use this to investigate how further technical improvements will impact on the quality of data obtained from such difficult systems.

$$\text{Signal} = \text{I} = \text{Number of photons in peak after background removal} \quad (2)$$

$$\text{Noise} = \sigma(\text{I}) = \sqrt{\text{Total number of photons in peak (including background)}} \quad (3)$$

For the actual experiment:

$$\text{I}/\sigma(\text{I}) = ((\Sigma_{j=1,N} C_j) - (\langle B \rangle * N)) / \sqrt{\Sigma_{j=1,N} C_j} \quad (4)$$

where $N$ is the number of pixels, $C_j$ the photon count in pixel $j$, and $B$ is measured background.

For a detector with minimal point spread and beam no bigger than pixel size:

$$\text{I}/\sigma(\text{I}) = ((\Sigma_{j=1,N} C_j) - (\langle B \rangle * N)) / \sqrt{((\Sigma_{j=1,N} C_j) - (\langle B \rangle * (N-4)))} \quad (5)$$

For a detector with minimal point spread function and doubled diameter:

$$\text{I}/\sigma(\text{I}) = ((\Sigma_{j=1,N} C_j) - (\langle B \rangle * N)) / \sqrt{((\Sigma_{j=1,N} C_j) - (\langle B \rangle * (N-4)) - ((\langle B \rangle * 0.75) * 4))} \quad (6)$$

### Effect of minimizing detector point-spread function

To fully integrate the observed spots, it was necessary to sum counts over 12 pixels (each 82 μm square ) since although the X-ray beam was collimated to $80 \times 80\,\mu\text{m}^2$ current generation CCDs have a substantial point-spread function, whereas a perfect detector would collect each spot in no more than four pixels. The effect of this (Eqs 4 and 5) would be to increase the signal-to-noise for one of the strongest 4 Å reflections from 2 (experimentally measured) to 3.2.

### Effect of detector size

Doubling the detector diameter (and keeping pixel size fixed) would then further increase the signal-to-noise from 3.2 to 5.4 (Eqs 5 and 6).

### Effect of illuminating a larger crystal volume and focusing the beam on the detector

If the beam size hitting the crystal is increased from $80\,\mu\text{m} \times 80\,\mu\text{m}$ to $160\,\mu\text{m} \times 160\,\mu\text{m}$, then if the crystal is still at least as large as the beam and if the beam is then focused at the detector such that the diffraction spots are still contained in no more than four pixels, the signal-to-noise is further improved to a final value of 10.9, this time simply by increasing the signal rather than decreasing the noise.

### Summary

Further technological improvements at synchrotron beamlines promise to have a substantial effect on data quality for weakly diffracting crystals, enabling the measurement of higher resolution data and the collection of useful data from more challenging problems.

which otherwise tend to bias the selection of data. For small viruses, depending on the lattice properties, an oscillation range between 0.3 and 0.5° is usually suitable for the collection of a 3 Å resolution data set and to minimize accumulated background scatter. For detectors incapable of continuous read-out there is obviously a compromise to be made between collecting as large a swept volume of reciprocal space as possible on each image, to maximize the number of fully recorded reflections, and the

**Table 16.2** Analysis of the American method of data collection

| | 5(2°) | 10 (4°) | 20 (8°) | 40 (16°) | 60 (24°) | 90 (36°) |
|---|---|---|---|---|---|---|
| 1 | 2.2 | 4.4 | 8.5 | 16.3 | 23.4 | 33.0 |
| 2 | 4.3 | 8.4 | 16.2 | 29.4 | 40.8 | 54.4 |
| 222 | 8.2 | 15.7 | 29.0 | 48.9 | 63.5 | 77.8 |
| 4 | 8.4 | 16.1 | 29.6 | 50.3 | 64.9 | 79.2 |
| 422 | 15.1 | 28.0 | 48.4 | 72.6 | 85.5 | 94.2 |
| 23 | 22.7 | 40.2 | 64.0 | 86.2 | 94.6 | 98.6 |
| 432 | 37.6 | 60.9 | 83.8 | 96.8 | 99.2 | 99.8 |
| 3 | 6.5 | 12.6 | 23.5 | 41.3 | 54.8 | 69.5 |
| 321 | 12.3 | 22.8 | 40.5 | 63.5 | 77.8 | 89.4 |

Data from randomly oriented images. The top row shows the angular range collected and the left-hand column the point group. The other figures shown are the percentage completeness to 3 Å resolution for each point group and angular range. These were calculated by predicting reflections for 90 images with randomly generated missetting angles, arbitrary cell, wavelength of 0.9 Å and oscillation range of 0.4°. Only reflections more than 50% recorded were accepted. These reflections were then reduced to the unique subset for each point group, merged, duplicates removed, and the percentage completeness calculated. To a first approximation, these figures scale directly according to the oscillation range used. The figures are essentially independent of resolution.

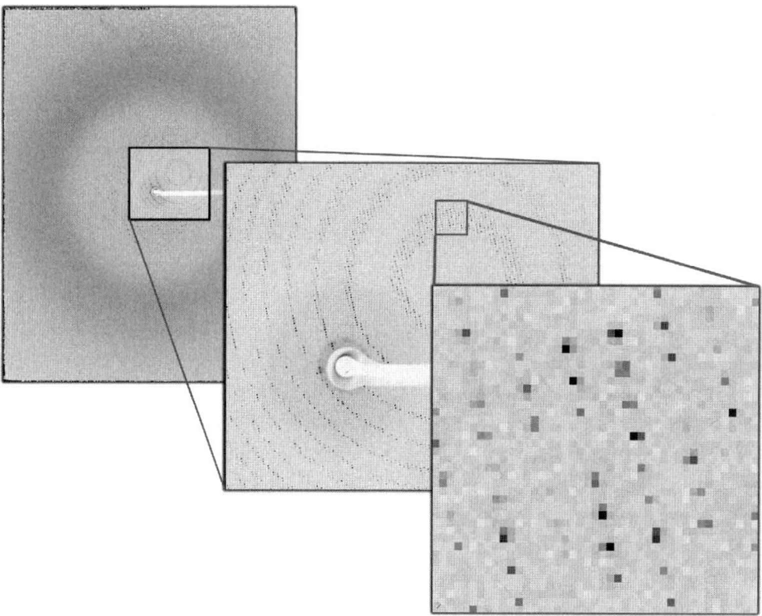

**Figure 16.1** Data recorded on a Marresearch (Marresearch GmbH, Hans Böckler-Ring 17, 22851 Norderstedt, Germany) amorphous Se detector from a crystal of BTV, shown in close up, to illustrate the low point spread and excellent separation of spots from the large cell ($796 \times 822 \times 753$ Å$^3$). X-ray source: ESRF ID14 EH1, wavelength 0.933 Å.

need to avoid reflection overlap at high resolution. This situation is ameliorated if it is possible to collect a substantial number of contiguous oscillations from a crystal (e.g. if the crystal is frozen). Nevertheless, the usually small mosaic spread of virus crystals coupled with the near parallelism of X-ray beams produced by undulators often permits a good proportion of fully recorded reflections, even with an oscillation range of 0.3°. Where it is not possible to collect more than one or two frames from each crystal position, incompletely recorded reflections can still be used (as will be discussed below). It is worthwhile to check the alignment of the beamline components, take any necessary steps to minimize divergence (for instance by reducing the width of the fan of radiation incident upon the focusing optics), especially if using an older (second generation) synchrotron, and to obtain an accurate measure of the direct beam position. We have found viral polyhedra (protein microcrystals which produce powder-like diffraction rings) to be very useful for deriving an accurate direct beam position (Anduleit *et al.*, 2005). Beam alignment is best checked using a scintillator and along-the-beam viewing. In the absence of this, X-ray sensitive paper can be mounted in place of the crystal and exposed whilst orthogonal to the beam direction and then rotated to be perpendicular to the direction of view of the camera. The beam position is noted and then virus crystals centred to that point and then the inverse rotation applied to crystals prior to exposure to X-rays. Minimizing the backstop shadow allows the measurement of low-resolution reflections; it is possible to routinely achieve 105 Å (Gouet *et al.*, 1999) with minimal changes to a standard beamline. Such low-resolution data are particularly helpful when starting phases are obtained from a low-resolution cryo-EM reconstruction or via *ab initio* methods and for visualizing less well ordered parts of the structure; 100 Å is adequate for these applications. Tsuruta *et al.* (Tsuruta *et al.*, 1998) report the measurement of reflections to about 300 Å from a single crystal. These very low angle measurements require long sample to detector distances and careful elimination of parasitic scattering at small angles and air scattering at intermediate angles. The limited dynamic range of the present detectors may require attenuation of the beam to prevent saturation of the lowest resolution data.

High pressure crystallography (>2 kbar) has been used to significantly enhance the diffraction quality of CPMV crystals, which undergo a phase transition to a crystal form exhibiting enhanced order (Fourme *et al.*, 2002). This technique, combining the use of a diamond anvil cell and ultra-short-wavelength undulator radiation, may have more general applicability and is rather well suited to virus crystals since the limited rotation range permitted by diamond cells means that complete data sets can be collected from a single sample only for high-symmetry space groups. Furthermore, there is a necessity to keep the sample at, or close to, room temperature to preserve hydrostatic compression.

The Laue method (using white radiation) has not yet been used for a *de novo* virus structure determination. Whilst this method continues to have potential for studying dynamic events it compounds the problem of spot resolution and will only be feasible with detectors with greater resolution than is furnished by current CCD detectors.

## 16.5 Data processing

Preliminary space group assignations are made by data processing programs, for example DENZO (HKL suite) (Otwinoski, 1993), MOSFLM (CCP4, 1979), XDS (Kabsch, 1988), and d*TREK (Pflugrath, 1999). The agreement of symmetry-related reflections and the intensities of possible systematic absences should be carefully checked, together with packing considerations, and sometimes the final space group assignment may rest upon the results of molecular replacement analysis. Pseudosymmetry may be apparent in the low-resolution reflections and this can hinder the correct space group assignment (Fry *et al.*, 2003). Since icosahedral viruses have innate 532 point group symmetry, structures can be correctly solved in the absence of the full complement of translational information (e.g. echovirus 1 (Filman *et al.*, 1998)).

Data processing is quite straightforward and we will illustrate this for the HKL suite. Initially, the images must be autoindexed such that the crystal orientation is known precisely relative to the camera axes. It is important to use even the very weak reflections at this stage and if using DENZO (Otwinoski, 1993), set the weak level appropriately (e.g. 1 sigma). Parameters such as the direct beam position and crystal-to-plate distance need to be known very accurately to avoid mis-indexing (given the high density of spots) and the spot shape and background templates tailored quite carefully to ensure the most accurate measurement of the crowded and often weak data. It is helpful to define a large background area, often encompassing several spots. Somewhat irritatingly, the overlap problem (Rossmann, 1979) is still not properly addressed with current software despite the fact that such a treatment would be relevant to many tricky crystals, including those of membrane proteins where the mosaic spread is often high, especially after freezing (S. Iwata personal communication). Knowledge of the unit cell parameters, mosaicity and beam characteristics, allows the correct assignment of reflections as either fully or partially recorded. The quality of the data (based upon the comparison of multiply recorded or symmetry related reflections) will tend to be worse than for crystals with smaller unit cells (and hence stronger reflections). This may be compensated for by the high degree of over-sampling of the molecular transform, so that even a shell of data for which the merging R-factor is 50% (and redundancy is low), can produce a good contribution to the electron density map with low phase errors. Owing to the isometric nature of virus particles, it is common for there to be an ambiguity in the relative orientation of the icosahedral and crystallographic axes. For example, in a cubic space group not possessing a four-fold axis, there are two alternative ways of indexing the reciprocal axis that will not normally be distinguishable until data from different crystals are merged. Incompatible indexing, illuminated by high merging R-factors, can be remedied by the appropriate index permutation. Occasionally this ambiguity crops up within a single crystal which is formed from a mosaic of alternatively orientated scattering blocks, culminating in twinned data. In this case it is feasible to use the underlying non-crystallographic symmetry to deconvolute even 50% twinned crystals, as described by Lea and Stuart (1995). Partially recorded reflections which cannot be combined with their complementary parts on successive images can be salvaged using the knowledge of the fraction of each partial reflection recorded to allow the observed intensity (and associated standard deviation!) to be scaled up to the value it would be if the reflection were fully recorded. The final merged dataset may contain millions of unique reflections, possibly necessitating the reconfiguration of software applications to deal with the sheer bulk of data.

In common with other crystal systems where the diffraction tends to fall off rapidly with resolution, we find that sharpening of the observed diffraction data by application of a negative B-factor improves the behaviour of subsequent phasing and refinement steps and the interpretability of the electron density (see e.g. Grimes *et al.*, 1998).

## 16.6 Phase determination

### 16.6.1 Initial phase determination

Armed with a reasonable set of structure factor amplitudes, the over-sampling of the molecular transform means that very poor starting phases can often be refined to great accuracy. Determining the precise orientation and position of the virus particle(s) in the unit cell may be trivial because the virus, with its very high symmetry, is required to lie at a special position, whilst packing considerations may place strong restrictions in other cases. Where conventional molecular replacement analysis is required then the precise particle orientation can be determined by putting the model structure into a locked rotation function. Generally, however, a self-rotation function (polar coordinates) will provide the precise particle orientation(s) where plots of sections at constant kappa will show the direction of non-crystallographic symmetry axes (since there are only 12 five-fold axes, the $\kappa = 72°$ section tends to be the easiest to interpret). The parameters used for a self-rotation function tend to be relatively standard, although the optimum maximum Patterson vector length may be longer than expected, thus as shown in Fig. 16.2, the best signal-to-noise can be achieved with a maximum vector length significantly longer than the radius of the virus particle. Translational searches can then be performed using standard packages, for example X-PLOR (Brunger, 1992), CNS (Brunger *et al.*, 1998), and MOLREP (Vagin and Isupov, 2001). Note that these calculations can be extremely time consuming and in the case of higher symmetry space groups we have found it useful to simplify the process by deconstructing the analysis into several one- (or two-) dimensional searches, each performed with the appropriate subset of symmetry operators. Phaser (McCoy *et al.*, 2005) is useful for handling low-resolution maps.

Once the position and orientation of the particle are known, phase refinement and extension can

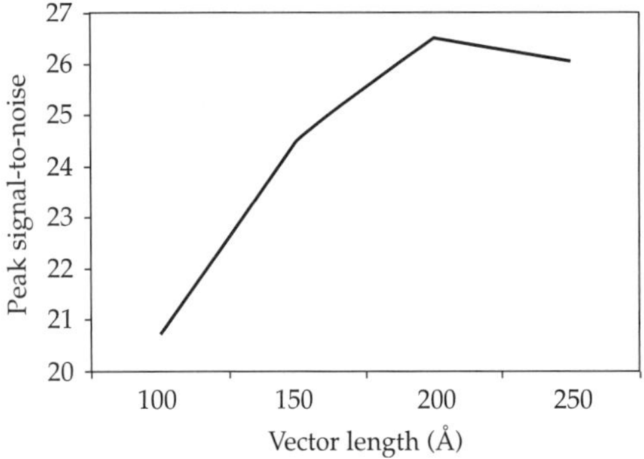


**Figure 16.2** A plot of peak signal-to-noise versus maximum vector length for the $k = 72$ section of a self-rotation function for a picornavirus (approx. 150 Å radius), space group $P22_12_1$, $a = 354$, $b = 372$, $c = 319\,\text{Å}^3$, calculated in XPLOR (Brunger, 1992) using 200000 Patterson peaks from data to 5 Å resolution.

proceed. Again, the internal symmetry provides strong constraints so that only the unique part of the structure requires consideration. One utility a virus crystallographer does not have at their disposal is the use of probabilistic direct methods since the sharpness of phase probability distributions is inversely proportional to the square root of the number of atoms. This method is currently limited to structures up to 1000 non-hydrogen atoms with atomic (conventionally better than 1.2 Å) data. Starting phases may, however, be derived indirectly via heavy atom or anomalous dispersion methods. Using Patterson maps to solve, by hand, the positions of atoms in a substructure becomes difficult above 10 atoms, but direct methods can be used to solve these substructures (even at lower resolution) thus theoretically up to 1000 atoms in a substructure can be solved, which should be adequate even for larger viruses. Grossly incomplete heavy-atom data can still be useful (for instance 5% completeness was sufficient in the study of CPV (Tsao *et al.*, 1992)).

Where the structure of one or more similar viruses is available, low-resolution starting phases can be calculated from the model virus (or summed structures) correctly placed in the cell of the unknown structure (Fry *et al.*, 1993). Cryo-EM reconstructions can provide adequate starting phases and envelope information for the determination of novel structures, for example BTV (Grimes *et al.*, 1998), Reovirus (Reinisch *et al.*, 2000) and solution scattering, analysed with icosahedral harmonics, may also provide suitable low-resolution starting models (Zheng *et al.*, 1995). *Ab initio* phasing has been used with native diffraction data to generate correct phases and electron density to 3 Å resolution (Miller *et al.*, 2001). In this case a genetic algorithm was used to determine coarse, low-resolution (20.5 Å) lattice models of the virus that obey the known non-crystallographic symmetry (NCS) constraints. It is imperative to know the precise particle position and orientation and to include all the low-resolution terms in the transform. Model bias can be removed from phases derived from such starting points by refinement and extension to higher resolution.

### 16.6.2 Phase refinement and extension

Iterative NCS averaging (even with minimal five-fold NCS) usually provides very accurate phases to the resolution limit of the data. NCS averaging can also break the centric symmetry of *ab initio* shell-based phases (Naitow *et al.*, 1999). The disadvantage of applying such constraints is that it can prevent the elucidation of features that do not obey the capsid symmetry. The viral genome cannot properly follow the capsid symmetry since it is usually a linear, non-repetitive structure. It therefore needs to be treated carefully when icosahedral symmetry averaging is utilized, for example by masking. In many crystallographic analyses, the RNA or DNA genome has been almost completely disordered, or occasional bases have been visible, stacking with aromatic residues on the interior of the virus capsid. In a few viruses, for example bean pod mottle virus (BPMV), the nucleic acid reflects capsid symmetry in the vicinity of particular axes (Chen *et al.*, 1989). The organization of the genome has been most clearly visualized in a crystallographic study of the dsRNA virus BTV (Gouet *et al.*, 1999) which allowed a model representing 80% of the genome to be constructed. A substantial amount of structure was also observed for the dsDNA bacteriophage PRD1 (Cockburn *et al.*, 2004).

The most efficient method for implementing symmetry averaging with larger, higher-resolution structures is in real space following the method largely devised by Bricogne (Bricogne, 1976). In outline, the electron density for the non-crystallographically related subunits within the viral envelope is averaged, then back transformed, the resultant phases are combined with the original observed structure factor amplitudes (Fobs) (with suitable weighting) and a new electron density map computed (Fig. 16.3). If phases are to be extended, the map is back-transformed to a slightly higher resolution than it was calculated at, providing phase estimates to this new limit which can then be fed back into the cyclical procedure. The region outside the molecular envelope, and thus beyond the limits of applicability of the local non-crystallographic symmetry, is flattened to represent disordered solvent. There are now a number of efficient computer programs to perform the real space part of this procedure (Kleywegt and Read, 1997). A classic analysis of the effect of some parameters on the outcome of the phase refinement process was made by Rayment (Rayment, 1983). To optimize the procedure (Fry *et al.*, 1993) (particularly important where redundancy is at a minimum), Fobs should be filled in by including calculated values from the back-transformed map with suitable weighting, reflections should be weighted according to the likely error in their phases, solvent flattening should be used with the maximum solvent volume consistent with no truncation of the protein, preferably with an automatic procedure (Fry *et al.*, 1993; Wang, 1985), and the envelope should be recalculated at intervals. Possible errors occur where the high-resolution phase solutions satisfy the non-crystallographic symmetry constraints but are anticorrelated with the low-resolution phasing solution. Such errors should not arise if the shell thickness for phase extension is sufficiently small. A suitable rate of phase extension can be calculated from the interference function, $G$, for the virus particle. If the virus is spherical (of radius $R$) the argument of $G$, is given by $(n/a)R$ where $n$ is a number of reciprocal lattice points and $a$ is the cell dimension. The largest values of $G$ will occur within the first positive portion of the function, corresponding to $(n/a)R < 0.725$, and thus for a given radius and unit cell, the shell thickness that is likely to contain reliable phase information can be determined (e.g. for $R = 150$ Å and $a = 400$ Å, $n$, the shell thickness for phase extension, is 1.9 reciprocal lattice units). Some authors have used a process of phase extension in steps of roughly this amount followed by a number of cycles of phase refinement (Arnold, 1987), whereas others have used a continuous phase extension procedure where each extension step is much smaller, combined with a special treatment of the outermost resolution shell (e.g. omission of calculated values for unmeasured data) to facilitate improved phase convergence (Fry *et al.*, 1993).

The process of phase refinement and extension is normally monitored by reference to averaging $R$-factors and correlation coefficients in reciprocal space which should be analysed as a function of resolution (although similar measures in real-space can also be useful). The correlation coefficient is a more trustworthy indicator of success than the $R$-factor. Poor agreement of the strong low-resolution terms may inhibit phase extension, for example where missing low-resolution data cause serious series termination ripples. One approach to solving this difficulty is to fill in unmeasured data by phase extension to lower resolution in an analogous fashion to that employed for increasing resolution. Always check carefully that reflections close to the back stop shadow have been correctly measured otherwise it may be necessary to impose additional cut-offs on the measured data. Similarly, series termination errors at high resolution can be reduced by phase extension a little beyond the limiting resolution of the data (allowing calculated data to soak up spurious ripples in the map).

## 16.7 Model building – refinement

Fitting of a chemically reasonable model into the electron density using a program such as O (Jones *et al.*, 1991) or Coot (Emsley and Cowtan, 2004) should be straightforward (although the relative instability of the viral genome means that care should be taken to ensure that an accurate and appropriate protein sequence is available). A model is normally constructed for a single icosahedral unit. If this icosahedral asymmetric unit contains multiple copies of a single protein, then this structure, once built for one subunit, can be rotated and adapted to fit the electron density for the NCS related subunits. This starting model is then refined (e.g. using XPLOR (Brunger, 1992), CNS (Brunger

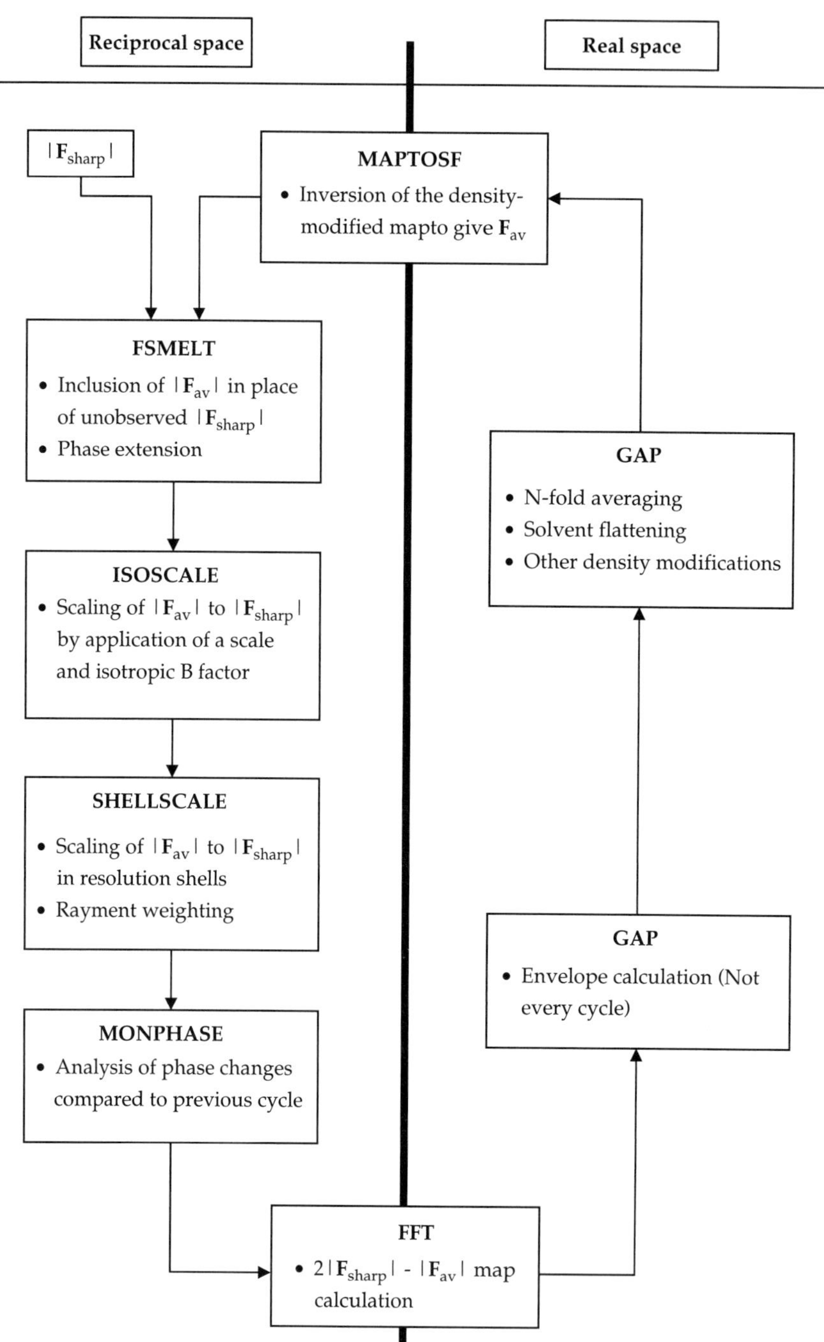


**Figure 16.3** Flowchart summarizing a general averaging protocol in reciprocal and real space (Diprose, 2000). (N depends on the level of non-crystallographic redundancy.) FSMELT, ISOSCALE, SHELLSCALE, MONPHASE, and MAPTOSF are in-house programs (D. I. S., J. M. Grimes, J. M. Diprose, unpublished data and unpublished routines of R. Bryan), FFT is a CCP4 program (CCP4, 1994). $\mathbf{F}_{sharp}$ are the sharpened $\mathbf{F}_{obs}$. The $\mathbf{F}_{obs}$ are scaled against the corrected $\mathbf{F}_{c}$ by application of a negative B factor in order to increase the weight of the high resolution terms, reducing domination by strong low-resolution terms. $\mathbf{F}_{av}$ are the calculated structure factors back-transformed from the averaged map.

*et al.*, 1998)), minimizing the differences between calculated and observed data whilst enforcing stereochemical restraints and non-crystallographic constraints and using an appropriate bulk solvent correction. We tend to use a rather explicit method to correct for the 'bulk solvent' scattering, a mask is first generated to define the solvent region in the unit cell to which an average electron density is assigned. Back-transformation of this region provides $F_{solvent}$ terms which can be added to the calculated structure factors for the model. The solvent structure factors are corrected using the formula:

$$F'_{solvent} = F_{solvent} * k_{solvent} \exp(-B_{solvent}S^2/4) \quad (7)$$

where the scale factor 'k' and B factor 'B' are determined by iterative trials measuring the agreement between $F_{calc}$ and $F_{obs}$. Note that it may be worthwhile to flatten separately the interior and exterior of the capsid since the mean electron density within the capsid may be elevated by the presence of disordered genomic nucleic acid.

Although the refinement and model building protocol is essentially standard, the presence of non-crystallographic symmetry means that the R-free behaves in an unreliable way (due to strong correlations between the working and test set of data). Thus, for virus structures the justification for using the R-free, namely to guide refinement strategy, is thrown into question.

The refined coordinates will correspond to either the icosahedral asymmetric unit or the crystallographic asymmetric unit, hence symmetry operations must be applied to generate the whole capsid. A useful repository of virus structure information is the website http:// viperdb.scripps.edu/ where portions of the viral capsid can be generated.

## 16.8 Use of cryo-EM as a complimentary method

Flash cooling of samples, the use of field-emission guns to generate more coherent electron beams, and improved software has radically extended the power of electron microscopy, so that it is now almost routine to provide structural information for viruses to better than 10 Å, and sometimes 7 Å resolution (Conway *et al.*, 1997). The major strength of this method is that it does not require crystals nor, in principle, particularly pure sample. This may be important for future studies of viruses interacting with cellular components which may not be readily crystallizable. Three-dimensional reconstructions from cryo-electron microscopy have thus far been used to study complexes between viruses and Fab fragments (Hewat *et al.*, 1997), whole antibodies (Hewat and Blaas, 1996), and receptor molecules (Xiao *et al.*, 2001) giving low-resolution information on these complexes (Gilbert *et al.*, 2003). By combining this information with X-ray data for the individual components that make up such complexes (Rossmann *et al.*, 2001), it is possible to derive more precise information about the interactions between the components. Combinations of these methods and solution small-angle X-ray scattering also show promise in determining both the dynamic and static aspects of virus assembly and maturation (Lee and Johnson, 2003). Recent examples of combining cryo-EM and X-ray crystallography are found in the structural studies of herpesvirus (Chiu and Rixon, 2002), BTV (Grimes *et al.*, 1998; Hewat *et al.*, 1994), and PRD1 (Abrescia *et al.*, 2004).

## 16.9 Case study 1 – BTV

BTV belongs to the family *Reoviridae*, viruses with double-stranded RNA segmented genomes. They contain two capsid layers; the internal capsid, or core, contains transcription complexes which transcribe the viral RNA. The BTV core is approximately 700 Å in diameter and is composed of two principal structural proteins; 780 copies of VP7 (38 kDa) arranged as trimers on a $T = 13$ quasiequivalent lattice form the bristly core surface and 120 copies of VP3 (100 kDa) the 'subcore' shell.

The structure determination proceeded in steps starting with separate structural analyses of the component protein VP7. Information from a 22 Å resolution cryo-EM reconstruction of the viral core (Grimes *et al.*, 1997) was then combined with the high-resolution atomic structure for the VP7(T13) trimer (Grimes *et al.*, 1995) to yield a model providing phase information to a higher resolution than the EM reconstruction alone. Infectious BTV-1 crystals (unit cell dimensions $a = 796$ Å, $b = 822$ Å, and $c = 753$ Å), containing half a particle in the asymmetric unit, were safely premounted in quartz capillary tubes at the containment laboratory of the Institute of Animal Health (Pirbright, UK) prior to transportation to the ESRF. The unique dataset (3,299,866 reflections to 3.5 Å, Rmerge = 22.9%) (where Rmerge = $\sum_j \sum_h |I_{j,h} - \langle I_h \rangle| / \sum_j \sum_h \langle I_h \rangle$ and $h$ are unique reflection indices, $I_{j,h}$ are intensities of symmetry related reflections, and $\langle I_h \rangle$ is the mean intensity) was collected at ID2 (a high brilliance undulator beamline at the ESRF). A self-rotation function provided the orientation of the virus and a translation search, using the cryo-EM derived model, located the particle. The resolution was extended in reciprocal space from 12 Å to 6 Å using rigid body refinement. Averaging and solvent flattening then refined these phases, and an averaged map at 3.8 Å allowed unambiguous chain tracing and sequence alignment (reciprocal space R-factors and correlation coefficients were 16.9% and 0.89 respectively). The structure was then refined at 3.5 Å using XPLOR (Brunger, 1992) (despite the limited completeness of the data at this resolution) (Grimes *et al.*, 1998).

Crystals of another serotype, BTV-10, crystallized with unit cell parameters $a = b = 1115$ Å, $c = 1584$ Å, with one core particle per asymmetric unit. Data were, again, collected on ID2 at the ESRF where the highly parallel beam and limiting apertures on the incident beam of 50–80 µm allowed the diffraction orders to be resolved at a wavelength of 1 Å using a 30 cm MAResearch imaging plate placed 1.05 m from the crystal. The BTV-1 structure was used as a search model in the determination of the space group and particle position but the complexity of this model necessitated the factoring of the symmetry operators to reduce the problem to 1-D and 2-D searches. Rigid body refinement and a grid search refinement of the unit cell dimensions provided model structure factors whose phases formed the basis for cyclic averaging and solvent flattening (Gouet *et al.*, 1999).

Beyond these structures, a wealth of information has emerged from crystallographic studies (Diprose *et al.*, 2001) examining the effect of soaking substrates and effectors (e.g. $Mg^{2+}$, nucleotides, oligonucleotides, and inorganic phosphate). This revealed multiple, distinct binding sites and an expansion of the capsid (brought about by the binding of oligonucleotides and $Mg^{2+}$) around the five-fold axes where the transcribed RNA would be translocated into the cytoplasm.

## 16.10 Case study 2 – PRD1

The PRD1 bacteriophage, though lacking a tail, has a host-derived lipid bilayer lying beneath the icosahedral pseudo $T = 25$ protein capsid, which encapsulates the dsDNA genome and is used as a genome ejection device. Sharing a common evolutionary ancestor with adenovirus, it has receptor-binding spikes protruding from the vertices of the capsid (vertex to vertex diameter 740 Å). The critical path for obtaining diffraction quality crystals necessitated the development of a very rapid virus purification and concentration method. However, the resultant crystals only diffracted to low resolution. Making the reasonable assumption that the flexible spikes on the vertices (protein P2) might interfere with crystallization, a mutant virus with a defect in the gene encoding P2 was selected. Mutant *sus539* virus particles were produced on *S. enterica* wild-type strain DS88 in liquid cultures at 310 K followed by purification in sucrose gradients and by a MemSep anion-exchange chromatography cartridge. It was verified that the particles had the remaining complement of protein but they proved metastable, resulting in poorly formed crystals. This was remedied by treating the virus with 0.025% (w/v) glutaraldehyde immediately after the final anion-exchange chromatography step and crystallizations were set up after 15 min incubation (other cross-linking agents such as DSS (disuccinimidyl suberate) and DSP (dithiobis-succinimidylpropionate) were tested and proven ineffective). Successful crystallization conditions were 3.4–4.4% (w/v) polyethylene glycol (PEG) 8000 in 100 mM potassium phosphate (pH 7.2), 400 mM NaCl, at 20°C. These crystals, grown by vapour diffusion either in hanging or sitting drops, were very fragile and did not diffract beyond 20 Å. An improvement was found by treating the crystals with cryoprotectants (preferably PEG 20 K), whereby approximately 10% of the crystals diffracted to almost 4 Å (presumably an annealing phenomenon arising from dehydration of the unit cell). To avoid mechanical stress, the crystals remained suspended in buffer during data collection thus requiring an alignment procedure incurring minimum movement of the capillary tube on the goniostat. Attempts to cool the crystals were unsuccessful. An advance on this approach involved growing the crystals in thin-walled quartz capillary tubes by vapour diffusion obviating the need for crystal handling (Cockburn *et al.*, 2003). A 3 μl volume of the virus and precipitant solution was injected into a thin-walled quartz capillary 15 mm × 0.5 mm, fixed to a sitting drop microbridge and set to equilibrate (vapour diffusion via the open ends) against reservoir solution in the well of a sitting drop tray. Crystals grown by this technique, were 'annealed' one week prior to data collection by addition of approximately 1 μl 20% PEG 20K injected into the capillary using a drawn-out pipette. Data were collected predominantly at the beamlines of ID14 at the ESRF. The capillaries (extracted from the trays) were sealed with vacuum grease, maintaining the mother liquor in place, and prealigned prior to exposure. The beam size was most usually 80 μm × 80 μm, oscillation range 0.25°, crystal-to-plate distance 440 mm and the wavelength 0.933 Å and most data were collected on ADSC Q4 CCD detectors. Strangely, crystals exhibiting a regular morphology diffracted far less well than those with a fern-like appearance. Data were collected from the latter, taking exposures at 100 μm intervals along the length of the 'fern', with 1° rotations about $\phi$ between exposures, permitted up to 20 images to be collected from a single crystal. These crystals were mostly isomorphous with those grown in sitting drops, belonging to space group $P2_12_12_1$, $a = 895$ Å, $b = 905$ Å, $c = 925$ Å (cell 1) and one particle in the asymmetric unit, a subset of crystals, however, belonged to the same space group but with a different unit cell (cell 2). The data were processed with DENZO and a version of SCALEPACK (Otwinoski, 1993) compiled to handle large numbers of partially recorded reflections. Data with fractional partiality above 0.7 were scaled to full intensity (using the program POST, D. I. S and J. M. Diprose unpublished).

Phase information was obtained by combining an X-ray structure of the major capsid protein with a cryo-EM reconstruction of the virus. The position and orientation of this model were refined against the experimental data, separately for the two crystal forms, using XPLOR3.1 (Brunger, 1992). An initial estimate of the particle orientation was obtained from a self-rotation function. Patterson correlation refinement was performed against 60–15 Å resolution data to optimize the orientation. The position and orientation of the particle were then refined at 30 Å using the E1E1 target function, giving an initial set of NCS operators for the two cell types. The set of trimers of the major capsid protein, P3, comprising one icosahedral asymmetric unit, were refined at 6 Å resolution and the particle orientation and position were refined at 8 Å resolution. Further refinement with data to 5 Å gave final correlation coefficients of 0.23 and 0.42 for cell 1 and cell 2 respectively.

A $2F_o - F_c$ map was calculated for each cell type (to 4.2 Å resolution) with the corresponding model phases, which were then refined by iterative solvent flattening and 60-fold averaging (using the program GAP, D. I. S, Jonathan Grimes and Jonathan Diprose, unpublished). Missing reflections were substituted with appropriately weighted $F_c$ values derived from Fourier transformation of the modified map.

SeMet data were used for the calculation of difference Fourier maps. For SeMet labelling the virus was grown on M9 glucose-thiamine medium and 60 mg/ml L-selenomethionine added 20 min after infection. SeMet substituted crystals were isomorphous with cell 1. The clear peaks for the selenium positions helped with the unequivocal identification and tracing of previously structurally unknown viral proteins (using O (Jones *et al.*, 1991)). The atomic structural information gathered on PRD1 has led to the proposal of a general assembly mechanism scaleable to very large viruses and a series of unexpected observations on the structure of the viral membrane (Abrescia *et al.*, 2004; Cockburn *et al.*, 2004).

## 16.11 Conclusions

There is no doubt that knowledge of the structure of viruses at atomic resolution has contributed to our understanding of virus biology, leading to the more rational design of antiviral agents and ultimately the development of new vaccines. In the future it should become possible to capture different functional states, assembly intermediates, and view the non-icosahedral components. More comprehensive studies such as those with BTV are beginning to unravel the complexity of these machines. Undoubtedly, more complex virus structures will be tackled by combinations of methods, especially crystallography and cryo-EM.

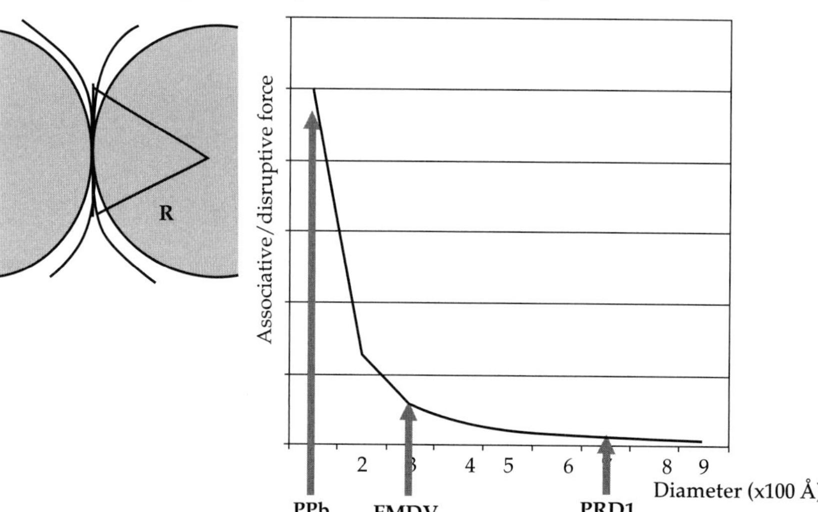


**Figure 16.4** A plot showing likely crystal stability with increasing particle size based on the observation that for a given surface roughness (e.g. 10 Å) the likely contact area is proportional to the particle radius $R$, whilst the dispersive force is proportional to $R^3$ (i.e. the particle mass). PPb = glycogen phosphorylase b, FMDV = foot-and-mouth disease virus; PRD1 = PRD1 bacteriophage.

Certain factors are likely to influence future analyses of more complex viruses. Crystal stability is governed by packing interactions and, as can be seen from Fig. 16.4, is, to a first approximation, inversely proportional to the square of the virus radius, presumably underlying the problems with crystal stability for analyses such as that of PRD1. Even assuming that well-ordered, stable crystals can be formed, technical considerations will place an upper limit on the unit cell size from which useful data can be collected. Nevertheless, with some improvements in beam and detector technology, we expect that data collection from cells up to 2000 Å should be feasible for even a primitive unit cell.

Phasing problems tend not to be limiting owing to the non crystallographic symmetry. Given reasonable quality data, even low-resolution starting phases (most likely from cryo-EM) will yield a good map given non-crystallographic symmetry averaging, solvent flattening, and phase extension. Such maps are often better than would be expected from their nominal resolution and often permit chain tracing of even β-sheet rich proteins at approximately 4.5 Å resolution. The crystal structure of PRD1 demonstrates the possibility of tackling lipidic viruses and hopefully some of the techniques devised for this analysis will be transferable to other systems and open up this important area of structural virology.

Finally, large scale, genomic approaches to the three-dimensional structure determinations of viruses and their complexes, using high-throughput methodology are already underway (Sutton *et al.*, 2004; SPINE, 2006) and it will be interesting to see how profound their impact on structural virology will be.

## References

Abrescia, N. G., *et al.* (2004). Insights into assembly from structural analysis of bacteriophage PRD1. *Nature* **432**, 68–74.

Anduleit, K., *et al.* (2005). Crystal lattice as biological phenotype for insect viruses. *Protein Science* **14**, 2741–2743.

Arnold, E., *et al.* (1987). Structure determination of a common cold virus, human rhinovirus 14. *Acta Crystallogr.* **A43**, 346–361.

Bamford, J. K., *et al.* (2002). Diffraction quality crystals of PRD1, a 66-MDa dsDNA virus with an internal membrane. *J. Struct. Biol.* **139**, 103–112.

Bamford, D. H., Grimes, J. M. and Stuart, D. I. (2005). What does structure tell us about virus evolution? *Curr. Opin. Struct. Biol.* **15**, 655–663.

Berman, H. M., *et al.* (2000). The Protein Data Bank. *Nucleic Acids Res.* **28**, 235–242.

Bernal, J. D. and I. Fankuchen (1941). X-ray and crystallographic studies of plant virus preparations. III. *J. Gen. Physiol.* **25**, 111–146.

Bricogne, G. (1976). Methods and programs for direct-space exploitation of geometric redundancies. *Acta Crystallogr. A* **32**, 832–847.

Brown, J., *et al.* (2003). A procedure for setting up high throughput nanolitre crystallization experiments II:Crystallization results. *J. Appl. Crystallogr.* **36**, 315–318.

Brunger, A. T. (1992). *XPLOR Version 3.1.* Yale University Press, New Haven and London.

Brunger, A. T., *et al.* (1998). Crystallography and NMR system: A new software suite for macromolecular structure determination. *Acta Crystallogr. D* **54**, 905–921.

Caspar, D. L. D. and Klug, A. (1962). Physical principles in the construction of regular viruses. Cold. Spring Harbour Symp. *Quant. Biol.* **27**, 12219–12223.

CCP4 (1994). The CCP4 suite: Programs for protein crystallography. *Acta Crystallogr. D* **50**, 760–763.

CCP4 (1979). *The SERC (UK) Collaborative Computing Project No. 4, a Suite of Programs for Protein Crystallography.* Daresbury Laboratory, Warrington, WA4 4AD, UK.

Chen, Z. G., *et al.* (1989). Protein-RNA interactions in an icosahedral virus at 3.0 A resolution. *Science* **245**, 154–159.

Chiu, W. and Rixon, F. J. (2002). High resolution structural studies of complex icosahedral viruses: a brief overview. *Virus Res.* **82**, 9–17.

Cockburn, J. J., *et al.* (2003). Crystallization of the membrane-containing bacteriophage PRD1 in quartz capillaries by vapour diffusion. *Acta Crystallogr. D* **59**, 538–540.

Cockburn, J. J., *et al.* (2004). Membrane structure and interactions with protein and DNA in bacteriophage PRD1. *Nature* **432**, 122–125.

Conway, J. F., *et al.* (1997). Visualization of a 4-helix bundle in the hepatitis B virus capsid by cryo-electron microscopy. *Nature* **386**, 91–94.

Crowfoot, D. and Schmidt, G. M. J. (1945). X-ray crystallographic measurements on a single crystal of a tobacco necrosis virus derivative. *Nature* **155**, 504–505.

Cullis, A. F., *et al.* (1962). The structure of haemoglobin. IX. A three-dimensional Fourier synthesis at 5.5A resolution: description of the structure. *Proc. R. Soc. London Ser. A.* **265**, 161–187.

Diprose, J. M. (2000). *Structural Studies on Orbiviruses.* D.Phil. thesis, University of Oxford.

Diprose, J. M., *et al.* (1999). Bluetongue virus: the role of synchrotron radiation. *J. Synchrotron Rad.* **6**, 865–874.

Diprose, J. M., *et al.* (2001). Translocation portals for the substrates and products of a viral transcription complex: the bluetongue virus core. *EMBO J.* **20**, 7229–7239.

Dokland, T., *et al.* (1998). Structure determination of the phiX174 closed procapsid. *Acta Crystallogr. D* **54**, 878–890.

Filman, D. J., *et al.* (1998). Structure determination of echovirus 1. *Acta Crystallogr. D* **54**, 1261–1272.

Fourme, R., *et al.* (2002). Opening the high-pressure domain beyond 2 kbar to protein and virus crystallography–technical advance. *Structure* (Camb) **10**, 1409–1414.

Franklin, R. E. (1955). Structure of tobacco mosaic virus. *Nature* **175**, 379–381.

Fry, E., Acharya, R. and Stuart, D. (1993). Methods used in the structure determination of foot-and-mouth disease virus. *Acta Crystallogr. A* **49**, 45–55.

Fry, E. E., *et al.* (2003). Crystal structure of Swine vesicular disease virus and implications for host adaptation. *J. Virol.* **77**, 5475–5486.

Geerlof, A., *et al.* (2006). The impact of protein characterization in structural proteomics. *Acta Crystallogr. D* **62**, 1125–1136.

Gilbert, R. J., Grimes, J. M. and Stuart, D. I. (2003). Hybrid vigor: hybrid methods in viral structure determination. *Adv. Protein Chem.* **64**, 37–91.

Gilliland, G. L., Tung, M. and Ladner, J. E. (2002). The Biological Macromolecule Crystallization Database: procedures and strategies. *Acta Crystallogr. D* **58**, 916–920.

Gouet, P., *et al.* (1999). The highly ordered double-stranded RNA genome of bluetongue virus revealed by crystallography. *Cell* **97**, 481–490.

Grimes, J., *et al.* (1995). The crystal structure of bluetongue virus VP7. *Nature* **373**, 167–170.

Grimes, J. M., *et al.* (1997). An atomic model of the outer layer of the bluetongue virus core derived from X-ray crystallography and electron cryomicroscopy. *Structure* **5**, 885–893.

Grimes, J. M., *et al.* (1998). The atomic structure of the bluetongue virus core. *Nature* **395**, 470–478.

Harlos, K. (1992). Micro-bridges for sitting drop crystallisations. *J. Appl. Cryst.* **25**, 536–538.

Harrison, S. C., *et al.* (1978). Tomato bushy stunt virus at 2.9 Å. *Nature* **276**, 368–373.

Hendrickson, W. A. (2000). Synchrotron crystallography. *Trends Biochem. Sci.* **25**, 637–643.

Hewat, E. A. and Blaas, D. (1996). Structure of a neutralizing antibody bound bivalently to human rhinovirus 2. *EMBO J.* **15**, 1515–1523.

Hewat, E. A., Booth, T. F. and Roy, P. (1994). Structure of correctly self-assembled bluetongue virus-like particles. *J. Struct. Biol.* **112**, 183–191.

Hewat, E. A., *et al.* (1997). Structure of the complex of an Fab fragment of a neutralizing antibody with foot-and-mouth disease virus: Positioning of a highly mobile antigenic loop. *EMBO J.* **16**, 1492–1500.

Jancarik, J., *et al.* (2004). Optimum solubility (OS) screening: an efficient method to optimize buffer conditions for homogeneity and crystallization of proteins. *Acta Crystallogr. D* **60**, 1670–1673.

Jones, T. A., *et al.* (1991). Improved methods for building protein models in electron density maps and the location of errors in these models. *Acta Crystallogr. A* **47**, 110–119.

Kabsch, W. (1988). Automatic indexing of rotation diffraction patterns. *J. Appl. Cryst.* **21**, 67–71.

Kendrew, J. C., *et al.* (1960). Structure of myoglobin. A three-dimensional Fourier synthesis at 2A resolution. *Nature* **185**, 422–427.

Kleywegt, G. J. and Read, R. J. (1997). Not your average density. *Structure* **5**, 1557–1569.

Lea, S. and Stuart, D. (1995). Deconvolution of fully overlapped reflections from crystals of foot-and-mouth disease virus $O_1$ G67. *Acta Crystallogr. D* **51**,160–167.

Lee, K. K. and Johnson, J. E. (2003). Complimentary approaches to structure determination of icosahedral viruses. *Curr. Opin. Struct. Biol.* **13**, 558–569.

McCoy, A. J., *et al.* (2005). Likelihood enhanced fast translation functions. *Acta Crystallogr. D* **61**, 458–464.

Miller, S. T., Hogle, J. M. and Filman, D. J. (2001). Ab initio phasing of high-symmetry macromolecular complexes: successful phasing of authentic poliovirus data to 3.0A resolution. *J. Mol. Biol.* **307**, 499–512.

Muchmore, S. W., *et al.* (2000). Automated crystal mounting and data collection for protein crystallography. *Structure Fold Des.* **8**, R243–246.

Naitow, H., *et al.* (1999). A low-resolution structure of rice dwarf virus determined by ab initio phasing. *Acta Crystallogr. D* **55**, 77–84.

Naitow, H., *et al.* (2001). Purification, crystallization and preliminary X-ray analysis of L-A: a dsRNA yeast virus. *J. Struct. Biol.* **135**, 1–7.

Nettleship, J. E., *et al.* (2005). Sample preparation and mass spectrometric characterisation of crystal-derived protein samples. *Acta Crystallogr. D* **61**, 643–645.

Otwinoski, Z. (1993). Oscillation data reduction program. In: *Data Collection and Processing*, Sawyer, L., Isaacs, N. and Bailey, S., eds, pp. 56–62. Daresbury laboratory, SERC, Warrington, UK.

Pflugrath, J. W. (1999). The finer things in X-ray diffraction data collection. *Acta Crystallogr. D* **55**, 1718–1725.

Pokric, M., *et al.* (2002). Large area high-resolution CCD-based X-ray detector for macromolecular crystallography. *Nucl. Instrum. Meth. A.* **477**, 166–171.

Rayment, I. (1983). Molecular replacement method at low resolution:optimum strategy and intrinsic limitations as determined by calculations on icosahedral virus models. *Acta Crystallogr. A* **39**, 102–116.

Reinisch, K. M., Nibert, M. L. and Harrison, S. C. (2000). Structure of the reovirus core at 3.6 A resolution. *Nature* **404**, 960–967.

Rossmann, M. G. (1979). Processing oscillation diffraction data from very large unit cells with an automatic convolution technique and profile fitting. *J. Appl. Cryst.* **12**, 225–238.

Rossmann, M. G. (1998). From the structures of simple salts to those of sophisticated viruses. *Acta Crystallogr. A* **54**, 716–728.

Rossmann, M. and Erickson, J. (1983). Oscillation photography of radiation-sensitive crystals using a synchrotron source. *J. Appl. Cryst.* **16**, 629–636.

Rossmann, M. G., Bernal, R. and Pletnev, S. V. (2001). Combining electron microscopic with X-ray crystallographic structures. *J. Struct. Biol.* **136**, 190–200.

Snell, G., *et al.* (2004). Automated sample mounting and alignment system for biological crystallography at a synchrotron source. *Structure* **12**, 537–545.

SPINE special issue (2006). Structural Proteomics IN Europe. *Acta Crystallogr. D* **62**, 1103–1285.

Stanley, W. M. (1935). Isolation of a crystalline protein possessing the properties of tobacco mosaic virus. *Science* **81**, 644–645.

Sutton, G., *et al.* (2004). The nsp9 replicase protein of SARS-coronavirus, structure and functional insights. *Structure* (Camb) **12**, 341–353.

Tsao, J., *et al.* (1992). Structure determination of monoclinic canine parvovirus. *Acta Crystallogr. B* **48**, 75–88.

Tsuruta, H., *et al.* (1998). Imaging RNA and dynamic protein segments with low-resolution virus crystallography: experimental design, data processing and implications of electron density maps. *J. Mol. Biol.* **284**, 1439–1452.

Vagin, A. A. and Isupov, M. N. (2001). Spherically averaged phased translation function and its application to the search for molecules and fragments in electron-density maps. *Acta Crystallogr. D* **57**, 1451–1456.

Verdaguer, N., *et al.* (2004). X-ray structure of a minor group human rhinovirus bound to a fragment of its cellular receptor protein. *Nat. Struct. Mol. Biol.* **11**, 429–434.

Walter, T. S., *et al.* (2005). A procedure for setting up high throughput nanolitre crystallization experiments. III. Crystallization workflow for initial screening, automated storage, imaging and optimization. *Acta Crystallogr. D* **61**, 651–657.

Walter, T. S., *et al.* (2003). A procedure for setting up high-throughput nanolitre crystallization experiments. I protocol design and validation. *J. Appl. Cryst.* **36**, 308–314.

Wang, B.-C. (1985). Resolution of phase ambiguity in macromolecular crystallography. *Method Enzymol.* **115**, 90–117.

Wien, M. W., *et al.* (1997). Structural studies of poliovirus mutants that overcome receptor defects. *Nat. Struct. Biol.* **4**, 666–674.

Wikoff, W. R., Schildkamp, W. and Johnson, J. E. (1997). Increased resolution data from a large unit cell crystal collected at a third-generation synchrotron X-ray source. *Acta Crystallogr. D* **56**, 890–893.

Wynne, S. A., *et al.* (1999). Crystallization of hepatitis B virus core protein shells: determination of cryoprotectant conditions and preliminary X-ray characterization. *Acta Crystallogr. D* **55**, 557–560.

Xiao, C., *et al.* (2001). Interaction of coxsackievirus A21 with its cellular receptor, ICAM-1. *J. Virol.* **75**, 2444–2451.

Zheng, Y., Doerschuk, P. C. and Johnson, J. E. (1995). Determination of three-dimensional low-resolution viral structure from solution x-ray scattering data. *Biophys J.* **69**, 619–639.

Zlotnick, A., *et al.* (1999). Separation and crystallization of $T = 3$ and $T = 4$ icosahedral complexes of the hepatitis B virus core protein. *Acta Crystallogr. D* **55**, 717–720.

CHAPTER 17

# Macromolecular crystallography in drug design

Sherin S. Abdel-Meguid

## 17.1 Introduction

Given the important role macromolecular crystallography has played and will continue to play in the discovery of new drugs, it is only fitting to end this book with a brief chapter entitled 'Macromolecular crystallography in drug design'. Among all the technologies in drug discovery, macromolecular crystallography is one of the most powerful. Table 17.1 lists the marketed drugs that have been designed using knowledge and analysis of protein crystal structures. Many others are currently in preclinical or clinical development (Hardy and Malikayil, 2003).

My first introduction to this topic was in 1979, when I met Drs Karst Hoogsteen and Peter Gund of Merck Research Laboratories (Rahway, New Jersey) while they were visiting Purdue University. At that time, I was a Postdoctoral Fellow in Professor Michael Rossmann's laboratory at Purdue working on one of the first determinations of the crystal structure of an icosahedral virus (Abad-Zapatero *et al.*, 1980). During our conversation, I was introduced to the idea of 'rational' drug design. Both Drs Hoogsteen and Gund were convinced that protein crystallography could play a critical role in speeding up drug discovery. I have to say that I was highly sceptical, given that that was a time when the determination of a *de novo* protein crystal structure, even one of modest size (20,000 dalton), took as long as a year, if not more. In such a time scale, it would have been impossible to make a timely impact on drug discovery. However, advancements in protein expression and purification, crystallographic data acquisition, crystal structure determination methodology, and computer speed in the late 1970s and early 1980s, and the continued advancement in these and other related areas in the 1980s and 1990s, have considerably speeded up the structure determination of macromolecules. Now, it is certainly possible for protein crystallography to make a significant impact on drug discovery; it is even possible to determine the crystal structures of large molecular assemblies such as the ribosome in a relatively short time (Ban *et al.*, 2000). Today, all large and most small pharmaceutical and biotechnology companies have invested in macromolecular crystallography, with protein crystallography now an integral tool of drug discovery.

During the last 20 years, the term rational drug design has slowly been replaced with the more precise term structure-based drug design (SBDD). Of course, an even better term for using three-dimensional structures to design novel molecules is structure-based ligand design. Although the latter term would cover not only the design of novel drugs, but also the design of novel herbicides or pesticides, it has never caught on.

## 17.2 The drug discovery and development process

Drug discovery and development is a high-risk-high-reward business that requires a long-term vision, considerable technical and strategic experience, and multifaceted expertise. It may take as long as 14 years and cost as much as \$880 million to discover a drug; this cost includes failures (Tollman *et al.*, 2001). The steps necessary to discover and develop a drug can be summarized and simplified as follows (Fig. 17.1):

1. *Identify an unmet medical need*: it is important to choose a disease or a medical condition for which either no therapies are available or current therapies have known liabilities.
2. *Identify a target (Target Identification)*: in this step one selects a target molecule (in most cases a protein) that is believed to be involved in a human disease process. This and the next step are the most important steps in drug discovery because they are the first in a long process. The wrong target selection will lead to considerable loss of time and resources that could have been directed toward other, more productive, avenues.
3. *Validate the target (Target Validation)*: this is a difficult step, and may not be achieved until a drug candidate is in clinical trials. In this step one ensures that the target identified in Step 2 (above) is involved in the disease process and that modulating its biological action by a drug will lead to desired results. This can be accomplished by one of several processes, such as gene knockout, knockdown or over expression, RNAi, testing drug candidates in animal disease models, or knowing that a drug already on the market functions by modulating that target.
4. *Identify a drug candidate (Lead Identification)*: in this step, often using a robotic system, one screens chemical libraries containing thousands or millions of manmade and natural compounds for a drug candidate that modulates the selected target. This step can be described as 'looking for a needle in a hay stack'. Usually the identified drug candidate is not suitable for introduction in humans and must be optimized to better modulate the selected target.
5. *Optimize the lead molecule (Lead Optimization)*: in modern drug discovery this step involves optimization of lead molecules utilizing an iterative SBDD process (see below). This process entails the determination of the three-dimensional, atomic structures of the selected target in the presence of ligands (identified from the previous step), the use of those structures to design new ligands using computational chemistry approaches, the synthesis of those ligands, and the *in vitro* testing of the designed ligands for affinity and selectivity (Fig. 17.2). The end product is an optimized drug candidate (lead) that is ready to be tested in animal disease models.

**Table 17.1** Marketed drugs designed using knowledge of protein crystal structures

| Trade name | Generic name | Molecular target | Indication |
|---|---|---|---|
| Agenerase | Amprenavir | HIV protease | AIDS |
| Crixivan | Indinavir | HIV protease | AIDS |
| Kaletra | Lopinavir | HIV protease | AIDS |
| Viracept | Nelfinavir | HIV protease | AIDS |
| Norvir | Ritonavir | HIV protease | AIDS |
| Invirase | Saquinavir | HIV protease | AIDS |
| Gleevec | Imitinab | BCR-Abl tyrosine kinase | Cancer |
| Tomudex | Raltitrexed | Thymidylate synthase | Cancer |
| Trusopt | Dorzolamide | Carbonic anhydrase | Glaucoma |
| Capoten | Captopril | Angiotensin I-converting enzyme (ACE) | Hypertension |
| Tamiflu | Oseltamivir | Neuraminidase | Influenza |
| Relenza | Zanamivir | Neuraminidase | Influenza |

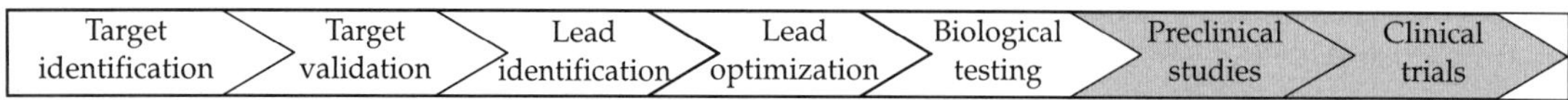


**Figure 17.1** Major steps in the drug discovery and development process. The non-shaded areas are steps in drug discovery, while grey areas are steps in drug development.

**6.** *Test the optimized drug lead in a disease animal model (Biological Testing)*: here the optimized drug lead is tested for efficacy in an animal disease model. If the drug lead is efficacious, it is advanced to the next step. If it is not, a different optimized drug lead is tested. This process continues until the ideal drug lead is identified.

**7.** *Undertake preclinical studies (Preclinical Development)*: this involves many different important steps, of which the most critical is safety evaluation of the efficacious drug candidate in animal models. Before introduction of any new drug in humans it must be shown to be safe in animals.

**8.** *Test the drug in the clinic (Clinical Trials)*: this is a long process consisting of several phases. It involves testing the drug in humans for safety and efficacy. The drug can be marketed and sold after the successful completion of this step and approval from government agencies such as the FDA, in the US.

- **(a)** *Phase I*: drugs are tested in healthy volunteers to determine safety and dosage.
- **(b)** *Phase II*: drugs are tested in patient volunteers to look for efficacy and side effects.
- **(c)** *Phase III*: drugs are tested in patient volunteers to monitor adverse reactions to long-term use.

Steps 1 to 6 are usually referred to as drug discovery, while steps 7 and 8 are referred to as drug development. The timeline for each of drug discovery and development can be as little as 4 years and as much as 7.

## 17.3 The iterative structure-based drug design (SBDD) cycle (lead optimization)

SBDD is an iterative process (Fig. 17.2), in which macromolecular crystallography has been the predominate technique used to elucidate the three-dimensional structure of drug targets (Qiu *et al.*, 2004; Babine *et al.*, 2004). Although both nucleic acids and proteins are potential drug targets, by far the majority of such targets are proteins. Given that many proteins undergo considerable conformational change upon ligand binding (Qiu *et al.*, 2004), it is important to design drugs based on the crystallographic structures of protein–ligand complexes, not the unliganded structure.

Crystallography has been successfully used in the *de novo* design of drugs, but its most important use has been, and will continue to be, in lead optimization (Step 5, above; Fig. 17.2). It is important to note that what is being optimized is the affinity

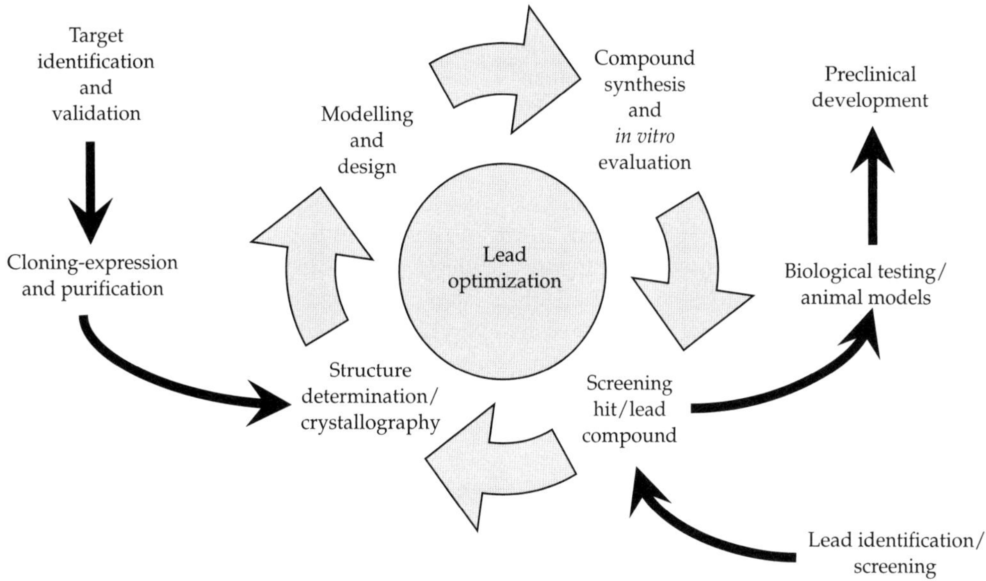


**Figure 17.2** The structure-based drug design (SBDD) or lead optimization cycle.

and specificity of compounds to their drug target. Lead optimization is a multistep process that can be summarized as follow:

**1.** The process starts with cloning, expression, and purification of the protein of interest. The protein is then crystallized in the presence of a ligand, which can be a non-hydrolysable substrate or can come from a biochemical or a cell-based screen. Ligands can also be low-affinity compound fragments or scaffolds (Card *et al.*, 2005). The latter are generally a collection of basic chemical building blocks, each with a molecular weight less than 200 Da (Erlanson *et al.*, 2004). It is important to note that if the screen identifies several promising ligands, each with a unique scaffold, one should try to determine the structures of the drug target with as many of these as possible.

**2.** Once one or more liganded structures have been determined and refined, analysis of each structure will reveal sites on the ligand that can be optimized to enhance potency to the drug target. This can be accomplished by redesigning the ligand with greater hydrophobic, hydrogen-bonding, and electrostatic complementarity to the molecular target. The design process can be simple and intuitive if one starts with a relatively high affinity lead. In this case, only minor modifications to the existing ligand are introduced. Many of these modifications can be proposed from previous personal knowledge, or can be derived by computer modelling. There are numerous commercial and academic computer programs to aid in the analysis and design of new ligands. A list of many of these programs can be found in Anderson (2003). However, it is important to note that computational methods are still not reliable in predicting binding modes and affinities of ligands, mainly due to inaccuracies in force fields, limitations in dealing with ligand and target flexibility, the lack of reliable scoring functions, as well as the difficulties in treating solvent molecules. Therefore, even for seemingly minor modifications of the leads, it is still necessary to confirm the binding mode experimentally; there are countless examples in which the mode of binding significantly changes upon introduction of minor modifications to the original ligand (Qiu *et al.*, 2004).

**3.** Now that ligands have been designed, they should be chemically synthesized. It is prudent, if synthetically feasible and relatively easy, to synthesize a small library of five to ten compounds around the proposed ligand to obtain structure–function relationship (SAR) data. Once the synthesized compounds are purified to greater than 80% purity, they are tested in a relevant biochemical or cell-based assay to determine whether or not the design was successful. Occasionally, a redesigned ligand will show less potency than the parent compound. Further cycles of structure determinations should reveal the reason.

**4.** The above three steps constitute one design cycle. It is often necessary to go through several iterations of the above cycle of structure determination, design, synthesis, and testing before a drug candidate emerges (Fig. 17.2).

## 17.4 Case study: structure-based design of cathepsin K inhibitors

Cathepsin K, a member of the papain superfamily of cysteine proteinases, is selectively and highly expressed in osteoclasts (Drake *et al.*, 1996; Bromme and Okamoto, 1995). It is secreted as a 314 amino acid proenzyme containing a 99 amino acid leader sequence (Bossard *et al.*, 1996). Cathepsin K plays an important role in bone resorption and is a potential therapeutic target for the treatment of diseases involving excessive bone loss such as osteoporosis (Veber *et al.*, 1997).

Cathepsin K was one of the first drug targets to be identified from analysis of human genome sequences. Its identification and characterization by SmithKline Beecham (now Glaxo SmithKline) and others led to a race to develop selective inhibitors of the enzyme (Abdel-Meguid *et al.*, 1999). Early in the study, the absence of sufficient pure cathepsin K for crystallographic structure determination compelled me and my colleagues at SmithKline Beecham to search for a protein that could serve as a suitable surrogate model. Papain was chosen because it is 46% identical in amino acid sequence to cathepsin K, it is commercially available in a pure form, and its crystal structure bound to inhibitors had been reported (Varughese *et al.*, 1989; Yamamoto *et al.*, 1991, 1992). At the time we initiated these studies, the inhibitors in all of the papain structures, including our structure of papain bound to leupetin (Leu-Leu-Arg-aldehyde), were found to bind on the non-prime side of the active site. Using these structures, we modelled a number of our di- and tripeptide aldehyde inhibitors into the non-prime side of the active site of papain and into a homology model of cathepsin K derived from papain. These modelling studies did not explain our SAR data, which showed strong preference for the presence of a Cbz or other aromatic moiety at the amino terminus of these peptides. Therefore, we undertook our own crystallographic studies with papain complexed to our inhibitors (LaLonde *et al.*, 1998). Surprisingly, the Cbz-Leu-Leu-Leu aldehyde inhibitor in our structure was found to bind on the prime side of the active site. A major point of interaction between the inhibitor and the protein was an edge-to-face interaction between the phenyl ring of the inhibitor and the indole ring of Trp181 (LaLonde *et al.*, 1998), a residue that is conserved between papain and cathepsin K. These observations led to the design of novel inhibitors spanning both sides of the active site (Abdel-Meguid *et al.*, 1999; Fig. 17.3). The prototype of this class of inhibitors was a symmetric inhibitor that resulted from an overlay of the crystal structure of papain containing the Cbz-Leu-Leu-Leu aldehyde and that containing leupeptin. The two inhibitors were merged computationally by replacing their aldehyde functions with a single ketone (Fig. 17.3). The resulting model of a ketone-containing inhibitor was further simplified by removal of the side chains on both sides of the ketone moiety. This was necessary since the arginyl and leucyl sidechains occupied the same region of space. Furthermore, a homology model of cathepsin K derived from the structure of papain suggested that Trp184 of cathepsin K (Trp177 in papain), a highly conserved residue within the papain superfamily, would form a better aromatic–aromatic interaction with the Cbz moiety. Thus, the hypothetical inhibitor was shortened by one Leu residue from the prime side (Fig. 17.3), resulting in a yet smaller molecule. A second Cbz moiety was introduced on the left side (Fig. 17.3) as a final step to make the inhibitor truly symmetric. This was done not to mimic any symmetry in the active site (there is none), but rather to simplify the chemical synthesis of this initial member of a new class of inhibitors. This Cbz group was also hypothesized to reach to Tyr67 on the non-prime side of the cathepsin K active site for additional aromatic–aromatic interaction. The resulting diacylaminomethyl ketone (1,3-bis[[*N*-[(phenylmethoxy)carbonyl]-L-leucyl]amino]-2-propanone) is shown in Fig. 17.3 (Abdel-Meguid *et al.*, 1999; Yamashita *et al.*, 1997).

By the time the diacylaminomethyl ketone inhibitor was synthesized, highly purified cathepsin K became available. This allowed us to obtain crystals and determine the structure of cathepsin K bound to the inhibitor; we showed that the inhibitor binds in the cathepsin K active site as predicted (Abdel-Meguid *et al.*, 1999; Yamashita *et al.*, 1997). It spans both sides of the active site and makes a number of key interactions with the enzyme. The phenyl groups on both ends of the inhibitor engage Trp184 and Tyr67 in a face-face and edge-face interaction, respectively. The crystal structure clearly shows the inhibitor covalently attached to the enzyme at the sulphur atom of Cys25 (the active site cysteine) as expected. The P2 leucyl side chain of the inhibitor fits snugly in the hydrophobic S2 pocket defined by residues Met68, Leu209, Ala134, Ala163, and Tyr67. Hydrogen bonding interactions are seen between ND1 of His162, NE2 of Gln19 and the backbone amide nitrogens of Cys25 and Gly66, each of which donate a hydrogen to oxygen atoms of the inhibitor. The remainder of the inhibitor interacts poorly or not at all with the enzyme, indicating potential for further optimization of this class of inhibitors. Spanning both sides of the active site allowed for enhanced potency and selectivity by taking simultaneous advantage of interactions on the non-prime and prime sides of the active site, and by allowing the use of a less reactive electrophilic carbon for attack at the cysteine. This novel diacylaminomethyl ketone proved to be a selective, competitive, reversible inhibitor of cathepsin K with a Ki of 23 nM (Yamashita *et al.*, 1997). It is a relatively poor inhibitor of papain, cathepsin L, cathepsin B, and cathepsin S, with Ki,app of 10,000 nM, 340 nM, 1300 nM, and 890 nM, respectively (Yamashita *et al.*, 1997; Table 17.2).

Additional cycles of SBDD were undertaken. They focused on separately optimizing each of the two halves of the symmetric diacylaminomethyl ketone inhibitor (DesJarlais *et al.*, 1998; Thompson *et al.*,

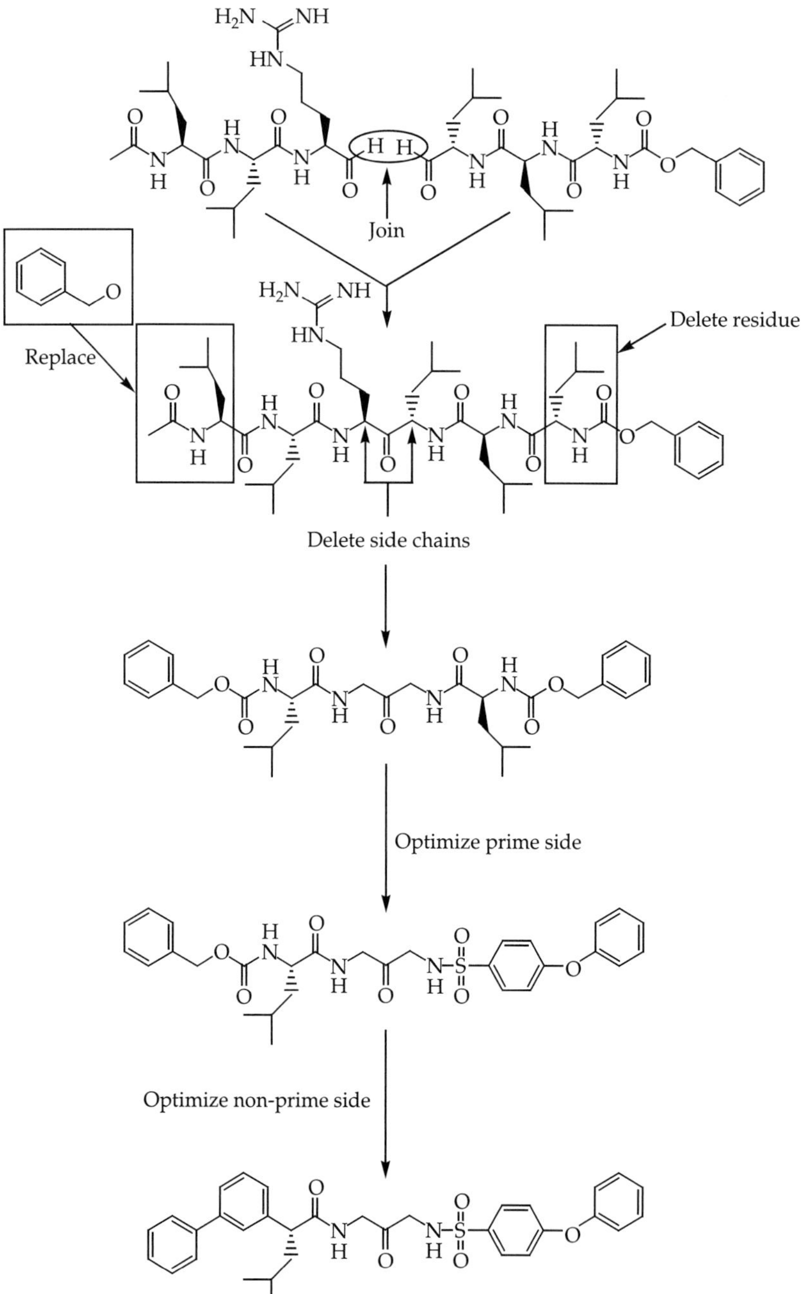


**Figure 17.3** Schematic representation of the design of the symmetric cathepsin K inhibitor diacylaminomethyl ketone (1,3-bis[[*N*-[(phenylmethoxy)carbonyl]-L-leucyl]amino]-2-propanone), based on the crystal structures of papain bound to leupeptin (Leu-Leu-Arg-aldehyde) and to Cbz-Leu-Leu-Leu-aldehyde, and an example of its further optimization.

**Table 17.2** Structure-based enhancement of cathepsin k inhibitor's potency and selectivity

| Inhibitor | Cathepsin K Ki,app (nM) | Papain Ki,app (nM) | Cathepsin L Ki,app (nM) | Cathepsin S Ki,app (nM) | Cathepsin B Ki,app (nM) |
|---|---|---|---|---|---|
| Symmetric ketone | 23 | 10, 000 | 340 | 890 | 1300 |
| Prime side optimized | 1.8 | – | 1400 | 80 | >10, 000 |
| Both sides optimized | 1.4 | – | >1000 | 910 | >10, 000 |

1997; Thompson *et al.*, 1998; Marquis *et al.*, 1999). Once each half was optimized, more SBDD cycles were needed to tweak the full molecule. This work resulted in numerous potent and selective cathepsin K inhibitors; an example is shown in Fig. 17.3. Improvements in potency and selectivity of the compounds shown in Fig. 17.3 are listed in Table 17.2. Much of this work was recently summarized in Veber and Cummings (2004).

## 17.5 Impact of structure-based drug design on drug discovery

Whether SBDD has a direct or indirect impact on the discovery and development of a particular drug has been, is, and will continue to be a debatable issue. To some extend this is a turf issue. It has to do with who designed the drug, the medicinal chemist, the computational chemist, or the crystallographer, and how much that individual *feels* the structural information has contributed to that design, if any. To minimize this problem, many drug companies have fostered a team spirit and have given credit to successful teams, instead of to individuals. Furthermore, now many chemists are trained in SBDD and are becoming comfortable with the use of structures in drug design. Regardless of whether SBDD has a direct or indirect impact on the design of new drugs, everyone today agrees that having access to a high resolution structure of a drug target, in complex with lead compounds, is extremely desirable, if not absolutely necessary, for the timely optimization of lead compounds.

Table 17.1 lists a number of successful drugs on the market that were designed using knowledge and analysis of protein crystal structures. Note that the list is dominated by HIV protease inhibitors, drugs for the treatment of AIDS. The speed by which these drugs were developed is a testament to the power of SBDD. Key was the availability of numerous crystal structures of HIV protease (Abdel-Meguid, 1993) shortly after the discovery that HIV protease is an aspartyl protease (Pearl and Taylor, 1987). Also critical was the extensive knowledge available on the development of inhibitors of the aspartyl protease, renin (Abdel-Meguid, 1993).

## 17.6 Experience with structure-based drug design

In the last 25 years, much has been learned from our experience with SBDD. Below, I will highlight some of the important points learned.

**1.** *The lack of crystals is detrimental to the process*: Although this is an obvious point, obtaining large single crystals that diffract to high resolution remains the primary bottleneck of protein crystallography. Therefore, it is critical to find ways to obtain suitable crystals. If the desired protein fails to crystallize, the classical approach is to switch to the same protein from a different species. Kendrew pioneered this approach when determining the structure of myoglobin by pursuing the sperm whale protein (Kendrew *et al.*, 1958, 1960). With advances in recombinant DNA technology, other approaches have been pursued (Jin and Babine, 2004). These include removing flexible regions and post-translational modifications and improving crystal packing by site-directed mutagenesis. The most general application of the latter is to replace surface residues having high conformational entropy, such as replacing lysines with alanines. When the protein structure is not known, this can simply be done by mutating each lysine in the protein, one at a time. Another approach is to form protein complexes with Fabs

(Ruf *et al.*, 1992) or in the case of serine protease with ecotin (Waugh and Fletterick, 2004), a macromolecular inhibitor of serine proteases. Fab- and ecotin-protein complexes also offer the advantage that these structures can be solved by molecular replacement.

2. *Design should be based on liganded structures*: As indicated above, many proteins undergo considerable conformational change upon binding to their ligands. Initiating ligand design based on an unliganded structure may be misleading if that structure is of a protein that will change its conformation upon ligand binding. To be on the safe side, one should always start ligand design based on a liganded structure of the target protein. An example of a protein that undergoes large conformational change upon ligation is EPSP (5-enol-pyruvyl-3-phosphate) synthase. The unliganded structure (Stallings *et al.*, 1991) shows a large cavity at the active site, much of which disappears upon ligation (Qiu *et al.*, 2004). Sometimes, different ligands may lead to different conformational changes of the protein target, making ligand design even more challenging.

3. *Aqueous solubility of ligands is important*: One of the bottlenecks in structure-based design is poor aqueous solubility of many ligands. If the ligands are insoluble in water, it is often difficult to form complexes under conditions of crystallization. Unlike the crystallization of small organic molecules, proteins must be crystallized from aqueous solutions or using solvents that are highly miscible with water. Therefore, it is sensible to introduce polar or charged groups to improve the inhibitor solubility, making structural studies more amendable. However, there are examples in which adding ligands in the solid form produces crystals of the protein–ligand complex that diffract to high resolution (Cha *et al.*, 1996).

4. *Use of surrogate enzymes can lead to important insights*: As described above in the Case Study, when the target enzyme is difficult to obtain or crystallize, a related enzyme can be used to provide insights in the design of novel ligands.

5. *Beware of crystal contacts*: In the crystal, it is possible for a ligand to make important contacts with residues from a neighbouring molecule, producing an artificial mode of binding that is not possible in solutions. Thus it is important to analyse all crystal contacts in the vicinity of the ligand, prior to proceeding with the design of new ligands.

6. *Allow for flexibility in the design of enzyme inhibitors to ensure optimal fit in an often rigid, active site cavity*: It is often very difficult to design a highly constrained ligand that complements and fits snugly in an enzyme active site. Although rigidity of the ligand is important to reduce entropy and to ensure greater affinity, it is often wise to initially introduce some flexibility to ensure proper fit in an often rigid active site. Much of this flexibility can be reduced considerably in later iterations of the drug design cycle. This can be achieved by designing molecules to present complementary electrostatic, hydrogen-bonding, and hydrophobic interactions to their drug target.

7. *Every water molecule is special*: Incorporation in ligand design of the position of water molecules that are firmly bound to the protein can impart affinity and novelty to the designed ligand. A prime example is the design of a class of HIV protease cyclic urea inhibitors that incorporates a water molecule known to bind to both flaps of the enzyme (Lam *et al.*, 1994). The crystal structure of the HIV protease–cyclic urea complex shows the urea carbonyl oxygen substituting for the position of the water molecule.

8. *Fill available space and maximize interactions*: A major goal of ligand design should be to fill as much of the space in the binding site as possible without rendering the designed ligand too large. Ligands greater than 500 Da have a lower probability of being orally bioavailable. It is also important to maximize both polar and non-polar interactions with the protein.

9. *Design of small molecules to interfere with protein/protein interaction requires knowledge of the structure of the complex*: Most protein–protein interfaces are large, hydrophobic surfaces. For example, the interface area between growth hormone (Abdel-Meguid *et al.*, 1987) and its receptor (Cunninghum *et al.*, 1991) is about 2100 Å$^2$. To rationally design a small molecule to interfere with such large surfaces is a considerable challenge requiring atomic details of the receptor surface, which may differ between unliganded and liganded forms. Generally, success in this arena is rare. Occasionally, protein–protein interactions may consist of only a small number of

contacts, such as the RGD (Arg-Gly-Asp) interaction with its receptors (Ku *et al.*, 1995). In such a case, the design task becomes essentially a small molecule-protein interaction problem and is more likely to succeed.

**10.** *Synthetic accessibility is essential*: It is important to design ligands that can be synthesized in a timely fashion using readily available or easy to obtain starting material. Given that many potential drugs fail for reasons that have nothing to do with their binding affinity, it is important that one go through a design cycle as fast as possible to obtain feedback on the suitability of the designed ligands as drugs.

**11.** *Iterative design is essential*: It is a rarity that the first ligand to be designed is the final one. As indicated above, it is common to go through several iterations of the structure-based design cycle before settling on the desired molecule that will be advanced to development.

**12.** *There is no substitute for experience*: Structure-based drug design is no different from most other areas; experience counts.

## References


Abad-Zapatero, C., Abdel-Meguid, S. S., Johnson, J. E., Leslie, A. G. W., Rayment, I., Rossmann, M. G., Suck, D. and Tsukihara, T. (1980). Structure of southern bean mosaic virus at 2.8 Å resolution. *Nature* **286**, 33.

Abdel-Meguid, S. S. (1993). Inhibitors of aspartyl proteinases. *Medicinal Res. Rev.* **13**, 731–778.

Abdel-Meguid, S. S., Shieh, H.-S., Smith, W. W., Dayringer, H. E., Violand, B. N. and Bentle, L. A. (1987). Three-dimensional structure of a genetically engineered variant of porcine growth hormone. *Proc. Natl. Acad. Sci. USA*, **84**, 6434–6437.

Abdel-Meguid, S. S., Zhao, B., Janson, C. A., Carr, T., D'Alessio, K., McQueney, M. S., Oh, H.-J., Thompson, S. K., Veber, D. F., Yamashita, D. S. and Smith, W. W. (1999). Rational approaches to inhibition of human osteoclast cathepsin k and treatment of osteoporosis. In: *Rational Drug Design*, ACS Symposium Series 719, Parrill, A.L. and Reddy, M.R., eds. American Chemical Society, pp. 141–152.

Anderson, A. C. (2003). The process of structure-based drug design. *Chem. Biol.* **10**, 787–797.

Babine, R. E. and Abdel-Meguid, S. S., eds. (2004). *Protein crystallography in drug discovery. Methods and Principles in Medicinal Chemistry*, Vol. **20**. Wiley-VCH.

Ban, N., Nissen, P., Hansen, J., Moore, P. and Steitz, T. (2000). The complete atomic structure of the large ribosomal subunit at 2.4 Å resolution. *Science* **289**, 905–920.

Bossard, M. J., Tomaszek, T. A., Thompson, S. K., Amegadzie, B. Y., Hanning, C. R., Jones, C., Kurdyla, J. T., McNulty, D. E., Drake, F. H., Gowen, M. and Levy, M. A. (1996). Proteolytic activity of human osteoclast cathepsin k. expression, purification, activation, and substrate identification. *J. Biol. Chem.*, **271**, 12517–12524.

Bromme, D. and Okamoto, K. (1995). Human cathepsin O2, a novel cysteine protease highly expressed in osteoclastomas and ovary molecular cloning, sequencing and tissue distribution. *Biol. Chem.* **376**, 379–384.

Card, G. L., Blasdel, L., England, B. P., Zhang, C., Suzuki, Y., Gillette, S., Fong, D., Ibrahim, P. N., Artis, D. R., Bollag, G., Milburn, M. V., Kim, S.-H., Schlessinger, J., Zhang, K. Y. (2005). A family of phosphodiesterase inhibitors discovered by co-crystallography and scaffold-based drug design. *Nat. Biotechnol*, **23**, 201–207.

Cha, S.-S., Lee, D., Adams, J., Kurdyla, J. T., Jones, C. S., Marshall, L. A., Bolognese, B., Abdel-Meguid, S. S. and Oh, B.-H. (1996). High resolution X-ray crystallography reveals precise binding interactions between human non-pancreatic secreted phospholipase A2 and a highly potent inhibitor (FPL67047XX). *J. Med. Chem.* **39**, 3878–3881.

Cunninghum, B. C., Ultsch, M., De Vos, A. M., Mulkerrin, M. G., Clauser, K. R. and Wells, J. A. (1991). Dimerization of the extracellular domain of the human growth hormone receptor by a single hormone molecule. *Science* **254**, 821–825.

DesJarlais, R. L., Yamashita, D. S., Oh, H.-J., Uzinskas, I. N., Erhard, K. F., Allen, A. C., Haltiwanger, R. C., Zhao, B., Smith, W. W., Abdel-Meguid, S. S., D'Alessio, K. J., Janson, C. A., McQueney, M. S., Tomaszek, Jr., T. A., Levy, M. A. and Veber, D. F. (1998). Use of X-Ray co-crystal structures and molecular modeling to design potent and selective, non-peptide inhibitors of cathepsin K. *J. Am. Chem. Soc.* **120**, 9114–9115.

Drake, F. H., Dodds, R. A., James, I. E., Connor, J. R., Debouck, C., Richardson, S., Lee-Rykaczewski, E., Coleman, L., Rieman, D., Barthlow, R., Hastings, G. and Gowen, M. (1996). Cathepsin K, but not cathepsins B, L, or S, is abundantly expressed in human osteoclasts. *J. Biol. Chem.* **271**, 12511–12516.

Erlanson, D. A., McDowell, R. S. and O'Brien, T. (2004). Fragment-based drug discovery. *J. Med. Chem.* **47**, 3463–3482.

Hardy, L. W. and Malikayil, A. (2003). The impact of structure-guided drug design on clinical agents. *Curr. Drug Discov.* **3**, 15–20.

Jin, L. and Babine, R. E. (2004). Engineering proteins to promote crystallization. In: *Protein Crystallography in Drug Discovery. Methods and Principles in Medicinal Chemistry*, Babine, R. E. and Abdel-Meguid, S. S., eds, Vol. 20, Wiley-VCH, pp. 209–216.

Kendrew, J. C., Bodo, G., Dintzis, H. M., Parrish, R. G., Wyckoff, H. W. and Phillips, D. C. (1958). A three-dimensional model of the myoglobin molecule obtained by X-ray analysis. *Nature* **181**, 662–666.

Kendrew, J. C., Dickerson, R. E., Strandberg, B. E., Hart, R. G., Davies, D. R., Phillips, D. C. and Shore, V. (1960). Structure of myoglobin: a three-dimensional fourier synthesis at 2 Å resolution. *Nature* **185**, 422–427.

Ku, T. W., Miller, W. H., Bondinell, W. E., Erhard, K. F., Keenan, R. M., Nichols, A. J., Peishoff, C. E., Samanen, J. M., Wong, A. S. and Huffman, W. F. (1995). Potent non-peptide fibrinogen receptor antagonists which present an alternative pharmacophore. *J. Med. Chem.* **38**, 9–12.

LaLonde, J. M., Zhao, B., Smith, W. W., Janson, C. A., DesJarlais, R. L., Tomaszek, T. A., Carr, T. J., Thompson, S. K., Oh, H.-J., Yamashita, D. S., Veber, D. F. and Abdel-Meguid, S. S. (1998). Use of papain as a model for the structure-based design of cathepsin K inhibitors: crystal structures of two papain-inhibitor complexes demonstrate binding to S′-subsites. *J. Med. Chem.* **41**, 4567–4576.

Lam, P. Y. S., Jadhav, P. K., Eyermann, C. J., Hodge, C. N., Ru, Y., Bacheler, L. T., Meek, J. L., Otto, M. J., Rayner, M. M., Wong, Y. N., Chang, C.-H., Weber, P. C., Jackson, D. A., Sharpe, T. R. and Erickson-Viitanen, S. (1994). Rational design of potent, bioavailable, nonpeptide cyclic ureas as HIV protease inhibitors. *Science*, **263**, 380–384.

Marquis, R. W., Ru, Y., Yamashita, D. S., Oh, H.-J., Yen, J., Thompson, S. K., Carr, T. J., Levy, M. A., Tomaszek, T. A., Ijames, C. F., Smith, W. W., Zhao, B., Janson, C. A., Abdel-Meguid, S. S., D'Alessio, K. J., McQueney, M. S. and Veber, D. F. (1999). Potent dipeptidylketone inhibitors of the cysteine protease cathepsin K. *Bioorg. Med. Chem.* **7**, 581–588.

Marquis, R. W., Yamashita, D. S., Ru, Y., LoCastro, S. M., Oh, H.-J., Erhard, K. E., DesJarlais, R. L., Smith, W. W., Zhao, B., Janson, C. A., Abdel-Meguid, S. S., Tomaszek, T. A., Levy, M. A. and Veber, D.F. (1998). Conformationally constrained 1,3-diamino ketones: a series of potent inhibitors of the cysteine protease cathepsin K. *J. Med. Chem.* **41**, 3563–3567.

Pearl, L. H. and Taylor, W. R. (1987). A structural model for the retroviral proteases. *Nature* **329**, 351–354.

Qiu, X. and Abdel-Meguid, S. S. (2004). Protein crystallography in structure-based drug design. In: *Drug Discovery Strategies and Methods*, Makriyannis, A. and Biegel, D., eds. Marcel Dekker, New York, p. 1–21.

Ruf, W., Stura, E. A., LaPolla, R. J., Syed, R., Edgington, T. S., Wilson, I. A. (1992). Purification, sequence and crystallization of an anti-tissue factor Fab and its use for the crystallization of tissue factor. *J. Crystal Growth* **122**, 253–264.

Stallings, W. C., Abdel-Meguid, S. S., Lim, L. W., Shieh, H.-S., Dayringer, H. E., Leimgruber, N. K., Stegeman, R. A., Anderson, K. S., Sikorski, J. A., Padgette, S. R. and Kishore, G. M. (1991). Structure and topological symmetry of the glyphosphate 5-enol-pyruvylshikimate-3-phosphate synthase: a distinctive protein fold. *Proc. Natl. Acad. Sci. USA* **88**, 5046–5050.

Thompson, S. K., Halbert, S. M., Bossard, M. J., Tomaszek, T. A., Levy, M. A., Meek, T. D., Zhao, B., Smith, W. W., Abdel-Meguid, S. S., Janson, C. A., D'Alessio, K. J., McQueney, M. S., Amegadzie, B. Y., Hanning, C. H., DesJarlais, R. L., Briand, J., Sarkar, S. K., Huddleston, M. J., Ijames, C. F., Carr, S. A., Garnes, K. T., Shu, A., Heys, J. R., Bradbeer, J., Zembryki, D., Lee-Rykaczewski, L., James, I. E., Lark, M. W., Drake, F. H., Gowen, M., Gleason, J. G. and Veber, D. F. (1997). Design of potent and selective human cathepsin K inhibitors that span the active site. *Proc. Natl. Acad. Sci. USA* **94**, 14249–14254.

Thompson, S. K., Smith, W.,W., Zhao, B., Halbert, S. M., Tomaszek, T. A., Tew, D.G., Levy, M. A., Janson, C. A., D'Alessio, K. J., McQueney, M. S., Kurdyla, J., Jones, C. S., DesJarlais, R. L., Abdel-Meguid S. S. and Veber, D. F. (1998). Structure-based design of cathepsin K inhibitors containing a benzyloxy-substituted benzoyl peptidomimetic. *J. Med. Chem.* **41**, 3923–3927.

Tollman, P., Guy, P., Altshuler, J., Flanagan, A. and Steiner, M. (2001). *A Revolution in R&D: How Genomics and Genetics are Transforming the Biopharmaceutical Industry*. The Boston Consulting Group.

Varughese, K.I., Ahmed, F.R., Carey, P.R., Hasnain, S., Huber, C.P. and Storer, A.C. (1989). crystal structure of a papain-E-64 complex. *Biochemistry* **28**, 1330–1332.

Veber, D. F. and Cummings, M. D. (2004). Structure-based design of cathepsin K inhibitors. In: *Protein Crystallography in Drug Discovery. Methods and Principles in Medicinal Chemistry*, Babine, R. E. and Abdel-Meguid, S. S., eds, Vol. 20, Wiley-VCH, p. 127–146.

Veber, D. F., Drake, F. H. and Gowen, M. (1997). The new partnership of genomics and chemistry for accelerated drug development. *Curr. Opin. Chem. Biol.* **1**, 151–156.

Waugh, S. M. and Fletterick, R. J. (2004). Crystallization and analysis of serine proteases with ecotin. In: *Protein Crystallography in Drug Discovery. Methods and Principles in Medicinal Chemistry*, Babine, R. E. and Abdel-Meguid, S. S., eds, Vol. 20, Wiley-VCH, pp. 171–186.

Yamamoto, A., Tomoo, K., Doi, M, Ohishi, H., Inoue, M., Ishida, T., Yamamoto, D., Tsuboi, S., Okamoto, H., Okada, Y. (1992). Crystal structure of papain-succinyl-Gln-Val-Val-Ala-Ala-p-nitroanilide complex at 1.7-Å resolution: noncovalent binding mode of a common sequence of endogenous thiol protease inhibitors. *Biochemistry* **31**, 11305–11309.

Yamamoto, D., Matsumoto, K., Ohishi, H., Ishida, T., Inoue, M., Kitamura, K. and Mizuno, H. (1991). Refined X-ray structure of papain.E-64-c complex at 2.1-Å resolution. *J. Biol. Chem.* **266**, 14771–14777.

Yamashita, D. S., Smith, W. W., Zhao, B., Janson, C. A., Tomaszek, T. A. Bossard, M. J., Levy, M. A., Marquis, R. W., Oh, H-J., Ru, Y., Carr, T. J., Thompson, S. K., Ijames, C. F., Carr, S. A., McQueney, M., D'Alessio, K. J., Amegadzie, B. Y., Hanning, C. R., Abdel-Meguid, S. S., DesJarlais, R. L., Gleason, J. G. and Veber, D. F. (1997). Structure and design of potent and selective cathepsin K inhibitors. *J. Amer. Chem. Soc.* **119**, 11351–11352.

# Index